LIVERPOOL PALS

LIVERPOOL PALS

A History of the
17th, 18th, 19th and 20th (Service) Battalions
The King's (Liverpool Regiment) 1914–1919

Graham Maddocks

Pen & Sword
MILITARY

First published in 1991 by Leo Cooper
an imprint of Pen & Sword Books Ltd
47 Church Street, Barnsley, South Yorkshire, S70 2AS

ISBN 978 1 84415 641 2

A CIP catalogue record for this book is available from the British Library.

Printed and bound in England by CPI UK

Pen & Sword Books Ltd incorporates the imprints of Pen & Sword Aviation, Pen & Sword Maritime, Pen & Sword Military, Wharncliffe Local History, Pen & Sword Select, Pen & Sword Military Classics and Leo Cooper.

For a complete list of Pen & Sword titles please contact
PEN & SWORD BOOKS LIMITED
47 Church Street, Barnsley, South Yorkshire, S70 2AS England
E-mail: enquiries@pen-and-sword.co.uk
Website: www.pen-and-sword.co.uk

Contents

This book is dedicated to all the Liverpool Pals, past and present, but especially to my great friend, 17518 Corporal E G Williams, of the 10th Battalion, who survived the Battle of the Somme, and my great-uncle, 16414 Sergeant A E Gray, of the 18th Battalion, King's who didn't.

Introduction

It has always been a source of wonder to me that the story of the Liverpool Pals has not been written in its entirety before. General The Hon F C Stanley wrote *The History of the 89th Brigade 1914–1918* in 1919, largely from his own and the official Battalion War Diaries, but apart from the fact that it did not include most of the exploits of the 18th Battalion, it had the many limitations that all books written so soon after a war will always contain. Nevertheless, Stanley's work has been quoted extensively in the text of this book, for, if nothing else, it gives a real flavour of the times, at least at officer level. More respect for actual dates and times has been exercised here, however, than was sometimes shown by the General! My wonderment that the story has not been told before lies chiefly in the fact that the Liverpool Pals who survived the Great War probably had more educated and articulate men amongst them than the survivors of any other four battalions of the New Army. One might have thought, therefore, that at least one of them, perhaps suffering an idle retirement in the 1960s, might have put pen to paper, but this did not happen. Thus, the early 1990s is probably the last time that this book could conceivably have been written, as there are now so very few survivors left – probably no more than a dozen worldwide.

I never set out in the first place to write the story of the Liverpool Pals. Had I done so, I would surely have begun the task twenty years earlier, when there were many Pals still alive and their memories were still crisp and sharp. Being a collector of militaria from an early age, my interest in the Pals was first stimulated as a schoolboy, when I bought, for two pounds ten shillings – in pocket money-sized instalments – a silver Pals cap badge. Part of my reason for the purchase was that the badge was hallmarked, and the rest was the erroneous story, probably told to me by the shopkeeper, that there were only ever 1,000 badges made, and that as a consequence, only the first 1,000 men to join the Pals actually received them! This popularly-held misconception is still believed today, even by some badge dealers. However, as the Eagle and Child badge was of beautiful design, and was not at all like the White Horse of Hanover, which I knew to be the cap badge of most of the battalions of the King's (Liverpool Regiment), I began to ask people what they knew about these Liverpool Pals. The response fascinated me. Everyone had heard of them, many had had relatives who had served with them, but no one knew a great deal about where they had actually fought, or what they had actually done! Throughout my conversations, however, everyone agreed that they were special, and positively different from the rest of the soldiers of the Great War. Thus, without any real aim in mind, I began to collect and collate material about them and tried to discover their history.

In 1972 I achieved a long-cherished ambition to visit the Somme, which turned out to be the first of over eighty visits to the Western Front I have made since, both as a serious student of the area and as a battlefield tour leader and guide. It is only when one sees the area fought over during those grim years that one begins to understand the nature of the suffering endured and the triumph celebrated by all the soldiers of the Great War. On my next visit to the Somme, in 1973, I carried with me a talisman – the original brass cap badge worn by 17122 Private Bill Gregory of the 18th Battalion, and last carried in that area in October 1916. Maybe it helped me to find the Glatz Redoubt at Montauban, or drew me to the grave of Colonel Trotter and other comrades in Peronne Road Cemetery at Maricourt, but that visit set me off on a desire to visit all the places where the men of the Liverpool City Battalions had fought. Since there was no point in simply travelling round the Western Front looking for cemeteries which contained Pals' graves, I began to research more fully the history of each Battalion, and where it had served. This in itself opened the floodgates of research, and after numerous appeals in the local press, I began to realise what a mammoth task it was. It was at this time that I learned just how kind people could be, and how eager they were to offer help and information. They were all anxious that the story of the Pals should be told, and that the endeavours of men of that generation should not be forgotten. It seemed a churlish waste not to attempt to record all the wealth of information I had been given. Thus, I eventually began to realise that, to do justice to the men of the Pals, I would have to write their story, if only to put the record straight and destroy the myth which I kept hearing, that all the Pals Battalions were wiped out on 1st July 1916, and none even reached their objectives!

The decision to write was the easy part. Little did I know how much of my life would be occupied by the task, albeit a rewarding one, and little did I realise just how single-minded I would have to become in order to satisfy the demands it would engender. There is no doubt that it would have been so very much more difficult had I not had the help and support of dedicated friends, whom I gratefully acknowledge elsewhere. Also, I can honestly say that, although I have wearied of the task many, many times, especially after five hours' solid work in the evening, perhaps following an already-too-full working day, there has never been one occasion when I have not been uplifted by researching and recording the deeds of such a special and splendid band of men. I realise fully, also, that their story will never be complete, and that the publication of this book will inevitably uncover new information that I would have loved to have included in the first place.

Another source of regret to me is that nowhere on the Western Front (or in the United Kingdom, for that matter) is there a memorial to the Liverpool Pals. Thus, it is my intention to use some of the proceeds of this book to place a brass plaque in St George's Memorial Church in Ypres in honour of all the Pals who fought and died in the Salient. Ultimately, however, I would dearly love to erect a stone memorial in the village of Montauban, on the Somme, to the memory of all the Liverpool and Manchester Pals who captured that village on 1st July 1916. This was the only truly successful attack of the day, anywhere along the Somme front, by any division. There is a small patch of grass not far from the centre of the village . . . ? If anyone would like to be associated with either of these two projects, in any capacity, I would be delighted to hear from them.

This, then, is the story of the Liverpool Pals. They were the first and last of a kind. They were the first Pals Battalions to be raised, and the last to be stood down, and in between they served with honour, valour, success and comradeship on the battlefields of the Western Front and North Russia. Their story is one of triumph and tragedy, but above all, one of the perseverance of the British character in adversity. Nothing has been hidden from their account, which shows cowardice as well as bravery, and stupidity as well as intelligence. Alongside great feats of endurance and heroism, there is the tragedy of a soldier who died under the strain of punishment, and another who was shot by firing squad for cowardice, and who finally had to be 'finished off' by one of his own friends. Most of all, though, this story tells of their dedication to an ideal, and a love of country the like of which has rarely been seen since. They were 'Pals' in the true sense. They had worked and played together in peace, and they fought and all too often died together in war.

Graham Maddocks
May 1991

Acknowledgements

It is impossible to write a work of this kind, so long after the events it portrays, without the help of many, many people. All those mentioned in this very long list have given their help freely and willingly, and without any thought of reward, and my very great thanks are offered to them all. I have also scoured many sources for reminiscences to add to those of the former soldiers I have interviewed personally, and in certain cases, the writers of those reminiscences were themselves casualties in the Great War. I pay tribute to them all. If I have forgotten to acknowledge anyone who has helped in any way, then I am very sorry, and the omission has certainly not been intentional. My first thanks must go to The Right Honourable The Earl of Derby for graciously writing the Foreword to this book which chronicles the fortunes of the four battalions of soldiers raised by his grandfather.

FORMER SOLDIERS THE KING'S (LIVERPOOL REGIMENT)

17th Battalion

Brigadier-General the Hon F C Stanley DSO, CMG 17th Battalion; Lieutenant-Colonel B C Fairfax; Lieutenant-Colonel G Rollo, DSO and Bar; Lieutenant E W Willmer; 15971 Lance-Corporal H Foster; 50752 Private J Grogan; 16006 Sergeant S Harris; 15910 CQMS E R W Potter; 15406 Private I Smith; 15917 Private J A Tudor; 21646 Private C J Wright.

18th Battalion

Lieutenant-Colonel E H Trotter, DSO; Lieutenant-Colonel W R Pinwill, Captain G Ravenscroft; 16224 Private T H Brown; 16813 Private H Foggo; 17212 Private W Gregory; 17054 Private G Groome; 21915 Lance-Corporal W Heyes; 17059 Private G E Jones; 300087 Private W R McLeish; 16319 Private J F Mills; 16197 Private J Oakley; 16400 Private S R Steele; 21913 Private J Stockley; 24644 Private W Stubbs.

19th Battalion

21515 Private C S Hawnaur; 25575 Private W B Owens; 17518 Corporal E G Williams; 17519 Private T S Williams; 17993 Private F Winn;

20th Battalion

22189 Lance-Corporal A S Askew; 22080 Private R J Fleetwood; 21956 Corporal G E Hemingway; 26540 Lance-Corporal J Quinn; 25561 Private H Redhead.

ROYAL FIELD ARTILLERY

Lieutenant-Colonel N Fraser-Tytler.

ROYAL GARRISON ARTILLERY

292261 Gunner T P Brennan.

THE RIFLE BRIGADE

360306 Private W Hill, 8th Battalion.

THE ROYAL ARMY CHAPLAINS' DEPT

The Reverend A H Balleine CF.

RESEARCH TEAM

Especial thanks are offered to the following six friends, who have worked tirelessly to help me produce this book. Without their help, it would never have been completed, at least not this decade! It is in all ways odious that their names have to come in any set order, as they were all part of a willing and impromptu team. Thus, I thank them in alphabetical order, although their massive contributions have been in all ways equal, each in his own different way. This book is as much theirs as mine! PAUL BYRNE has undertaken long and tiring research in many libraries to discover dates of births of officer casualties, and reports of deaths in local newspapers. He has also offered his encouragement over many pints of beer in our local Rugby Club and secured valuable reference works for me to study. FRANK DAVIES has also undertaken many hours of research in the Public Record Office and accompanied me on research trips to the Western Front, offering his enthusiasm and encouragement over many glasses of French beer. TOM McDONOUGH has undertaken local research in Liverpool libraries on so many occasions and turned up vital information amounting to many pages of photocopy. He has also used his encyclopaedic knowledge of Liverpool to do much research and walk many miles looking for the family graves of former Pals, and helped Frank and myself on our trips to the Western Front, both with encouragement and beer. DEREK SHEARD has given me support and friendship over many years, in England and on the Western Front, and has been a great help with his constructive criticism of the written word. He has also drawn accurate maps and diagrams, compiled the 'Time Line' Appendix, and has always been there when I needed help. CHAS THOMPSON has not only helped with the processing of the many photographs included in the book, but has entered on computer the records of some 4,000 original Pals, and nearly 3,000 casualties (the later list has been included in Appendix II). PETER THRELFALL has tirelessly worked for the past five years to uncover relatives of former Pals, and any references, past or present, to Pals in the press. Like Tom, he has spent many, many hours scouring local cemeteries, and also many more on the Western Front. Many of the original postcards used are from his extensive collection. Perhaps his most unbelievable achievement, however, has been to scour all the cemetery registers of all the graveyards and memorials looked after by the Commonwealth War Graves Commission to find and record the burial sites or memorial locations of nearly 3,000 Pals casualties.

I also offer thanks to all of the following: Miss J P ADAMS of the Lord Chancellor's Department; Mrs Maureen ARTHUR for Grantham research; Tim ASHLEY; Pauline ATKINSON; Stuart BARR; Miss Dorothy BELLAMY; Miss A BIRD; Robbie BONNER; Mary BOSWORTH; Bill BLUNDELL; Joy BRATHERTON; Josephine BROWN; Ceri BYRNE; Bob CARTER; Alex CHADWICK; Jean CLARK; Ann CLAYTON; Annette CLIFFORD; Edith COOK; The late Rose COOMBS; Jean CRESSWELL; Peter CROSBY; Geoffrey CRUMP; Ben CUMMINGS; Richard DAGLISH; Ken DAVIES of Vauxhall Motors; Richard DA VIES; Mrs Rosa M E DERRICK; Mrs M DRAPER; Tom and Janet FAIRGRIEVE; Keith FORD; George and Ian of FOTOKAM, Birkenhead; Ian FOULDS; Margaret FINN; Julian FINCH; Mr A E EVANS; David EVANS of the Liverpool Scottish; Major J McQHALLAM; Graham GATHERCOLE; Hal GIBLIN; John GILL; John GIMBLETT of the Commonwealth War Graves Commission; Mr J S GLEN of The Liverpool Cotton Association; Mrs J GRIFFITHS; Mr R A HAMILTON; Jane HARROP; Mick HEHIR; Betty HOGARTH; Ian HOOK; Malcolm HOPSON; Mr N HOUGHTON; Ken HUGHES; Margaret HUGHES; Colin JACKSON of C J Studios, New Ferry; Adrian JERVIS of Merseyside Museums; Majorie E JONES; Rhys JONES; Ken LAMBERT; Mrs J F LANE; Chris LAWLER; Peter LIDDLE; Roger LEAT; Annie LEWIS; Albert LOVELADY; John MADDOCKS for encouragement and proofreading; James MAGGS; the late Colin McCRERY; Steve McGINTY; Tom McKEAVENEY; Helen McMAHON; Steve MANNION of Merseyside Museums; Laurie MILNER; Graham MONEY; Mr A MOORE; Tony MOORE of The King's Regimental Association; Jim and Neil MURPHY; David NASH; Bill NELSON; Brian OULTON; Miss G E OWENS; Alan PAWSON for the loan of postcards; Alf PEACOCK; Major J H PETERS; Miss Eileen QUINN; Miss D P RICHARDS; George ROBERTS; Graham ROBERTS; Charlie ROSE of Grantham; Peter SAUNDERS *of the Liverpool Daily Post*; Jeff SCULLY; Sheena SHANAHAN; Arthur STIRRUP; Mr G G STEPHENSON; Julian SYKES; Bill TAGG of Liverpool Militaria; Eileen TAPERT; Ron TAYLOR of Bickerstaffe; Ken TOFT, bookbinder; Marie TOOHER; Miss Nicola J WHATTON, of Grantham; Miss Florence M WHITE, of Grantham; George WILSON and Iris YOUNG of the Derby Estate Library.

LIBRARIES
J HILTON of the Atkinson Library, Southport; Richard SHAW, Local History Librarian of Battersea Library; Roger DIXON, Irish & Local Studies Librarian, Belfast Public Libraries; Miss R BROWN, Principal Librarian/Archivist Borders Regional Library; Mrs Janet GRIFFITHS of the Chester Local History Library (Mormon Church); The Reference Librarian, Chester Public Library; Marian O'DWYER,

Librarian Cork County Library; J G EVANS, Director of Education, Derbyshire Library Service; Area Archivist, Dolgellau Area Record Office; Gavin WRIGHT County Library Dover Division; Mr P HOWARD of Dublin Public Library; E A HOLLINGDALE, Team Librarian Eastbourne Central Library; Mary HOBAN, Reference Librarian at Hartlepool Public Library; B K C SCOTT, Team Librarian, Hastings Central Library; Miss P MARTIN of Humberside Libraries and Arts Unit; Mrs S M CAWS of Isle of Wight County Reference Library; B R CURLE, Local Studies Librarian of Kensington & Chelsea Central Library; Mrs A HEAP, Local History Librarian at Leeds City Library; Mr J P SELLARS, Librarian, Luton Central Library; The Librarian of Minchinhampton Library; Dianne SPENCE, Branch Librarian of Morley Branch Library, Leeds; Mena WILLIAMS, District Librarian of Preston District Central Library; Mr G MOORE of the North East of Scotland Library Service; Miss I J HUCKSTEP of Ramsgate Library; Miss N BEALES of Sale Library; Tim ASHWORTH of Salford Local History Library; The Librarian, St John's Wood Library; R A PRESTON of Shrewsbury Library Local Studies Department; Sue BATES of the Solihull Central Library; D B TIMMS, District Librarian, St Annes-on-Sea District Library; Mrs V L Hainsworth of St Helens Library; Howard BLACK of Stratford Reference Library; Malcolm HOLMES, Local Studies Librarian of Swiss Cottage Library; Local History Librarian of Taunton Local History Library; Area Librarian of West Devon Area Central Library; Miss M SANDERS of Worcester City Library; Mrs M P MACDOUGAL of Worthing Divisional Library.

INSTITUTIONS & ORGANISATIONS

I offer many thanks to the staff of the Commonwealth War Graves Commission, both in Maidenhead, Berkshire, and Beurains, near Arras, France, for the willing and patient help both have given, in tracing the burial and memorial places of Pals all over Europe. Grateful thanks are also offered to the Commission's gardening staff in many countries, for tending the graves of fallen Pals so beautifully and so well. I am sure that the men would be proud to know how well they are looked after! *Many thanks, also, to the following:* Liverpool City Council; The Imperial War Museum; The Taylor Picture Library, Barnsley; The Liddle Collection, The Library, The University of Leeds; The Public Record Office; The Royal Air Force Museum; 208 Battery, 103 Air Defence Regiment, Royal Artillery (Volunteers); Secretariat D'Etat Charge des Anciens Combattants, France; Volksbund Deutsche Kriegsgrberfrsorgee.V., Germany.

Foreword

With the Great War now over 70 years ago, it is inevitable that the story of the Pals Battalions is but a distant memory in the minds of people. It is an excellent thing that their history should be written as a tribute to the great patriotism and loyalty of the citizens of Liverpool at that time.

My Grandfather's call to form the Pals Battalions was answered by quite a staggering enthusiasm and the numbers far surpassed expectations. That the cream of Liverpool's sons should have enlisted in such numbers was his theme for recruiting over the rest of the County. He became known as the army's best recruiting 'Sergeant'.

THE RIGHT HONOURABLE THE EARL OF DERBY

Bibliography

PUBLISHED SOURCES

Allison J E *The Mersey Estuary* (Liverpool University Press, 1949); Babington A *For the Sake of Example: Capital Courts Martial 1914–18; The Truth* (Leo Cooper, 1983); Baker A *Battle Honours of the British and Commonwealth Armies* (Ian Allan Ltd, Shepperton, Surrey, 1986); Banks A *A Military Atlas of The First World War* (Heinemann, 1973); Bagley J J *The Earls of Derby 1485 –1985* (Sidgwick and Jackson, 1985); Becke A F *Major Order of Battle of Divisions* (four volumes) (HMSO1935–1945) (republished by Sherwood Press [Nottingham] 1987–1989); Berton P *Vimy* (McClelland and Stewart, Toronto, Canada, 1986); Bishop W A *Winged Warfare: Hunting the Huns in the Air* (Hodder & Stoughton, 1918); Boumphrey I & M *Yesterday's Wirral – Wallasey and NewBrighton* (Boumphrey, 1986); Bradley J *Allied Intervention in Russia 1917–1920* (Weidenfeld & Nicolson, 1968); Brice B *The Battle Book of Ypres* (republished by Spa Books with Tom Donovan, Military Books, Stevenage, Herts, 1987); Buchan J *The Battle of the Somme—First Phase* (Nelson, 1927); Bullock D L *Allenby's War: The Palestine-Arabian Campaigns 1916–1918* (Blandford Press, 1988); Brown M *Tommy Goes To War* (Dent & Sons, 1978); Burrows J W *The Essex Regiment: 2nd Battalion (56th) (Pompadours)* (Burrows, Southend-on-Sea, 1937); Chandler D (Editor) *A Guide to the Battlefields of Europe* (Hugh Evelyn Ltd, 1965); Chappell M *British Battle Insignia (1): 1914–18* (Osprey Publishing, 1986); Chappell M *British Infantry Equipments 1808–1908* (Osprey Publishing, 1980); Chappell M *British Infantry Equipments 1908–1980* (Osprey Publishing, 1980); Churchill W C *The World Crisis: The Aftermath* (Thornton Butterworth Ltd, 1929); Clark A *The Donkeys* (Hutchinson, 1961);Cooksey J *Pals: The 13th and 14th (Service) Battalions (Barnsley) The York & Lancaster Regiment* (Wharncliffe Publishing Ltd, Barnsley, 1986); Coombs R E B *Before Endeavours Fade* (Battle of Britain Prints International Ltd, 1976); Coop J O, The Reverend, *The Story Of The 55th (West Lancashire) Division (Daily Post,* Liverpool, 1919); Conan Doyle Sir A *The British Campaigns in Europe 1914–1918* (Bles, 1928); Cox & Company (Compilers) *List of Officers Taken Prisoner in the Various Theatres of War Between August 1914, and November 1918* (London Stamp Exchange, 1988); Edmonds Sir J E *Military Operations of the British Army in the Western Theatre of War in 1914–1918* (sixteen volumes) (HMSO 1922–1949); Edwin Gibson T A and Kingsley Ward G. *Courage Remembered* (HMSO 1989);Ellis J Eye *Deep in Hell* (Fontana, 1976); Enser A G S A *Subject Bibliography of the First World War* (Andre Deutsch 1979); Eyre G E M *Somme Harvest* (Jarrolds, 1938); Falkenhayn E G S, General von, *General Headquarters, 1914–1916, and Its Critical Decisions* (Hutchinson, 1919); Farrar-Hockley A H *Death of an Army* (Barker, 1967); Farrar-Hockley A H *Goughie* (Hart-Davis, MacGibbon, 1975); Farrar-Hockley A H *The Somme* (Batsford, 1964); Fosten D S V & Marrion R J *The British Army 1914–18* (Osprey Publishing, 1978); Fraser-Tytler N, Lieutenant-Colonel *Field Guns in France* (Hutchinson, 1922); Gaylor J *Military Badge Collecting* (Leo Cooper, 1971); Gibbs P *The Battles of the Somme* (William Heinemann, 1917); Gibson & Oldfield *City (Sheffield) Battalion, The 12th (Service) Battalion The York and Lancaster Regiment* (Wharncliffe Publishing Ltd, 1988); Giles J *The Somme: Then and Now* (Bailey, Folkestone, Kent, 1977); Giles J *The Ypres Salient: Flanders Then and Now* (Picardy Publishing Ltd, 1979); Gliddon G *When The Barrage Lifts* (Gliddon Books, Norwich, Norfolk 1987); Hankey M P A *The Supreme Command, 1914–1918* (Allen & Unwin, 1961); Hogg IV & Thurston L F *British Artillery Weapons and Ammunition 1914–1918* (Ian Allan, 1972); James E A *British Regiments 1914–1918* (Samson Books, 1978); James E A *A Record of the Battles and Engagements of the British Armies in France and Flanders, 1914–1918* (reprinted by the London Stamp Exchange, 1990); Kipling R *Rudyard Kipling's Verse: Definitive Edition* (Hodder & Stoughton, 1940); Lloyd George D *War Memoirs - Volume II* (Odhams Press Ltd, 1936); London Stamp Exchange Ltd (Publishers) *Histories of Two Hundred and Fifty-One Divisions of the German Army Which Participated in the War (1914–1918)* (London Stamp Exchange, 1989); Macdonald L *They Called lt Passchendaele* (Michael Joseph, 1978); Macvicar C A, The Reverend (Compiler) *Memorials of Old Birkonians Who Fell in the*

Great War 1914–1918 (Henry Young & Sons, Liverpool, 1920); MacGill P *The Red Horizon* (Jenkins, 1916); McGilchrist A M *The Liverpool Scottish 1900–1919* (Young & Sons, Liverpool, 1930); Midwinter E *Old Liverpool* (David & Charles, Newton Abbot, 1971); Middlebrook M *The First Day On The Somme* (Allen Lane, London, 1971); Middlebrook M *The Kaiser's Battle 21 March 1918: The First Day of the German Spring Offensive* (Allen Lane, London, 1978); Ministry of Pensions (Compiler) *Location of Hospitals and Casualty Clearing Stations, British Expeditionary Force 1914–1919* (Ministry of Pensions, 1923); Muir R *A History of Liverpool* (Williams & Northgate, 1907); Nash D B *German Artillery 1914–1918* (Almark Publishing, 1970); Nash D B *German Infantry 1914–1918* (Almark Publishing, 1971); Nash D B *Imperial German Army Handbook 1914–1918* (Ian Allan 1980); Norman T *The Hell They Called High Wood* (Kimber, 1984); Nicholls J *Cheerful Sacrifice – The Battle of Arras 1917* (Leo Cooper, London, 1990); Pearson C A (Publishers) *The Western Front: Then and Now* (C Arthur Pearson Ltd, 1928); Putkowski J & Sykes J *Shot At Dawn* (Wharncliffe Publishing Ltd, Barnsley, 1989); Purves A A *The Medals & Orders of the Great War* (J B Hayward & Son, 1975); Rankin R H *Helmets and Headdress of the Imperial German Army 1870 –1918* (Flayderman, New Milford, Connecticut, USA, 1965); Roberts E H G *The Study of the 9th King's in France* (Northern, Liverpool, 1922); Simkins P *Kitchener's Army: The Raising of the New Armies, 1914–16* (Manchester University Press, 1988); Slowe P & Woods R *Fields of Death* (1986 Standard Art Book Company, Ltd); *The Roll of Honour* (The Standard Art Book Company, Ltd); Stanley F C *The History of the 89th Brigade 1914–1918* (*DailyPost*, Liverpool, 1919); Stephens F J & Maddocks G J T*he Uniforms and Organisation of the Imperial German Army 1900–1918* (Almark Publishing, 1975); Stephens P (Publishers) *British Vessels Lost at Sea, 1914–1918* (Patrick Stephens, Cambridge, 1977);Talbot-Booth E C Lieutenant *The British Army, Its History, Customs, Traditions and Uniforms* (Colonn & Company, 1940); Tapert A *Despatches From The Heart* (Hamish Hamilton, 1984); Taylor P and Cupper P Gallipoli, *A Battlefield Guide* (Kangaroo Press, Kenthurst, Australia, 1989); Thompson E (Editor) *Liverpool's Scroll Of Fame* (Quills, Liverpool, 1920); *The Times* (Newspaper) *The Times Diary and Index of the War, 1914–1918* (*The Times*, London, 1923, reprinted by J B Hayward & Son, 1985); Turner W *Pals: The 11th (Service) Battalion (Accrington) The East Lancashire Regiment* (Wharncliffe Publishing Ltd, Barnsley, 1987); Westlake R *The Territorial Battalions* (Spellmount, Tunbridge Wells, 1986); Windrow M & Mason F K *A Concise Dictionary of Military Biography* (Purnell Book Services 1975); Winter D *Death's Men: Soldiers of the Great War* (Allen Lane, London, 1978); Wolff L *In Flanders Fields: The 1917 Campaign* (Longmans, Green and Company Ltd, 1958); Wurtzberg C E T*he History of the 2/6th Battalion, The King's (Liverpool Regiment) 1914–1919* (Gale & Polden, Aldershot, 1920); Wyrall E *The History of the Kings Regiment (Liverpool) 1914–1919* (three volumes) (Edward Arnold & Co., 1928)

UNPUBLISHED SOURCES
Heywood R P *Rough Diary of Day to Day Events in France and Flanders Throughout the War*; Quinn J, Lance-Corporal *Letters from the Front* (published privately, for the family, Liverpool, 1918); Battalion War Diaries of the 17th, 18th, 19th and 20th Battalions of the King's (Liverpool Regiment).

NEWSPAPERS AND JOURNALS: PAST AND PRESENT
Belfast Evening Telegraph; Belfast Northern Whig; Berwickshire Advertiser; Birkenhead & Cheshire Advertiser; Birkenhead News; Barrows Worcester Journal; Bystander; Cheshire Observer; Crosby Advertiser; Crosby Herald; Dover Express; Derbyshire Times; Eastbourne Chronicle; Eastbourne Gazette; East Kent Times; Eccles Journal; Evening Standard; Grantham Journal; Gun Fire; Hansard; Kelso Chronicle; Kensington News & West London Times; The Kingsman; Liverpool Courier; Liverpool Daily Post; Liverpool Daily Post & Mercury; Liverpool Echo; London Gazette; Luton News; Morley Advertiser; Northern Daily Mail; Die Nurnberg Nachrichten; Ormskirk Advertiser; Preston Guardian; Salisbury & Winchester Journal; St Marylebone & Paddington Record; Shrewsbury Chronicle; Somerset County Gazette; Stand To!; Stroud News; The Times; Wallasey Chronicle; Wallasey News; The War Illustrated; Western Evening News; Wirral Globe.

Chapter One

'Nothing would ever be the same again'

Liverpool in 1914

The Liverpool of early summer 1914 was a very different place from that of today. It was prosperous, proud, alive with trade and bustling with people, engaged in all types of commerce. Some 30,000 dock workers along seven miles of docks were kept busy loading or unloading ships of all nationalities and cargoes of all types. The River Mersey itself was just as full of ships awaiting either the turn of the tide, or their own turn to dock on either side of the river, and the arterial routes leading away from the seaport were similarly bristling with raw materials for Britain's factories, or finished goods destined for the consumers of Empire and New World. The people of the Mersey themselves were a curious blend of nationalities, seamen passing through, indigenous descendants of Anglo-Saxon, Norse and Celt, and the large ethnically Irish population, whose ancestors had come to the port either in search of work, or on the first stage of the trail of emigration to America or Australia.

Merseyside's prosperity had originally been founded on three main commodities, manufactured goods, cotton and slaves. Vast fortunes had been made in the eighteenth century and the early years of the nineteenth by entrepreneurial merchants who were prepared to risk capital on the long and arduous routes of the triangular trade. This started in Liverpool and then moved to Africa, where slaves could simply be taken, or bartered for very cheaply in exchange for manufactured goods such as cooking utensils and tools. The unfortunates were then transported to the West Indies or the United States, where they were sold for high profit, many of them to work in the cotton plantations. The profit gained was then used to buy sugar, molasses, rum and raw cotton, which was taken back to Liverpool in the now empty ships. To give some idea of the huge profits that could be made, in just eleven years at the end of the eighteenth century, 303,737 slaves brought in £15,186,850 for the slave traders. Even though this was a gross figure, it does not take into account the double crop of profits that would have been realised when the return cargoes were sold in England.

By this stage in British economic history, the textile factories of Lancashire and Yorkshire were clamouring for all the raw cotton they could get. The new inventions of the Industrial Revolution, and the damp Pennine climate, so suitable for spinning and weaving cotton yarn, made Liverpool the obvious port through which cotton and vast profits would pass. By the time that the last slave ship left Liverpool on 1 May 1807, the markets were well established and Britain's growing empire had created its own consumers. Because the Industrial Revolution had begun in Britain at least fifty years before it began to develop anywhere else, Liverpool's geographical position as the nearest large port to the Americas gave it an incredible advantage.

By 1900 British ships carried more of the world's goods than all the other maritime nations put together, and a large proportion of all world trade passed through Liverpool. Cargoes of raw products vital to the nation's prosperity were handled on Merseyside, from cotton, wheat and oil seed, to timber, wool and fruit. In return, manufactured goods of all kinds left the river on journeys to all the world's markets.

By 1914, not only did Liverpool ships carry the vast proportion of goods to and from the United States, but the growing transatlantic passenger trade also used the Mersey as well. Two of the major companies in this field were the well known Cunard

The slave trader Mary *out of Liverpool, being intercepted by two British warships whilst illegally carrying negro slaves in 1806.*

Line and White Star Line, (owners of the ill-fated *Titanic*), both based in Liverpool.

By the end of the nineteenth century all this trade and prosperity had created a new generation of middle-class businessmen, who were tough, efficient and self-reliant. Apart from establishing lots of small independent family businesses, to Liverpool's obvious commercial advantage, they had educated their sons well in order that they would carry on the concerns and continue to make them more prosperous. Many of these businesses would, however, evaporate after 1918, when there were no sons left to carry them on.

Towards the end of the nineteenth century, these people began the move out of the centre of Liverpool, to establish vast residential areas on the city outskirts. A highly efficient ferry service, (with a history dating back to 1330), and a rail link under the Mersey first opened in 1886, and electrified in 1903, made possible a fairly speedy crossing to the city. As a result, many decided to make their homes on the Wirral peninsula, and Birkenhead and Wallasey also grew rapidly in size, initially as dormitory towns, but later in their own rights, as manufacturing and commercial areas. The world-famous ship building firm of Cammell Laird & Co., for instance, was based in Birkenhead, and not only provided jobs for thousands of workers, but also well built ships for Britain's Royal Navy and Mercantile Marine.

There was still, however, the grinding poverty and horrific slum conditions in Merseyside, which existed in any industrial city in post-Victorian Britain. Many lived in squalor, just above, and frequently just below, the breadline, in an age when there was only charity or religion to alleviate their social distress.

Wallasey ferry boats at the Landing Stage just before the Great War.

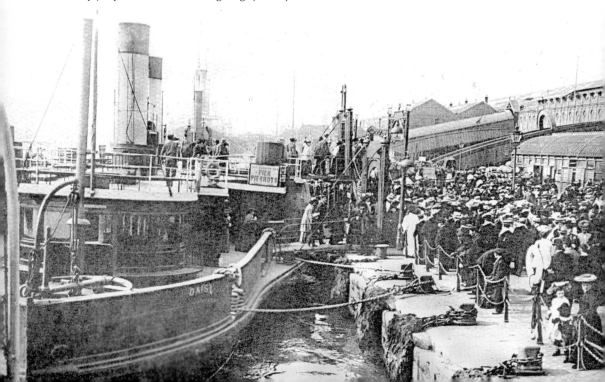

Hamilton Square Birkenhead before the outbreak of war.

Perhaps, however, a greater pride in the community, and a closer identification with the achievements of the region than is seen today helped make these terrible conditions seem more bearable.

Entertainments were different from today, also, and most people were content with what they made for themselves. With public houses open all hours, trade was fairly brisk, and without doubt, there were some main thoroughfares in the city where there literally was a pub on every corner. Although the cinema was beginning to creep into the range of popular entertainments, the music hall still provided fairly good

Central Underground Station Birkenhead, boasting 'Electric Trains Every Few Minutes, under the Mersey to Liverpool.

Church Street Liverpool, the main shopping centre of the city, in about 1900.

quality light entertainment, and those desiring more serious or more dramatic productions could visit one of the many theatres in the area.

For those seeking more cultural activities, the art gallery, the library or the museum, all established in 1851, by a special Parliamentary Act, and all housed in magnificent buildings, offered free exhibitions, the wealth of which were rarely seen outside the metropolis. The very city centre buildings themselves, including the magnificent St. George's Hall, completed in 1854, which would play a part in the Pals' story, reflected the confidence, prosperity and wealth of the city before the war. The growing popularity of football, both as a player or a spectator sport, had already made its mark on Merseyside by that time, Everton Football Club having been founded in 1878, and Liverpool Football Club following in 1892.

St George's Hall, Liverpool, where the first recruits 'took the King's shilling', on 31 August 1914.

Holiday makers on the golden sands of New Brighton, enjoying one of the last summers of peace. New Brighton Tower, in the background, was 100 feet taller than the one at Blackpool.

Over on the Wirral shore more entertainments could be found. In 1914 there were seven ferry landing stages, providing frequent, quick and cheap crossings to and from Liverpool. In the summer months an average of three million crossed the river to visit the seaside resorts of New Brighton, Hoylake and West Kirby. A further eight million came by rail from further afield to swell the throng of holidaymakers and day trippers. New Brighton itself was a very popular resort, with a long promenade, a menagerie, a circus, several funfairs, and a tower of its own, which, at 621 feet high, was 100 feet taller than the more famous Blackpool Tower. Holidaymakers looking across the Mersey would not merely have seen the passage of the huge transatlantic liners, and the smaller cargo ships, but after 1911, the Liver Building.

This magnificent structure, famous the world over, was completed in that year, its twin towers topped by twin bronze Liver Birds, the mythical cormorants which are always associated with Liverpool. The huge clock on the front tower, was first set in motion on 22 June 1911 at the exact time that George V was crowned king. The building itself was opened on 19 July. The Liver Building, perhaps more than anything else, would become a symbol of home for the

The opening to the Albert Dock, Liverpool, with the Liver Building in the background.

thousands of Mersey seamen and soldiers who would pass that way through the next four years of turmoil.

The eve of war

Few people in Merseyside could have realised at the end of June 1914, that the assassination of the heir to the Austrian throne and his wife by a gang of Serbian fanatics, could have led to the obscenity of a world war that would change their lives for ever. Merseysiders were perhaps more tolerant of foreigners than most British people, if only because they came into contact with more foreigners than most, but nevertheless they felt themselves to be superior to the rest of the world. It wasn't that they hated foreigners, rather that they felt slightly sorry for them, and regarded them with the same kind of benign tolerance usually reserved for children. They also believed that the British way of life was the most civilised that the world had ever seen, and was the model for the rest to follow. That the murder of Archduke Franz Ferdinand at Sarajevo would end British and European primacy, perhaps for ever would have been a preposterous thought – even had anyone thought of it!

The first weekend in August 1914, and the last weekend of peace for over four years was unusually hot. Monday 3 August was Bank Holiday Monday, and although the businesses in the city had begun to realise that war was not only likely but imminent, (the Bank Rate had shot up to an almost unheard of 10%, the highest figure since May 1866), most people had no inkling of what lay ahead. Those who had even contemplated the possibility of war were not unduly concerned by the thought, as Britain was at her peak, and any enemy would surely be dispatched with consummate speed. The thoughts in the minds of most dwelt simply on the holiday ahead, and the bonus of glorious weather. For many, the Bank Holiday would be spent on the sea shores of the Wirral, or picnicking in one of the many parks and open spaces provided by far-seeing councils.

New Brighton before the war. In the background is Fort Perch Rock, built to deter sudden attacks on the Mersey. It only fired its guns 'in anger' twice, on the first day of World War I and on the first day of World War II. Both were false alarms.

For those seeking ready-made entertainment, however, the Liverpool 'Empire' theatre, which promised 'a cool place for real enjoyment', was offering 'straight from the London Palladium', light entertainment entitled 'Dora's Doze'. The 'Royal Court' offered 'The Arcadians', which it advertised as 'a musical comedy, always merry and bright', whilst at the 'Royal Hippodrome', Jimmy Learmouth, billed as 'The Inimitable Comedian', was starring, with his famous sketch featuring 'Colonel Cobb the Red Hussar'. Jimmy had just taken over top billing from the previous week's star, the unlikely 'Linga-singh, the Mysterious Hindoo Sorcerer'! The cinema, in its infancy as an art form, was represented that weekend, in several theatres, including the Lime Street Picture House, which was showing 'The Crucible of Faith', billed as 'A

Lord Street and Church Street, Liverpool.

Vitagraph Drama', and 'The Hall, – A Keystone Comedy'.

Across the River Mersey, the famous 'Argyle Theatre' in Birkenhead was featuring 'Bert Le Mont – The Singing Comedian', whilst 'The Tivoli' in New Brighton offered different types of 'High Class Vaudeville'. Further down the Wirral Peninsula, Hooton Park Racecourse advertised its Bank Holiday race meeting, with the cheapest entry to the course being offered at one shilling, (five new pence). Curiously, this was the exact amount that could be earned for one day's service by a Private in the infantry of 1914. More curious still, well over a thousand of the weekend's holidaymakers would, within six weeks, get free and continuous entry into Hooton Park for a very different reason, and also earn one shilling a day!

Cruises from the River Mersey that weekend would take the holidaymaker on a day trip to Llandudno for three and sixpence, (17 new pence), or to the Isle of Man for four and sixpence steerage class, (22 new pence), or six and sixpence saloon class, (32 new pence). Anyone who wished to go further afield for a longer holiday could have have sailed to the Canaries with the Elder Dempster Line for only fifteen guineas (£15.75).

However, despite the obvious signs of public enjoyment that Monday, there were other obvious signs of a more foreboding nature, should anyone have wished to look for them.

> *I got the quarter to midnight train to Preston, and I had the heck of a job to get on the station at Liverpool, – this was on the Monday and war hadn't been declared, – August Bank Holiday. It was packed with Navy reservists who had been called up, and they were all saying goodbye to their wives and sweethearts, and we had to push our way in, even to get on our own train! There were three long trains there, and from what I could gather, there was one for Portsmouth, one for Falmouth and one for Chatham. All the Navy's reservists had been called up, and that was the day before they declared war.* **360306 Private W. Hill, 8th Bn. The Rifle Brigade**

August and after

At 11.00 p.m. on 4 August Great Britain declared war on Germany and the whole machinery of Empire began to turn itself, albeit at first without direction, towards the task in hand. Earlier on the previous day, 'The Liverpool Daily Post & Mercury' had published an advertisement on its front page informing all Cammell Laird's workers that their traditional August week's holiday had been cancelled, 'owing to urgent work', and that

Magazine Promenade, New Brighton, clearly showing the tower in the background. Neglected through the war years, it was declared unsafe when peace came, and was pulled down in the early 1920s.

they should return to work the following morning. Part of that 'urgent work' would be to help build over 30 warships for the Royal Navy during the course of the war. The holiday was truly over!

"War was declared on 4 August 1914. This was the Bank Holiday weekend, which was then on the first Monday in August, not the last, as now, and I was away on holiday in Wales. Since the assassination of an Archduke, in a place called Sarajevo, events had moved swiftly, and with dark threats, but to many young men, – and I was one – these threats seemed remote and not likely to affect me personally, though I had begun to feel concerned when I went away that fateful weekend. When I returned to Birkenhead, the full impact hit me. Many of my friends had already joined the forces, and it quickly came to me that this was my war, and that I must get into it as soon as possible. I have found out since, that I was one of the vast majority who did not perceive the full impact of the war as soon as I should have done. But I do not think it is surprising, as for almost a hundred years, Britain had been at peace." Lieutenant E. W. Willmer 17th Bn. K.L.R.

Despite its more evident maritime role of crewing many of the ships of Britain's Mercantile Marine, Merseyside was nevertheless well supplied with volunteer soldiers, especially infantrymen. There was a long tradition of voluntary service in the city itself, dating back to the Napoleonic Wars. During the international crisis of 1859, for instance, Liverpool was one of the first areas to raise a unit of rifle volunteers, and this, The 1st Lancashire Rifle Volunteer Corps was the forerunner of all the Territorial infantry battalions in Liverpool. By 4 August 1914, the Territorial Force, introduced in 1908, was well represented in the city, by the six Territorial battalions of the King's (Liverpool Regiment).

The 5th Battalion was the direct descendant of the 1st Lancashire Rifle Volunteers, and its members wore blackened badges, in memory of its Rifle Volunteer associations. The 6th Battalion had also been raised in 1859 as the 5th Lancashire Volunteer Rifles, and was still known as The Liverpool Rifles'. Its cap badge was the slung bugle horn traditionally worn by rifle regiments, surmounted by the rose of Lancashire, once again in blackened brass.

Originally rifle regiments had been raised during the forest fighting of the American War of Independence, when muskets were of limited use and accuracy. To help camouflage soldiers amongst the trees, dark green uniforms had been worn, and all badges and buttons had been blackened. As ordinary commands could not be seen or heard, they were given by bugle.

First steps of war: Prussian curassiers entering the Belgian border village of Mouland on their way to Liege, in August 1914.

Belgian lancers on their way to meet the invaders.

As a result, traditionally, most rifle regiments wore blackened badges and buttons, and incorporated a bugle horn in the design of their badges.

The 7th Battalion, first raised in 1860, as the 15th Lancashire Rifle Volunteers, wore the same pattern cap badge as the regular battalions, the White Horse of Hanover, over a scroll labelled 'THE KINGS', but in all white metal. The 8th (Irish) Battalion was raised from men of Irish descent, not a difficult feat in Liverpool, and was the direct descendant of the 64th Lancashire Volunteer Rifles. Its cap badge was the Irish Harp in blackened brass. The 9th Battalion had originally been the 80th Lancashire Rifle Volunteers, and its members also wore a white metal version of the regulars' White Horse. The 10th (Scottish) Battalion had originally been raised in 1900, from men of Scottish descent, and its badge was a white metal version of the regular badge, superimposed on a St. Andrew's cross and a sprig of thistles. All of these battalions had sent volunteers to fight in South Africa during the Boer War.

Just outside Liverpool, but still recruiting (within the area, were the 4th Battalion The South Lancashire Regiment, based at Warrington, and the 5th Battalion, The South Lancashire Regiment, based at St. Helens. Across the River Mersey, in Birkenhead, was the 4th Battalion, The Cheshire Regiment.

Men joining the Territorial Force before the war did so under the terms that they could not be ordered overseas, and were meant as a home defence force. However, if a man was willing to serve overseas if required, he could indicate this by volunteering for Imperial service, in which case he was rewarded with the issue of a white metal badge in the form of an oblong box bearing the words 'IMPERIAL SERVICE', and surmounted by a crown. This was then worn on the right breast. Those who did volunteer for such service were meant, in time of war, to replace regulars on garrison service anywhere in the Empire. They were not, at first anyway, meant to take the field alongside regular troops.

On 5 August, the famous hero of Khartoum, Lord Kitchener, after some persuasion, accepted the position of Secretary of State for War, and at once began to organise recruitment for the war which he at

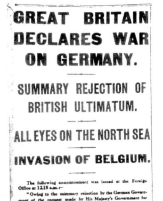

GREAT BRITAIN DECLARES WAR ON GERMANY.

SUMMARY REJECTION OF BRITISH ULTIMATUM.

ALL EYES ON THE NORTH SEA

INVASION OF BELGIUM.

The following announcement was issued at the Foreign Office at 12.15 a.m.:—

"Owing to the summary rejection by the German Government of the request made by His Majesty's Government for assurances that the neutrality of Belgium would be respected, His Majesty's Ambassador in Berlin has received his passports, and His Majesty's Government has declared to the German Government that a state of war exists between Great Britain and Germany as from 11 p.m. on August 4."

Huge crowds in Whitehall and Trafalgar Square greeted the news with round after round of cheers.

11 p.m. London time is midnight Berlin time, the hour at which the British ultimatum expired.

The King held a Council at midnight to sign the proclamation of war.

Great Britain had sent an ultimatum to Germany which

The blackened brass cap badge of the 6th Battalion KLR, sometimes known as The Liverpool Rifles.

The White Horse of Hanover cap badge worn in slightly different forms by the regular battalions of the King's (Liverpool Regiment) and the 5th, 7th and 9th Battalions of the Territorial Force.

The blackened brass cap badge worn by the 8th (Irish) Battalion KLR.

The white metal cap badge of the 10th (Scottish) Battalion KLR.

The brass and white metal cap badge worn by the 4th and 5th Battalions of the South Lancashire Regiment.

The brass and white metal cap badge worn by the 4th Battalion The Cheshire Regiment.

Field Marshal Earl Kitchener of Khartoum, appointed Secretary of State for War, on 5th August 1914.

least knew would not be over by Christmas, despite the popular theory! Kitchener's decision not to call upon the Territorial Force to make up the gaps in the army is well known, and several reasons have been put forward to explain this. One theory is that he had a deep-seated mistrust of part-time soldiers, picked up observing the French during the Franco-Prussian War, and his own side, during the Boer War. Another is that he believed that it would be easier to train men for active service who had no ideas about soldiering at all, rather than to retrain those whose military knowledge might be un-suitable or out of date. (Kitchener's attitude to the Territorial Force is very well chronicled by Peter Simkins in his book 'Kitchener's Army The Raising of the New Armies, 1914–16.')

As a result, although some units of the Territorial Force did go to the Western Front before the end of 1914, most would not arrive there until the following year, when the escalating scale of the war meant that earlier considerations had to be pushed aside. Curiously, the Liverpool Scottish, perhaps one of the more professional, and better trained pre-war units, was one of the first Territorial Force infantry battalions to be sent to Flanders. It arrived in the front line trenches near Kemmel, on 27 November 1914, and thus just failed to qualify for the campaign bar '5th AUG. – 22nd NOV.1914' which was later to be worn on the ribbon of the 1914 Star by men of the original British Expeditionary Force.

However, all this was in the future. On 7 August 1914, Kitchener made his first appeal for 100,000 men to serve in the army for the duration of the war. His famous 'pointing finger' poster, drawn by Alfred Leete, was to follow later. The initial hope was that each existing County infantry regiment would raise one battalion, extra to its Territorial Force commitment, which would be numbered in sequence and designated a 'Service' battalion. Thus, in the city, a newly raised battalion would be called The 11th (Service) Battalion The King's (Liverpool Regiment), as the Liverpool Scottish, the 'youngest' Territorial unit in the regiment, was numbered the 10th Battalion.

On 19 August 1914, in a gesture which well portended the future, Lord Derby wrote a letter which was published in the local press, asking for volunteers for this first Service battalion. By 23 August, the new battalion, based at Seaforth Barracks, had its full quota of recruits, and Lord Kitchener was informed that it was ready for duty. The 11th Battalion was, therefore, the first Service battalion in the country to be raised, and it would serve throughout the war as a pioneer unit. Other local Service battalions were to follow, three more in the King's (Liverpool Regiment), not including the Pals Battalions, six in The South Lancashire Regiment and eight in the Cheshire Regiment.

By this time, the Battles of Mons and Le Cateau had been fought, and despite rousing accounts of the heroism of the army in the national press, most people were able to read between the lines, and realised that the military situation was, at best, quite grave. However, this also brought about a new and sterner mood amongst the young males of the country, and the recruiting offices began to bulge with new volunteers. On 28 August, Kitchener called for his second 100,000 men, and by the end of the month daily recruiting figures had broken all records.

In the main, however, the early recruits for the New Armies were from the agricultural, unemployed, unskilled or semi-skilled workforce, and in the metropolis, and the large industrial cities of the north, there was a huge, as yet untapped work force of better educated people who ran the offices and businesses which contributed greatly to the country's prosperity. They were certainly no less patriotic than those who had rushed to join in the early days of the war, but perhaps because they still saw the profession of soldiering with pre-war eyes, as being rough, course and brutish, or maybe because their lives were more settled and tied in to a regular structured routine, they just needed a different kind of encouragement.

The Pals are born

The 17th Earl of Derby is usually given the credit for the concept of the Pals battalions, and it was certainly he who brought the whole idea into fruition, but it is likely that the original idea came from the War Office in mid-August 1914. Peter Simkins, in his excellent book 'Kitchener's Army', indicates that Sir Henry Rawlinson acting upon Kitchener's instructions, was instrumental in initiating the raising of the 10th (Service) Battalion, The Royal Fusiliers, on 19 August 1914. Although not officially accorded battalion status until 21 August, and then unofficially known as 'The Stockbrokers' Battalion', it was, to all intents and purposes, a Pals battalion, in as much as it was made up from men who worked in the offices of the City of London, and wished to serve together as comrades. This idea was to embody the concept of the Pals battalions, but without doubt, it was from the industrial towns of the north that the idea grew into a massive reality.

Certainly, on 24 August 1914, perhaps encouraged by the success of the Stockbrokers, Kitchener discussed the whole concept with Lord Derby, who gained the former's permission to raise a battalion of men from the business houses of Liverpool. Lord Derby by this stage had earned himself the unnofficial title of 'England's best recruiting sergeant', particularly for his work with the West Lancashire Territorial Association, and by his efforts in raising the 11th (Service) Battalion of the King's. Certainly, the day after his meeting with Kitchener, he informed his brother, Ferdinand Stanley D.S.O., then a Captain in the 3rd Battalion The Grenadier Guards, that he intended to raise a battalion of comrades in Liverpool, with Ferdinand, promoted to the rank of Major, in command. In September 1919, Lord Derby described those early days:-

> "I can claim but little credit for the formation of the 89th Brigade. The desire to serve their country in an hour of need was a predominant feeling amongst Liverpool men, and when I proposed the formation of what came to be known as the Pals' Brigade, I merely voiced the wish expressed to me by many would-be recruits that they should be allowed to serve with their friends. The appeal was, therefore, likely to be a great success before it was even made." The Rt. Hon. Earl of Derby

Derby's basic idea was that men who worked together in the close confines of a business, and who met together socially as 'pals', might well respond to a call to serve together, and if necessary fight together, so long as they were not separated, and made to serve with people of different circumstances whom they did not know, and to whom they could not relate. If the original idea was not Derby's, then he was the perfect innovator, and it was certainly he who coined the title 'Pals', which, though purely unofficial, was to stick thereafter, in the minds of the soldiers themselves, and the public in general

Derby's idea was first put forward in the Liverpool press, on 27 August 1914, and suggested that business people who might wish to join a battalion of comrades, to serve their country together, might care to assemble at the headquarters of the 5th Battalion The King's (Liverpool Regiment), at St. Anne Street,

The Rt. Hon. The Earl of Derby.

at 7.30, the following evening. The Earl also wrote to the larger business institutions explaining the country's need, and suggesting that they encourage their workforce to enlist at once! Obviously at this stage even Derby was unsure of the exact response to his suggestion, but seemingly his call to arms was all the business world of Liverpool needed!

Long before 7.30, on the evening of 28 August, St. Anne's Street was crowded with young eager men trying to get into the drill hall. Those inside found that the hall itself was packed to capacity, and men were standing in the aisles, the doorways and even on the stairs. So great was the crush, that another room below also had to be opened to take all those who wanted to enlist. When Lord Derby arrived and stepped onto the platform to address the multitude, his welcome was tumultuous, and this was only matched by the cheering and the throwing of hats in the air which accompanied the news that Derby's brother Ferdinand was to command the new battalion when it was formed. It was obvious to Lord Derby even then, that there were more than enough men present to form one battalion, and he spoke accordingly.

I am not going to make you a speech of heroics. You have given me your answer, and I can telegraph to Lord Kitchener tonight to say that our second battalion is formed. We have got to see this through to the bitter end and dictate our terms of peace in Berlin if it takes every man and every penny in the country. This should be a Battalion of Pals, a battalion in which friends from the same office will fight shoulder to shoulder for the honour of Britain and the credit of Liverpool. I don't attempt to minimise to you the hardships you will suffer; the risks you will run. I don't ask you to uphold Liverpool's honour; it would be an insult to think that you could do anything but that. But I do thank you from the bottom of my heart for coming here tonight and showing what is the spirit of Liverpool, a spirit that ought to spread through every city and every town in the kingdom. You have given a noble example in thus coming forward. You are certain to give a noble example on the field of battle.

The Rt. Hon. Earl of Derby.

Private J Oakley, 18th Battalion

Thus, Lord Derby, in that short speech, not merely made the first use of the term 'Pals' to what had been previously called a 'Comrades Battalion', he also embodied the whole concept and nature of all the Pals battalions, from all the cities of the north. He then went down to the room below, and repeated his speech, with similar response, inviting all would-be recruits to assemble on the following Monday morning, 31 August 1914, at St. George's Hall on Lime Street, for attestation.

We joined because we were all young chaps together, and we had this 'For God, King and Country' idea, and we meant it, and I think I came out of it the other end still believing it, despite all the adversities, and despite the fact that we knew that nothing would ever be the same again.

16197 Private J. Oakley 18th Bn. K.L.R.

By 8.00 am, on 31 August, the area outside the hall, St. George's Plateau, was even more packed than St. Anne's Street had been, with men waiting patiently to enlist. Anticipating that most men would come from the offices and businesses of Liverpool, separate tables for attestation in the hall were set aside for each of the main areas of commerce in the city. These were:

The Cotton Association, The Corn Trade Association, General Brokers and The Stock Exchange, The Provision Trade, The Seed, Oil and Cake Trade Association, The Sugar Trade. Fruit and Wool Brokers, The Cunard Line, The White Star Line and Steamship Companies, The Timber Trade, The Law Society and Chartered Accountants and Bank and Insurance Offices. Some concerns, like Cunard and The Stock Exchange, actually formed up their men first, and then marched them to St. George's Hall en masse to enlist. A similar gesture was made in Wallasey, across the Mersey, so that men wishing to enlist from the Wirral Peninsula could all arrive together at what was to become Liverpool's largest recruiting office.

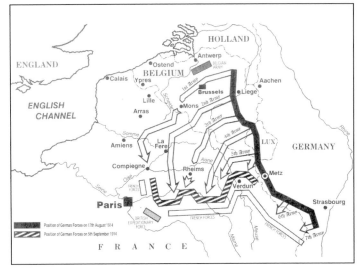

The invasion of Belgium and France.

Some German troop pause on their march towards Brussels to eat a meal of ham and black bread.

In Liverpool, at 9 a.m. on August 31st, the all male office staff of the Liverpool Gas Company assembled as usual for duty in the Duke Street Head Office. Before mid-morning, those of military age (18–35 years), were asked to go to the Board Room on the first floor, as the Chairman, Sir Henry Wade Deacon, wanted to talk to them. The Chairman said that he had received a letter from the Earl of Derby, who had been in touch with the Prime Minister, Mr. Asquith. The Premier had outlined to the Earl the grave military situation, and urged the vital necessity for the recruitment of all men who were willing to enlist in the Army. 'The Gas Company', said Sir Henry, 'would grant leave of absence with half-pay to all those who enlisted'. Lord Derby's plan, explained Sir Henry, was for the en masse recruitment of Liverpool city men into new battalions which would consist of friends or 'pals', with whom they had already been associated in their working lives. Lord Derby sent similar letters to Liverpool's banks, insurance firms, exchanges, shipping and other commercial offices, and no doubt, on that August morning, similar scenes to that at Duke Street were being

The road from Tirlemont to Brussels thronged with refugees fleeing before the rapid German advance.

Answering Lord Kitchener's call: Would-be Pals marching up Dale Street in Liverpool on 31 August 1914, to enlist at St George's Hall.

enacted throughout the city.

The desks being locked, the embryonic warriors left in a steady stream for St. George's Hall, where enlistment was to take place. The final word of farewell was spoken by the Chief Clerk, who, no doubt rather rattled by the prospect of operating gas accountancy with a greatly depleted staff, encouraged the recruits with the remark 'Well, I think you are all going to have a nice holiday'.

At St. George's Hall, it was certainly a man's world, with hundreds of city-garbed young men directed to rooms overlooking St. John's Gardens, rooms in which were clerks ready to take down recruits' personal details, name, address, age, religion and so on; and magistrates with bibles on which each man swore allegiance to the King, his heirs and successors. These formalities over, the next ordeal was in other nearby rooms, where doctors were medically examining each recruit. The men in these rooms were ordered to remove all their clothes, after which the doctors took over. Firstly a long steady visual appraisal, and then a more detailed examination in the usual manner. A slight element of drama was possible at this stage of enlistment. I can recall that whilst awaiting examination, a young man in front of me was visually examined and then told by the doctor 'Sorry, old man, we can't take you, you've got a hernia'. Volubly protesting, the would-be soldier said that his three friends had been accepted, and what could he do about it? 'An operation would probably put it right', said the doctor. This man had the operation, and several weeks later, by one of those odd coincidences with which army life seems to abound, he managed to enlist and be posted to my platoon in which his friends were already becoming competent infantrymen. Later, he was killed!

16006 Sergeant S. Harris 17th Bn. K.L.R.

By 10 o'clock, Lord Derby had passed a full battalion of 1,050 recruits, and decided that as they all had to be medically examined, and processed into the army, he would not be able to take any more that day. Reluctantly, all those waiting outside were told to return on Wednesday 2 September, when the process would begin again. In fact, it continued well after the Wednesday, into the following week, and by the following Monday, Lord Derby had over 3,000 recruits, enough in fact to raise three battalions of Pals.

On the Saturday, September 5th, my brother and I went into Liverpool's St. George's Hall to join the Comrades Battalions being raised by Lord Derby. It was a fine sunny afternoon, and we joined the crowds of naked youths milling around the corridors and rooms of that classical architectured hall, on its plateau overlooking the city and the Mersey and its docks and riversides. Teams of doctors with their stethoscopes and cards for eyesight testing soon passed us as fit. Some of us with our unequally sighted eyes took the opportunity in waiting our turn, to memorise some of the lines of the smaller letters on the card, so as not to give away the fact that in civilian life we wore glasses. It was more difficult for those with flat feet, for these could not be disguised, to the disappointment of those so afflicted.

17518 Corporal E. G. Williams 19th Bn. K.L.R.

The magnificent interior of St George's Hall, scene of much military activity in 1914.

At this stage, he decided to call a halt to the recruiting campaign, for the time being at least, and try to turn the 3,000 men he did have into some kind of efficient military unit. The response from Merseyside had been truly remarkable, – three battalions of men had been raised in just over a week, from an area which had already contributed many men to other Service Battalions, and many more to the Royal Navy and Mercantile Marine. By mid – October, Lord Kitchener gave Derby permission to raise a fourth battalion of Pals, to bring the men so far recruited up to Brigade strength. On 19 October, he utilised once more the city's press, to call for recruits for this fourth battalion, which would serve alongside the other three, which were then undergoing training. By 11 November this battalion was also full, and the excess recruits were formed into two reserve Pals battalions.

Although the four main Pals battalions were officially titled the 17th, 18th, 19th and 20th, Service Battalions of the King's (Liverpool Regiment), they were also called the 1st, 2nd, 3rd and 4th City Battalions of the K.L.R., although they were universally known as 1st Pals, 2nd Pals, etc., etc. The two reserve battalions, designated 21st and 22nd Battalions of the K.L.R., never left Liverpool, but were used as feeder and training units for the other four, for the rest of the war, the 21st serving the 17th and 18th, and the 22nd serving the 19th and 20th.

Above: *Would-be recruits forming up on Exchange Flags, Liverpool before marching to St George's Hall to enlist, on 31 August.*

Below: *Pals still in their civilian clothes, marching down Aigburth Road, Liverpool, led by Lieutenant Colonel F C Stanley, in September 1914.*

Pals recruits marching down Ullet Road, Liverpool, led by Lieutenant Colonel F C Stanley, in September 1914.

Men of the 17th Battalion in their mess hall at the former watch works barracks at Prescot.

Knowsley Hall 1914: Lord and Lady Derby are on the top of the steps, and Lord Derby's brother, Lieutenant Colonel F C Stanley, later Commanding Officer of the Pals Brigade, is seated on the dapple horse.

Chapter Two

'A motley crew, and very unsoldierlike!'

Early days

It was one thing raising four active battalions of troops, with two more in reserve, it was a completely different thing to billet and train them. One of the reasons that the War Office was so keen to allow cities and towns to raise their own battalions was that they would also take initial responsibility for their feeding, clothing, quartering and training. In Liverpool, these resources had already been severely stretched before the Pals concept had become a reality.

Nevertheless, everyone was keen to get the job done, and even on the afternoon of 31 August 1914, after they had just attested, squads of men could be seen setting out for Sefton Park, about three miles away, to begin drilling under the supervision of any of their comrades who had previous experience in the Territorial Force, the Police, or even the Boy Scouts or the Boys Brigade. For the most part, however, the new recruits were told to go home and wait until they were called for duty.

This call was to come fairly quickly, in the form of a postcard from the "Officer Commanding Liverpool Commercial Battalion", at St George's Hall, sent to each recruit, inviting him to parade at the West Lancashire Riding School in Aigburth, at 2.30 pm, on the following Saturday, 5 September 1914. Some 2,400 men out of the nearly 3,000 already enrolled turned up, and after being sorted out into companies within each battalion, they were given some rudiments of drill, were formed up in column of four, and told they had to march to the city, led by the Colonel of the 17th Battalion, Lieutenant Colonel F C Stanley DSO.

Naturally, this march was met with much enthusiasm by the local populace, who cheered every inch of the way from Aigburth to the review stand outside St George's Hall in Lime Street. There, on the platform, to witness this historic march past, was General Sir Henry McKinnon KCB KCVO, the officer responsible for the Western Command area, and other senior military staff. Although Lord Derby himself was not present, due to ill health, Lady Derby was, with her daughter Lady Victoria Stanley. Although

Men of the 19th Battalion training in Sefton Park in September 1914.

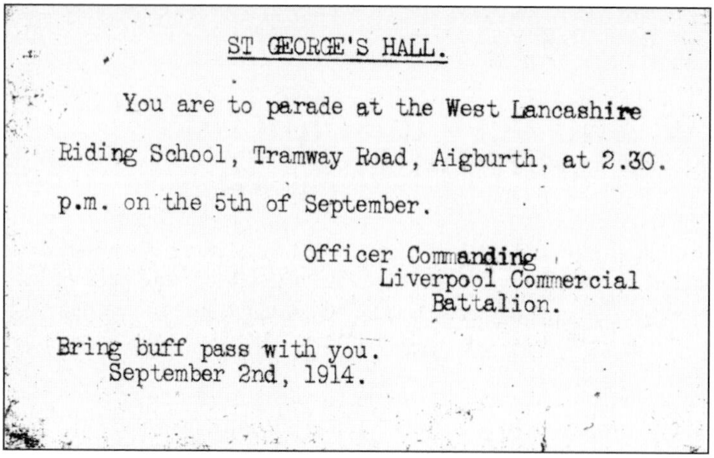

Postcard sent from St George's Hall calling new recruits to train at Aigburth.

ST GEORGE'S HALL.

You are to parade at the West Lancashire Riding School, Tramway Road, Aigburth, at 2.30. p.m. on the 5th of September.

Officer Commanding
Liverpool Commercial
Battalion.

Bring buff pass with you.
September 2nd, 1914.

the march past was generally hailed to be a great success, not everyone was delighted by it!

The battalion was then told that it had to form fours and would march into the City. Following completion of the task of forming the platoons into fours and explaining the meaning of the orders 'Quick March', and 'Halt', the 1st Pals set off for the City. My father, mother and sister watched the march-past in Lime Street. Their verdict: – 'a motley crew and very unsoldierlike'.

16006 Sergeant S Harris 17th Bn KLR

After the march past was over, the recruits were once more sent home to await further instructions, being granted money as a 'living out allowance', until they could be found proper accommodation.

It was at this point that Lieutenant-Colonel Stanley realised that the three battalions so far raised, would need more than just accommodation. Apart from himself, and some recruits with previous military training, there was no one who had much idea at all about soldiering. However, Stanley solved part of this problem quite simply, with a visit to the Orderly Room of his regiment, the Grenadier Guards, in London, and simply recruited six sergeants who had been called back from the Reserve for war service. They were all offered enhanced rank if they joined the Pals, and their transfer proved to be an excellent move, and one which was to give the battalions some much needed backbone.

It was from the Grenadier Guards, also, by tradition, the family regiment of the Stanleys, that the commander of the 18th Battalion was found. Stanley was able at this time, to secure the services of Lieutenant Colonel E H Trotter, DSO, to command the 2nd City Battalion. From the moment that he took command of the Battalion, Trotter endeared himself to his men, and although many accounts of the Great War speak of officers who were adored by their men, in Trotter's case, it was certainly true. As soon as he took command, he began to devote himself to the welfare and training of his charges. Colonel E F Gosset, who had served for many years in The King's, then agreed to take command of the 19th Battalion, and although he was not to see the Battalion go to France, his experience and enthusiasm in the early days helped to mould it into the efficient fighting unit it would become. At this time, the 20th Battalion had still to be raised.

The first problem of accom-

Grenadier Guards sergeants teaching drill to 18th Battalion recruits at Hooton Park Racecourse in 1914.

The barracks of the 17th Battalion in the former Prescot Watch Works in 1914.

modation was solved, in a fashion, following a suggestion by Canon Mitchell, the Vicar of Prescot, a small township not far from the Derby estate at Knowsley. Within Prescot itself was the empty and abandoned remains of an old watch factory. This had been built in 1899 for the Lancashire Watch Company who had tried to establish a British watch-making concern there to rival the Swiss. At its peak, the firm had employed 1,400 people, but it was not a commercial success and had closed in 1906. Canon Mitchell thought that it might make a good barracks for a Territorial unit, and suggested this to Lord Derby, who knew at once who he ought to put there! Following inspection, it was found to be filthy and dilapidated in parts, but sound enough to house a battalion of men. As the 17th Battalion was technically the senior, to it would go the first right of permanent accommodation.

Before anyone moved in, the place was completely whitewashed throughout, the white wash itself being given jointly by the Cunard Line, and The White Star Line, many of whose former employees in the 17th Battalion would benefit personally from the gesture. On Monday 14 September 1914, the 1st City Battalion marched in, and took over, 'A' Company occupying the single-storey building, and 'B', 'C' and 'D' Companies taking over the three-storied block. The total area of the accommodation was 41,048 square feet.

The following day Mr C Dale and Mr A Carruthers, the Chairman and Hon Secretary respectively of the Hooton Park Club Committee offered Lord Derby the use of the racecourse on the Wirral and all its grounds and facilities for training another battalion of troops. The next in 'seniority' was the 18th Battalion, and it was duly allotted this venue for training, and marched in there on 23 September 1914. This still left the 19th Battalion without a 'home', and as Lord Derby realised that there was a very strong possibility, even at this stage, that a fourth Pals battalion might be raised, he decided that a permanent camp would have to be built in the grounds of his own home, Knowsley Park. Until this could be accomplished, however, the men of the 19th Battalion would have to rely on being billeted in their own homes, or homes near to Sefton Park, where they would train each day.

The Prescot Watch Works in August 1914. The building is still standing today as the Prescot Trade Centre, and the Post Office shown in the photograph is now a Chinese take away.

Tented accommodation of the 18th Battalion pitched on Hooton Park Racecourse in the autumn of 1914.

The silver Eagle and Child crest of Lord Derby presented as a cap badge to the original recruits of all four battalions.

The Eagle & Child and other insignia

Whilst the problems of training and equipping over three thousand men for war began to make themselves apparent, the fairly pressing need for uniforms brought forward another problem, that of the badges the Pals would wear.

On the surface, there was no problem at all, because the Pals were Service Battalions of The King's (Liverpool Regiment), whose members had worn the insignia of the White Horse of Hanover since it was first granted to them in 1716 by King George I. Even in 1914, all but three of the Territorial Force battalions of the King's wore the White Horse. However, the unique quality of the Pals was that they had been raised in person by Lord Derby, and his crest, on the family achievement, was well known in Liverpool – an eagle perched on a cradle containing a baby. There were in 1914, as there are today, more than a few public

Hooton Hall, Cheshire, where officers of the 18th Battalion were billeted during training at the racecourse.

houses in the Liverpool area named 'The Eagle and Child', whose inn sign bore the Derby crest. Thus the idea began to take root, that the Derby Battalions ought to wear the Eagle and Child of the Derby family.

The origins of the crest itself are a little obscure, but it was almost certainly derived from a legend involving the Lathoms, a famous Lancashire family in mediaeval times. Separate versions of this legend exist, but all allude to a baby mysteriously 'found' in the grass below an eagle's eyrie, and then accepted as an heir to the Lathom estates. It is probable that the mysterious finding and acceptance of the child was a way of gaining and assuring the line of succession for an illegitimate offspring! Whatever the reason, by the middle of the fourteenth century the eagle and child device was accepted as the Lathom family crest. The Stanley family, later Earls of Derby, adopted it later in the century, following the marriage of Sir John Stanley and Isabel Lathom.

It is not clear who first officially suggested that the Derby crest should become the cap badge of the Pals battalions, but it is probable that it was Lord Derby's brother, Lieutenant-Colonel F C Stanley, himself. The idea was certainly a popular one, however, and realising that its acceptance would need the sanction of the War

Knowsley Hall, the home of the Earls of Derby, where all four battalions trained in 1915.

Office, and King George V himself, it was put forward for approval. On 14 October 1914, Stanley received word from Buckingham Palace that this approval had been granted, largely in appreciation of all Lord Derby's work for recruitment in West Lancashire, and throughout the country.

This was greeted with great delight by the recruits, who now really knew what they already felt, that they were something special. Lieutenant Colonel Stanley actually travelled in person to Prescot, Hooton Park and Sefton Park, where the three battalions were in training, to tell them the good news. (The 20th Battalion [4th City], had still not been formed at this stage.) He also informed them that Lord Derby had claimed the privilege to present each recruit with a solid silver version of the badge, as a memento of the unique occasion.

The reverse of the silver badge presented to 17167 Pte JR Bird, 18th Battalion, which has been made into a brooch to be worn by a lady. Pte Bird, from Hoylake in Cheshire, was wounded during the attack south of Montauban on 1st July 1916, and invalided out of the Army later in the year.

> *This will be our badge, and in times of hardship and danger may it recall to us that we are Lancashire men. I have informed Lord Derby of His Majesty's mark of favour, and he has claimed it as his privilege to present to each man his badge in silver. It only rests with us now to prove to His Majesty that we have been worthy of conveying this compliment to Lord Derby by fighting without flinching to retain it, and hand it down with pride and honour to our successors and to theirs after them.* **Brigadier General F C Stanley CMC DSO Officer Commanding, 17th Bn KLR**

The Liverpool jewellers Elkington and Company were given the order for the silver badges, and by early November, had several examples on display in the windows of their shop in Lord Street. Each one was hallmarked, some with the Chester Assay Office mark, but most with that of Birmingham, and they all bore the date letter for 1914. By the end of the year, Lord Derby had, in fact, presented one to each individual recruit, at a private ceremony outside his home at Knowsley.

> *I was that young looking that he stopped me, Lord Derby, as he gave me the badge, and he said 'How old are you?'. I looked at him, and I said 'Nineteen, Sir', and he said 'You liar – but best of luck, son!' It was a good badge, it was solid silver, you know, and we were very proud of it. It must have cost him a packet!* **16400 Private S R Steele, 18th Bn KLR**

Some confusion still exists as to who was actually presented with the silver badge. The intention was that only the original 3,000 recruits should receive it, and after that, no more were to be issued under any

The presentation of silver badges to the 19th Battalion at Knowsley Hall, in December 1914. As Lieutenant Colonel F C Stanley looks on, Lady Victoria Stanley gives each badge to her mother, Lady Derby, who then hands it to Lord Derby for presentation. The soldier marching away, having just received his badge, is 21515 Pte C S Hawnaur. Note that Lady Victoria is also wearing a badge.

Lieutenant Colonel E H Trotter, DSO, Commanding Officer of the 18th Battalion and genuinely much loved by his men.

Celluloid badge worn during the early war years by relatives of those serving in the Pals.

circumstances. However, it would appear that when Lord Derby raised the 4th City Battalion in November 1914, he presented its members also, with the silver badge. Certainly evidence exists of men who joined the 20th Battalion in 1914 who did receive the silver badge. Very few Pals, if any, ever wore the silver badge in their caps, for apart from the possibility of it being stolen, it would seem likely that brass versions of the Eagle and Child were issued at about the same time as the silver ones were presented, and these were worn from that time on. Most of the silver versions were destined to be converted into brooches, and given to mothers, wives or sweethearts, but nonetheless treasured. Certainly the Derby badge was a great mark of distinction for those who wore it, although it was not always referred to with due reverence! It was known, variously as 'Derby's Duck', 'The Bird', 'The Bird and Bastard', 'The Duck and Bastard', 'The Ruptured Duck', 'The Constipated Duck', and even worse!

Lord Derby also provided each man with a solid silver regimental cap badge in the shape of the family crest. This was an eagle standing over a child in the nest, with the motto 'Sans Changer'. There were voiced doubts as to the child's legitimacy, and the crest came to be known as 'Derby's (sometimes qualified) Duck!' **17518 Corporal E G Williams, 19th Bn KLR**

If the badge was popular amongst the Pals, it was equally unpopular amongst the members of the other battalions of the King's who wore the White Horse of Hanover. This may have been esprit de corps, or just plain jealousy, but it was a fact nevertheless.

On one occasion I was returning to the Battalion, and arrived at the base depot somewhere in France. One peacetime sergeant who noticed my cap badge screamed at me, 'What's that bloody bird doing in your hat? You're in the King's, so get an 'orse up there where it should be!'. Despite my protests, it was only the intervention of an officer that saved the day. **21515 Private C S Hawnaur, 19th Bn KLR**

The officers of the four battalions also wore the Eagle and Child badge on their caps, but in bronze, not brass, in keeping with King's Regulations governing officers' service dress. They wore the same sized badge on their collars, but in non-matched pairs. In other words, although two badges were worn, both the eagles faced the same way, to the wearer's right. Officers' buttons, however, were the regulation 'White Horse', King's' pattern, in brass, the same as worn by the Regular battalions. Other ranks, of course, wore the universal pattern button, bearing the royal coat of arms.

Some time after the cap badges were issued the other ranks in all four battalions were given distinctive shoulder titles. These were in the pattern later followed by other Pals battalions, in the form of a double scroll, with a numeral in the centre. The Liverpool ones had 'CITY BATN.' on the top scroll, 'THE KINGS', on the bottom scroll, and the numerals '1', '2', '3' or '4' in the centre, to denote each individual battalion. These were proudly worn whilst the battalions were in England, but do not seem to have survived in France much after the Battle of the Somme. This was

One of Pte J R Bird's identification discs, made from red fire proof fibrous material. The other one was green and octagonal. If a soldier was killed, the green one was left on the corpse, for future identification, and the red one was sent to Battalion HQ, as proof of the soldier's death.

probably for two reasons. The first was that they were large and cumbersome and tended to get caught in the rifle sling when sloping arms, and were thus either broken, or ripped off. Secondly, they were an obvious security risk in the front line, where they could give away the exact identification of each battalion, to an inquisitive German raiding party.

Nevertheless, they must have been worn in the front line at least until the summer of 1916, as the author has one in his collection, once worn by a soldier of the 20th Battalion, that was found on the old battlefield at Guillemont, on the Somme, as recently as 1987. It was discovered along with the remnants of some leather equipment, and a pair of boots, but no human remains, so perhaps the original owner survived, or more likely his remains had been found after the Great War and buried in the nearby Guillemont Road Cemetery, the rest of the equipment being discarded. It is possible, also, that only one set of 'CITY BATN.' shoulder titles was ever issued per man, because certainly after 1916, the Regular pattern curved shoulder title 'KINGS' was worn exclusively.

Later in the war, in 1917 and after, the three Pals battalions of the 89th Brigade, wore a square of coloured cloth on each upper arm to identify themselves. The 17th Battalion patch was black, the 19th patch was white, and the 20th patch was yellow. The 18th Battalion was no longer a part of the 89th Brigade at this time. The 18th Battalion left the 89th Brigade in France, in December 1915, for the 90th Brigade, which was still part of the 30th Division, at a time when it was considered that all New Army Brigades needed the 'stiffening' of a Regular Army unit before they saw action on a large scale. It was replaced by the 2nd Battalion The Bedfordshire Regiment, who served with the 89th Brigade until February 1918.

These coloured patches may well have had their origin on the Somme in 1916, as identification ribbons, although it is not known exactly where on the uniform they were worn. In a letter written home after July 1st 1916, a Private in the 17th Battalion made reference to such identification.

Original shoulder titles of the four battalions. The bottom one, of the 20th Battalion was discovered on the old battlefield at Guillemont on the Somme in 1987.

> *We have been dubbed the 'Black Kings' owing to our black ribbon for identification purposes.*
> **15406 Private I Smith, 17th Bn KLR**

At some stage, also, the Eagle and Child badge, adopted by the 30th Division as its own divisional insignia, was worn, as a single badge, in white weave, on a circle of black cloth, on the back of the tunic, below the collar, probably for the purpose of instant identification during trench raids.[1] In 1918 it was worn, by members of the 30th Division on each shoulder.

Prescot, Hooton and Sefton Park

Because of the prompt action of Lieutenant Colonel Stanley, as soon as recruiting began, the Liverpool Pals were clothed in khaki much sooner than most battalions of the New Army. On 31 August 1914, he actually sent a telegram to his wife, asking her to secure as much khaki cloth as she could find, and arrange for it to be made up into uniforms. As a result, men of the 17th Battalion were wearing khaki uniform in a matter of a few weeks, although the other battalions had to wait a little longer.

> *Within a couple of weeks, we were switched to an old watch factory at Prescot, which had been fitted out as a barracks. We had Grenadier Guards sergeants to train us in discipline, footdrill and musketry; only, at the start, we hadn't any uniforms and we hadn't any muskets!* **15917 Private J A Tudor, 17th Bn KLR**

Men of the 18th Battalion at Hooton in 1914. The soldier second from the left on the middle row, is 16414 Pte. A E H Gray, the author's great uncle, who as a sergeant, was wounded on the Somme on 8th July 1916, and died of his wounds at Le Havre on 31 August 1916.

We had about a week's rest before being summoned to Sefton Park for drilling, then we went from there to Hooton Park. We were under the grandstand. I went through two suits because we didn't get our uniforms until the end of October, and then they gave us all one pound allowance for wear and tear, and two pairs of shoes and boots. **17121 Private W Gregory, 18th Bn KLR**

They gave me a uniform that was too big, so I took it into Prescot and had it altered more to suit me. There was some trouble, though, over getting it back, because there was another chap called Redhead, and they had delivered it to him, and when I went for it, it wasn't there! **25561 Private H Redhead 19th Bn KLR, later 20th Bn KLR**

Supplies of boots and leather equipment were purchased equally swiftly for the use of the Pals, before other units could get them, and the early greatcoats were of such high quality that the troops in later times refused to swap them for officially made army issue ones!

Rifles, however, remained a great problem. The sudden expansion of Britain's army had left a great shortage of rifles. Virtually no New Army unit in England had the Short Magazine Lee Enfield, which was being used by the regulars on the continent, and not even the considerable influence of the Derby family could change this. Each battalion, however, was issued with one hundred obsolete Lee Metfords, which were in such a deplorable state that they were not allowed to be fired and could only be used for drill purposes. However, even 100 rifles did not go very far amongst 1,000 men, and so this essential part of the training of every soldier was sadly neglected. It was to be virtually a year before each man was issued with his own modern rifle and given sufficient training to learn how to use it properly!

The 17th Battalion, billeted in the old watch factory at Prescot, were the first to begin military training. Under the watchful eyes of their former Grenadier Sergeants, their routine was established virtually as soon as they moved in. They would be roused at 6.30 am, and then go for a three-mile run before breakfast, which was prepared and served by civilian caterers, who were not too popular, because of the standard and quality of some of their fare! Then, some of the recruits would set about the inevitable chores of running the barracks, such as loading and unloading wagons, guard and sentry duties and keeping the place clean. At the same time, others would be practising rifle drill, and then they would all swap around, so that

Men of the 18th Battalion and their Grenadier Guards instructors, at Hooton in 1914, posing with some of the precious Lee Metford rifles.

everyone had a chance to learn drill, with the old Lee Metfords. In the afternoons they would leave the barracks and go on route marches in the immediate area, to learn section, platoon and company formations. Once a fortnight they marched to St Helens where they were allowed to bathe in the municipal swimming baths. Many of the things they were taught, however, such as semaphore signalling, were to be totally useless to them in the future, but their instructors, often wearing Boer War medal ribbons, were more skilled in fighting in the veldt than in the mud of the Western Front!

Because Prescot was on the tram route into Liverpool, those who came from the city had no trouble getting home in the evenings, and at free weekends, virtually as often as they wanted. For those from further afield, the locals often laid on entertainment, in the form of concerts, and local sports, and provided billiards halls and reading and writing rooms. Many threw open their homes to the new citizen soldiers, and local theatres and cinemas provided reduced price seats. Many new recruits did complain, however, that if they tried to eat in some of the bigger restaurants in the city, as was their habit in civilian life, they were refused service, which was reserved for officers only! This was particularly galling, as most of them, by definition of profession, could have gained commissions anyway, but chose to follow Lord Derby's call and serve as private soldiers.

In early November 1914, the Second-in-Command of the Battalion, Major Beeman, was presented with a bloodhound, as a Battalion mascot, by a Mrs Armitage of Chorley, and it often marched at the head of the fife and drum band which had been formed.

Later, when I heard a band was being formed, I volunteered, because I could read music. Besides, I had been tipped off that if I were in the band, I would be excused fatigues, and I wanted to get home on Saturday, to play football. In due course, I became nicknamed as one of the snake charmers because I played a flute. We had to accompany the men on the route marches, supplying the music to keep them in step.

All the tunes we played became the subject of parodies by the men, and these became better known than the originals. The men fitted words to the Regimental march, which ran:

'What did you join the Army for? Why did you join the Army? What did you go to Prescot for? You must have been bloody-well barmy'.[2]
15917 Private J A Tudor 17th Bn KLR

A group of men from the 17th Battalion inside the ex-watch factory at Prescot in November 1914.

NCOs and officers of the 17th Battalion at Prescot in November 1914. The officers in obviously new uniforms, are wearing the cap and collar badges of the regular battalions of the KLR. Although the NCOs are wearing Pals cap badges, their shoulder titles have obviously not yet arrived. The tall Grenadier Guards' sergeant in the rear rank is wearing Boer War medal ribbons, and has, as yet retained his Grenadiers' insignia.

Across the River Mersey at Hooton Park Racecourse, the 18th Battalion was training in similar pattern, but as most of the Liverpool men were further away from their homes and couldn't return there so easily, there was a stronger feeling that they were real soldiers, almost as if they were on campaign. They occupied the stables and barns under the old grandstand, only recently vacated by the racehorses, and at first they drilled on the polo field which was in the centre of the racetrack. They were roused each morning at 6.00 am, and then sent on a cross-country run for fifteen or twenty minutes. Their Commanding Officer, Lieutenant-Colonel Trotter was very keen on running as a form of training, and the 18th Battalion became known as 'Trotter's Greyhounds', as they always won any subsequent inter-battalion running competitions. After breakfast, they drilled until midday, and undertook route marches of the Wirral in the afternoons. Tea was at 5.00 pm, and then the evenings were free until 'lights-out' at 10.15 pm. Once again, the locals made them very welcome, and organised frequent concerts and other entertainments, as well as taking them into their own homes. Mrs Rosa Derrick was a little girl in 1915, living in Clematis Cottage, Hooton Green, where she still lives today. She well remembers helping to play host to men of the 18th Battalion.

The Pals came, I think, in the September of 1914, and left in 1915 and mother opened the house to them. I was between nine and ten. There was a coal fire over there, and mother used to make bread. She used to buy the big sacks of flour, and I've seen them sit there and toast it all, and there was always a big jar of home-made jam, and mother made them tea. Of course they didn't all come at once, but we reckoned we had about 72 altogether that used to come in periodically.

Then they used to have inoculations, and some of them used to get mother to dress them at night. And then, after they went, some of them came to see us when they were home on leave, and I remember this fellow Fred Westmoreland, he came back from the gate and he put his arms round mother and said, 'We'll never come back', and he didn't. The boys used to write to mother, and mother used to make a cake, we used to call it the trench cake, a lovely big fruit cake, and she'd wrap it up and send it for them, and they used to share it in the trenches.

A smiling group of men from the 17th Battalion at Prescot in November 1914.

Officers and NCOs of No.2 Company of the 17th Battalion pictured at Prescot in 1914. The officer seated second from the right in the centre row is Second-Lieutenant D H Scott, who died of wounds on 2 July 1916, the only officer fatality from the Battalion of the attack on July 1st.

I've got the names of four that I can remember, Wilf Ashcroft; Short, I can't think of his Christian name, he gave me a beautiful doll, and its not long ago since I passed it on to a great friend of mine; then there was John Bradley, he was a Corporal, and he used to come quite often, and then there was Fred Westmoreland. The girl that was engaged to him, Dora, only died recently, she was 90 odd, and she used to come out here when Fred died, and we'd been great friends after that. **Mrs RME Derrick Hooton Green, Wirral.**[3]

On 27 November 1914, General Sir Henry McKinnon went to Hooton to inspect the Battalion and note its progress. Although he was very impressed by what he saw, he agreed that it was 'lamentable' that they had so few rifles, especially as they didn't work, but he promised to 'badger' the War Office to rectify the situation. Nevertheless, his 'badgering' was no more successful than that of Lord Derby!

The 19th Battalion, meanwhile, without a permanent home, had been drilling each day in Sefton Park, on an area traditionally known as 'The Review Field'. Its members lived at home, or by choice, in nearby lodgings, for which they received an allowance, and travelled to and from the park each day.

We drilled in Sefton Park, (going up by tram each morning), from 9 o'clock onwards, dressed in civvies. We learned squad drill, platoon drill, company drill and – a great moment, battalion drill. We skirmished on our stomachs and hands and knees, in open order, over the wide grass plains of the park. Then we went home at night. **17518 Corporal E G Williams 19th Bn KLR**

Meanwhile, the work of constructing the camp in Knowsley Park had been surging ahead. It was natural that given the talent which abounded in the ranks of the Pals and the organising skills of the Derby family, it would not take long at all.

In fact, within three days of Lord Derby deciding to use the Park, a

'Trotter's Greyhounds'. Some of the 18th Battalion at Hooton about to set out for their early morning cross- country run.

Lieutenant H C Watkins in charge of a platoon of No.3 Company, 18th Battalion, at Hooton in 1914.On the far left is 16920 Pte (later Sergeant) J Milne who was killed in action on 1st July 1916.

No.4 Platoon, No. 1 Company, of the 18th Battalion on the steps of the grandstand at Hooton Park Racecourse, with Grenadier Guards NCOs during the autumn of 1914.

former architect had drawn up the plans for the new camp. Within a week, the contract to build had been placed with a civilian contractor, and within just over a month the camp was ready to take its first troops! Thus, in early November 1914, the 19th Battalion abandoned its daily visits to Sefton Park and moved into new, purpose-built accommodation, at a time when many Regular and Territorial troops were shivering in tents all over the country. It was Lord Derby's wish that all the Pals Battalions should train together as quickly as possible, and realising that the 18th Battalion could not stay at Hooton indefinitely, he ordered that the construction work should continue, so that the 18th, too, should have a permanent home.

Shortly after this the 20th Battalion was raised, and this meant that the four battalions could be formed into a brigade. Lieutenant-Colonel Stanley was appointed in command of this brigade, at first numbered the 110th, and his place as Commanding Officer of the 17th Battalion was given to a very capable officer from the Durham Light Infantry, Lieutenant-Colonel B Fairfax, who had

Civilian kit laid out for inspection in the former stables at Hooton Park Racecourse, where the men of the 18th Battalion were billeted.

Some 'high jinks' at Hooton with these 18th Battalion men. On the extreme right (kneeling) is 16744 Pte J Redhead, who died at Boulogne on 20 July 1916 of wounds received on 1 July 1916 at Montauban.

A fatigue party of No.4 Company of the 18th Battalion at Hooton in 1914. They refer to themselves as 'Swabs', as they have each just washed and dried over 300 dishes.

No.4 Platoon, No. 1 Company of the 18th Battalion proudly display early entrenching skills at Hooton.

General Sir Henry McKinnon GOC Western Command, inspecting troops of No. 1 Company of the 18th Battalion at Hooton, on 27tNovember 1914. Second from the right is the author's great-uncle, then 16414 Pte A E Gray.

The newly built hutted accommodation at Knowsley, and the old mess tents, late autumn 1914.

recently returned from duty in France. Command of the 20th Battalion was at first given to Lieutenant-Colonel W Ashley, a local Member of Parliament and former soldier, but ill health meant that he soon had to stand down, and command passed to Lieutenant-Colonel H W Cobham, who was to earn well deserved promotion to Brigadier General in France.

Until accommodation could be built for this 4th City Battalion, it was quartered at the Tournament Hall in nearby Knotty Ash, and trained each day in Knowsley Park, as did the 17th Battalion which was now fairly comfortable, in the old watch factory.[4] Eventually, on 3 December 1914, the 18th Battalion left Hooton Park for good, and having marched through Liverpool, took up quarters at Knowsley Park Camp. On 29 January 1915, the 20th Battalion took over huts in the Park next to the 18th and 19th Battalions, and for the first time, the Liverpool Pals could train together as an infantry brigade.

NOTES
1. This information was given to the author in an interview with former 2I515 Private C S Hawnaur, 19th Battalion KLR, in 1973.
2. The Regimental quick march of the King's was 'Here's to the Maiden of Bashful Fifteen', and was generally sung on the march as: 'Why did we join the infantry? Why did we join the Army? Why did we join the infantry? We must have been bloody-well barmy.' Another Pals version of the parody went 'Why did we heed Lord Derby's call? Why did we join the Army? Why did we go to St George's Hall? We must have been bloody-well barmy.'
3. Of the four soldiers, all from Liverpool, mentioned by Mrs Derrick, only one, 16285 Private Wilfred Ashcroft, survived the war. The others, 16639 Lance-Corporal John Bradley, 17090 Private William Leo Short and 16343 Private Frederick J Westmoreland were all killed in action together, on 1 July 1916, during the Battalion's assault on the German trenches south of Montauban.
4. Although the 17th Battalion trained as part of this brigade, it never actually moved into the camp, as Prescot Barracks was only a couple of miles away from Knowsley Park, and near enough for the men to march to each day.

Chapter Three

'Will we win? – Rather!'

Knowsley

When the first drafts of the 19th Battalion took over the new camp at Knowsley in November 1914, hardly surprisingly, not all the huts were finished. Some of them lacked complete roofs, some had no glass in the windows, and no stoves, and some of them lacked beds or bedding material. However, they were actual huts, and they immediately made their occupants feel that they were real soldiers at last. Before long all the faults were put right and apart from the perennial problem of mud which beset any army camp in winter, it became obvious to the recruits that as their camp was purpose built, they were far better off than many Regulars or Territorials were at the same time!

> *We occupied huts beautifully built, divided into half-huts, by a central entrance. There were 32 men and an NCO to each half, making a platoon of 64 men, as it was in those days. They were good huts, good companions mostly, and good NCOs, mostly.* **17518 Corporal E G Williams, 19th Battalion KLR.**

It did come as quite a shock to some recruits to find that they were fed by civilian caterers, whose standards were far lower than most had been used to in civilian life. It was an equal shock to discover that their canteen was run by the same caterers, and thus they were not even able to buy different or decent food to augment their meagre and often sub-standard supply! However, wash houses and drying rooms were built into the camp structure, which made life easier for the soldiers, after long cold wet route marches. Similarly, their entertainment was assured, because, apart from the now well-established scene in nearby Prescot, a theatre was built on camp, (for the princely sum of £600).

By the end of January 1915, all four battalions were available for training together as a Brigade, and for the first time, the men, and particularly those of the 20th Battalion, who had the disadvantage of having been raised a clear two months after the others, began to shape up to the needs and rigours of military service. They still lacked rifles that could fire, and their training was not really commensurate with the needs of modern warfare. In order to give them some modicum of self reliance, Brigadier-General Stanley initiated a scheme by which parties of men, sometimes a battalion at a time, and sometimes in smaller units, were sent off into Lancashire, often billeted on a small village, to fend for themselves for three or four days at a time. These 'forays' were quite popular amongst troops, who were becoming bored with Knowsley, but not always for military reasons!

> *We went to Upholland for a week's toughening, and we had to sleep out in the open. Then the medical officer said that he wouldn't be responsible for the blokes if we didn't get them in somewhere, and so they stuck them all in pubs. Well, I had to see all the fellows out of the bar, and so on, I had a little room of my own, but a few minute later they were all out in the blinkin' bar again!* **16006 Sergeants Harris, 17th Battalion KLR.**

A taste of things to come – duckboards covering the 'awful mud of Knowsley'.

You can't do without cooks' runs the hand-written caption on this photograph. These were the men responsible for the culinary efforts at Knowsley. They had to put up with much abuse – good natured and otherwise.

The interior of one of the Knowsley huts.

Some of the 19th Battalion inside a hut at Knowsley in January 1915.

By early 1915 news was beginning to filter back across the Channel from France and Belgium that the system of trenches, albeit hurriedly dug at first, were there to stay, and that the whole style of training for warfare would have to be rethought.

> *From all that we had heard from France, it was very clear to us that there was a portion of the training which required a very great deal of attention, and that was digging. The country all around Knowsley was very highly cultivated, and it was impossible for us to secure ground for digging outside the Park; so it resolved itself into our having to make arrangements with Lord Derby to make a mess of some of his property.* **Brigadier-General F C Stanley CMC, DSO.**

This digging-up of Lord Derby's land is one of the most often remembered events amongst all the survivors of the Pals, and virtually all of them think that they were unfairly duped into digging a great portion of Lord Derby's estate for him, for next to nothing! On reflection, this is probably not the case, but the whole issue certainly provided a talking point which lasted for many years! The 'property' chosen for the Pals to 'make a mess of', was a high bank of earth quite near to the house at Knowsley, which apparently obscured the view towards the lake. Stanley was of the opinion that actually digging trenches was not so important as getting office workers, many of whom had never handled physical toil, to get used to handling picks and shovels. It is difficult to argue with his logic, although actual trench-digging might have given the troops better practice for the future.

Whatever the reasoning, each battalion in turn, once or twice a week, was ordered to set to and remove this great bank of earth, and despite their efforts, the task was still not complete when they finally left Knowsley in late April 1915. Nevertheless, many of the troops, with a typical Merseyside sense of humour, made the best of their task, managing, at the same time, and much to Lady Derby's dismay, to invade the house, the kitchens and even the stables, where they frequently rode the noble Peer's horses! Others found that the earth they were moving was virtually clay, and moulded, and then added, large appendages of truly mythical proportions, to many of the classical statues that surrounded Knowsley Hall. Others still, were content to play games with the

Four members of No.6 Platoon, No.2 Company, of the 18th Battalion at Knowsley. Seated on the fence are brothers 21868 Pte T S Edwards and 21867 Pte J L Edwards; standing below them are Pte T Holman and 21899 Pte J W Smith, who was killed in action on 1st July 1916.

A digging party from the 19th Battalion at Knowsley in 1915. One of the men in the front row is obviously so keen on his unit, that he is wearing a soft cap with 3rd Pals embroidered on the front. The sergeant on the left, probably a former Grenadier Guardsman, is wearing a regular 'King's' cap badge and Pals shoulder titles.

The pioneer section of one of the battalions at Knowsley.

light railway which carried away the earth.

It was a measure of the lack of knowledge of the arts of war, that we were not instructed in the digging of trench systems or dugouts. Instead, we removed mounds of clay, in levelling and landscaping the grounds, and had lots of fun with trucks and light rail tracks, and with the clay too! Some of it adorned the classical statuary, and not always in the best of taste, giving rise to hilarity among the troops, but not among the rightful owners! **17518 Corporal EG Williams, 19th Battalion KLR.**

We did hate that job. Once or twice a week, we had to parade particularly early, and walk to the Waterworks in Prescot Road, to pick the shovels and spades up. The stuff that we dug away went into a big hollow in the park, so they laid a railway down with tipping trucks. I was a sergeant, so I didn't have to dig, and I just stood at a certain spot where there was a dip, and counted the trucks as they went past. The fellows used to let them go down this slope, and the damn things would turn over, making a hell of a mess. **16006 Sergeant S Harris, 17th Battalion KLR.**

The measure to which this earth-moving made an impression on the soldiers is illustrated in the fact that they parodied a popular melody of the time, 'Moonlight Bay', with their own version, 'Derby's Clay', and later on, even sang it on the march, in France and Belgium. It went:

We were digging all day, On Derby's clay. The picks and shovels ringing, They seemed to say. If you don't do any work, You'll get no pay. So we dug dug dug like hell, For a bob a day.[1]

It was even being sung at Brigade and Battalion reunions some fifty years later!

The truth of the matter concerning all the excavation work, is that Lord Derby from the start of the work, was fully aware of the possible 'cheap labour' accusation that might be levelled at him, and insisted that the work should be surveyed by an independent valuer. Once it was completed, he promised to pay the amount

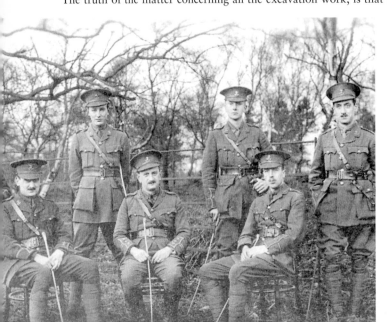

Some of the officers of No.2 Company of the 19th Battalion at Knowsley in 1915. The seated officer in the centre is Captain R K Morrison, who, as a major, was dismissed from the service and sent home to England in 1916. (See page 187).

of the valuation into a special Comrades' Fund', to provide comforts for the troops. Even though the work was never actually completed, and was independently valued at between £600 and £700, Lord Derby actually paid £1,000 into the fund. Writing about the affair after the war, Lord Derby's brother stated this case quite unequivocally.

A few people who did not know the facts of the case, and who would have had a grievance over anything, and who were only too anxious to have a dig at a public-spirited man, did not hesitate to say that he had employed the men of the City Battalions to carry out work for him. This libel has long been set at rest, but it is just as well, in these pages, to restate the whole case, and to show that, so far from him getting any advantage out of what we had done, he was distinctly the loser, because not only did he have to pay 1,000 to our Comrades' Fund, but it has entailed considerable further expenditure to him afterwards.[2] **Brigadier-General FC Stanley CMC, DSO.**

Despite the evidence to the contrary, the old adage that 'mud sticks' could almost literally be applied to this story, as the belief still persists today that somehow the Derby family got cheap labour, and suffered accordingly after the war!

Lord Derby's son, Lord Stanley, who died, and never got the title, put up for Parliament just after the war, in one of the Liverpool

Above: *Mess waiters and the officers' mess of the 19th Battalion at Knowsley in 1915.*

Below: *Contrasting with the above, men of the 19th Battalion enjoying dinner in the open air at Knowsley Park Camp.*

constituencies, and he lost it. They reckon he lost it because of the Pals telling these stories about the digging they had to do. **16006 Sergeant S. Harris, 17th Battalion KLR.**

Lord Kitchener and the leaving of Liverpool

As 1915 progressed, and with the limited resources at their disposal, the Pals continued to reach a high standard of military efficiency as a Brigade. In February Lord Kitchener asked Lord Derby if he could raise a couple of brigades of Royal Field Artillery from the area, to supplement the Pals infantry. This Lord Derby was able to do eventually, although the recruiting was slow, and these men and others who were to join the Royal Garrison Artillery later served in the 30th Division in France and Flanders as The County Palatine Artillery. Although they wore the traditional gun and carriage cap badge of artillerymen, and not the Eagle and Child, many of them considered themselves to be 'Artillery Pals'.

On 20 March 1915, as a boost to military and civilian morale, Lord Kitchener reviewed all the New Army troops who were training in the area, as they marched past him on the heights of St George's Plateau, outside St George's Hall, in Lime Street. He was at the head of many dignitaries, including Lord Derby and his family, local military commanders, Members of Parliament, the Mayors and Mayoresses of all the local districts, and esteemed local business men. Upwards of 100,000 people crowded into the immediate vicinity of the Hall, to watch some 12,000 soldiers march past, which took from noon until about one o'clock. First to march past were General Abdy and twelve men representing the newly raised County Palatine Artillery, and they were followed by nearly five thousand men of the four Pals Battalions, in order of precedence, each Battalion Colonel at the head of his men. Each man was in full battle order, although the elusive new rifles had still not arrived, and only the leading one hundred men in each battalion carried the old Lee Metfords. There is no doubt, that the military manner of these men, compared most favourably with the last time they had marched past the Plateau!

Lord Kitchener was the guest of Lord Derby at Knowsley Hall the night before the review, and the Pals provided a guard for him.

> *There were twelve of us, all six-footers. I was on sentry on the terrace at the front of the Hall. It was about 8.00 am. All was quiet, and then suddenly I heard a click. The door opened and out stepped the famous man, with a small black dog. I was scared stiff. Later, the guard was drawn up in two ranks and inspected by him. As he walked slowly along, he glared right through each one of us, just as depicted on that famous recruiting poster with the pointing finger. But I suppose we must have done the job all right, as we were dismissed for the rest of the day, and did not have to foot-slog from Knowsley Hall to St George's Hall where Lord Kitchener was to review troops stationed in the area.* **17054 Private G. Groome, 18th Battalion KLR.**

Kitchener's reaction to the men who had passed by him was certainly one of approval. He commented, 'They are splendid – we must have more of them'. This prompted Lord Derby to throw his Pals recruiting net wider still, and on 23 March he appealed in the press for even more recruits for the Pals

J E Rayner, The Lord Mayor of Liverpool, Lord Kitchener and Lord Derby about to ascend St George's Plateau to review the Pals, on Saturday 20 March 1915.

The 17th Battalion marching down Lime Street. Liverpool, on 20 March 1915.

Battalions. This time, he targetted the shopkeepers and shopworkers of Liverpool to come forward and form themselves into a company. In a move which may well have been prophetic, he asked employers to let shopworkers go and fill the vacancies with women instead! In the same appeal, he quashed the widely circulating rumour that the Pals would never go to war, but would be used only for home defence. He was adamant that this notion was total rubbish! However, despite his words, there was still a disgruntled feeling amongst a certain element on Merseyside that the Pals were cosseted and pampered, and would never see action.

It is not difficult to see why this rumour had arisen. By March 1915, it was beginning to dawn on the the British public that the Germans were not a band of murderous, inefficient, sausage eating buffoons, which was the way the popular press had portrayed them since the outbreak of war. This had been done in an attempt to rouse anti-German feelings amongst a nation who had regarded the French as Britain's natural enemy for the last 1,000 years, and the Germans as Anglo-Saxon cousins! Similarly, people realised that the war had not been finished by Christmas, and Britain was not victorious everywhere, as they had been encouraged to believe would be the case. Instead, lists of casualties had begun to point to the harsh realities of war, and even the Territorial Force, at first kept in England had begun to be filtered across the Channel to meet the ever increasing need for more men. When Territorial Force casualties began to appear in the newspapers, it was easy to make comparisons between them and the well turned out and very clean Pals Battalions training in safety and comparative luxury, at Knowsley, and very much in evidence all over Merseyside. Taunts of 'Derby's Lapdogs' began to be made at Pals walking in the streets, and some, even in uniform, were offered the iniquitous white feather of cowardice, normally reserved for those who would not join up, by sanctimonious young ladies who imagined it the best way that they could serve the war effort!

All was not well with the public image of the Pals in Liverpool, however. The Territorial battalions had been in France for months and they had been having casualties and reinforcements, whilst we were having

Lord Kitchener stood for three-quarters of an hour as 12,000 locally recruited men of various sections of the New Army marched down Lime Street, to be reviewed as they passed back in front of St George's Hall, on 20 March 1915.

Fatigues at Knowsley, in 1915.

Inside one of the wash houses at Knowsley in 1915.

The interior of the 19th Battalion recreation room at Knowsley in 1915.

a good, safe, time at home. Epithets like 'Derby's Lap-dogs' etc., were heard, and white feathers and scorn were the lot of some who went into Liverpool itself! **17518 Corporal E G Williams, 19th Battalion KLR.**

However, the problem was solved when, at the end of April, orders were received that the Brigade, (still numbered the 110th Infantry Brigade, at this time,) was to leave Knowsley, and depart for Belton Park Camp at Grantham in Lincolnshire. Apart from the fact that this would toughen the men by taking them away from home, where creature comforts were all too easily available to them, it would also allow them to train with the other brigades of the 37th Division. On Friday 30 April, with an atmosphere more normally associated with a carnival than a military oper-ation, the Pals marched to Prescot Station, battalion by battalion, to entrain for Grantham.

Lord Derby himself was one of the first to arrive at the station, accompanied by his brother, Brigadier-General Stanley. General McKinnon turned up later in the morning, as did a huge crowd of wellwishers, who were not allowed in the station, but were nevertheless content to wait outside the station gates, and cheer each company, as it arrived. The 18th Battalion had left Knowsley at 5.00 am and it was the first to arrive at the station, to catch the first and second of twelve special trains that would be needed to move all four Battalions. Two companies or just over 500 men from each Battalion, were assigned to each train. The trains themselves had

Some of the 17th Battalion at Knowsley. On the extreme left is 30101 Pte T J Blincow, later taken prisoner in 1918. (See page 205).

Swedish drill at Knowsley Park in 1915.

19th Battalion footballers at Knowsley in 1915.

The Liverpool Pals also trained their own officers in a specially raised Officers Training Corps, at Knowsley Park Camp.

Part of the Transport Section of the 19th Battalion in 1915.

No.1 and No.2 Companies of the 17th Battalion at Knowsley Park. Some of the statues that the soldiers loved to 'deface' can be seen in the background.

at least twelve carriages, and were longer than the platform at Prescot! The 20th Battalion was the next to leave, followed by the 19th, and finally, at one o'clock, the last two companies of the 17th left Prescot for ever.

All throughout the morning the holiday atmosphere had persisted, as more and more people arrived by tram and car from Liverpool. There was even a cinema photographer present, whose film would be shown the following week in Liverpool.[3] The men responded well to the encouragement of the crowds, waving caps and handkerchiefs back at them, and laughing and joking with families, and friends alike. Either unaware, or more likely uncaring, of what might lie ahead of them, they sang and hummed tunes and exchanged 'Good Luck' and 'Hope to see you soon' with all the well-wishers. At one point, as was the habit at that time, a single voice shouted out from the mass the question, 'Will we win?'. With one accord, two or three hundred voices gave him their reply – 'Rather!'

Grantham

The journey from Liverpool to Grantham, took about four hours, and at 10.00 a.m., on 30 April, the 18th Battalion arrived at the military platform of the Springfield Road Goods Depot, at Grantham Station, about four miles from the camp. It was met there, by the bugle and drum band of its Mancunian counterpart, the 18th Battalion The Manchester Regiment, (3rd Manchester Pals) who had arrived in Grantham earlier in the week. It then marched behind the band to its new accommodation in the camp, to be followed throughout the day, by the other three Liverpool Battalions. It is recorded in the Grantham Journal, that sing songs kept them 'merry and bright' on their first night at the camp, until after midnight!

Belton House, home of the Brownlow family, Belton Park Grantham, in 1915.

In many ways there were similarities between Belton Park Camp, and Knowsley Park Camp. Both were built on the estates of lords, on land generously lent to the War Office at no charge, for the duration of the war, and in both cases the camps were purpose built. Belton House was the family seat of Lord and Lady Brownlow, and it was built at the north end of the park. The part occupied by the Liverpool Pals was on a gentle slope, and opinions differed amongst the troops as to its practical usefulness. Certainly, it had nothing like the facilities of Knowsley, but could accommodate some 12,000 men. This meant, however, that the supply of water for drinking, washing and sanitation, which was only fed through a one-inch pipe, was desperately inadequate! The four City Battalions, who were quartered at the far end of the camp, fared worse than most, and at best, their water supply was sparse, which prompted Brigadier-General Stanley to say, with unaccustomed levity, 'Ours not to reason why, ours but to wash – or try.'

Another problem that the Pals encountered was that some of their predecessors had left the camp in a terrible state, especially those who had solved their lack of decent sanitary arrangements in a time honoured if unhygienic way!

> *By god it was awful when we got there, a sea of mud, it was bad in all respects. The militia had been there before us, and they were a dirty lot. They'd just stood on the steps of the huts, and piddled onto the mud, and the first thing that we were ordered to do was clear a couple of feet of soil. The M.O. said that he wouldn't be responsible if we didn't, so we had to clear a couple of feet of soil from all over the place, and cart it away in railway trucks, but after Knowsley, we were used to that!* **16006 Sergeant S Harris, 17th Battalion KLR.**

Obviously, Sergeant Harris' experience was not common to all of the four battalions' areas, and some troops were quite satisfied, at first, anyway, with their new quarters. This is the first impression of one soldier, written home the day after his battalion had arrived. Some of his ideas would change, as the weeks went by!

"The camp is beautifully situated at the foot, (or rather on the slope), of a hill, which rises sharply behind us to

Belton Park Camp, Grantham, where the four Pals battalions trained from May until September 1915.

The 17th Battalion marching through the streets of Grantham in July 1915.

some height, and is crowned by extensive woods. There is a tower on the top, and a splendid view is obtained on a clear day.[4] The park is quite different to Knowsley, – no bracken, but with lovely splendid old trees. The surrounding country, more advanced than around Liverpool, is very pretty and of course, we are seeing it in the present lovely weather. Two Companies, (500 men), mess in one large hut, and the babel at meal times is indescribable. We get Army rations here, in fact all arrangements are carried out by the ASC,[5] and we find the difference from Knowsley, I can tell you. The food of course is good, but nothing like so plentiful as before. Our beds are pretty much the same as at K., only laid on three planks supported by trestles about six inches from the ground. The camp is planned on better lines, – the canteen and the bath houses are a great improvement, and altogether I think we are likely to be comfortable here. **25576 Private W B Owens, 19th Battalion KLR.**

Once the troops had settled into camp life, they began to explore the surrounding countryside, and particularly the town of Grantham itself, which was only a three mile walk from Belton village, dominated as it was, by the imposing Belton House, home of Lord and Lady Brownlow. At first, the locals were not always friendly, as they had not always been treated with respect and courtesy by previous occupants of the camp. Also, the sudden arrival of thousands of men into the area brought an equally sudden but most unwelcome throng of prostitutes into Grantham, which naturally outraged its post-Edwardian sensibilities, and made the troops even more unpopular! This problem was solved by Grantham's Watch Committee, who recruited and appointed Britain's first uniformed policewoman in 1915, giving her the same powers of arrest enjoyed by her male colleagues. Her name was Edith Smith, and it is recorded that she was almost solely responsible for ridding Grantham of its prostitution problem, between 1915 and 1918. Exactly how the troops regarded her success, however, is not recorded!

The people of Grantham and its surrounding countryside were soon bowled over by the charm of the Pals, as had been the people of Prescot, Hooton and Knowsley, and before long they opened up their homes and their hearts to them and treated them like their own families.

Officers and Warrant officers of the 20th Battalion at Grantham in the summer of 1915. Seated second from the left is Lieutenant-Colonel H W Cobham, who commanded the Battalion.

The town was very nice, and there were a lot of houses where we could go and get baths for sixpence, because there were no proper facilities at the camp. Of course, we had to overcome the awful reputation of these militia fellows when they were there, but I think we managed that all right, and got on well with the people. Of course all the pubs there were 'blue', – instead of 'The Red Lion', it was 'The Blue Lion', and so on.[6] The pubs were quite nice, and we used to go in

'Bayoneting at Belton' during the summer of l915. The original caption to this photograph was, 'If only they were Germans'.

them at night, and on the mantlepieces in them, were adverts for cows being served by bulls, and that kind of thing, which used to amuse us! **16006 Sergeant S Harris, 17th Battalion KLR.**

I remember walking along Wharf Road in Grantham, and the people there were giving us a heck of a gay time, and all the time that we were in Belton Park, they treated us like Kings of England. They couldn't have treated us better if we had been. They treated us very well indeed." **24644 Private W Stubbs, 18th Battalion KLR.**

The local people were also very generous in offering accommodation to families from Merseyside who wished to visit their loved ones at Belton Park Camp, and many lasting friendships were forged between families whose common link was a member of the City Battalions.

However, the Pals were not in Grantham to enjoy the local hospitality and they soon got down to serious training as a division, which put them one step nearer to the war. Shortly after they arrived, they changed their Brigade and Division numbers, but still remained within the same structure. This happened because brigade and division numbers were originally given to men from reserve and extra reserve battalions of the Regular Army who had been taken back on strength in the autumn of 1914. In April 1915, the War Office decided to convert these battalions instead, to the training role, and re-allocated their numbers to the New Armies Divisions.[7] The four Liverpool Pals Battalions became the 89th Infantry Brigade, and, together with the Manchester and Oldham Pals, of the 90th and 91st Infantry Brigades, formed the 30th Division. At that stage of the war there were four battalions to a brigade, and three brigades to a division. Thus, an infantry division was composed of twelve battalions of men, each of which numbered just over a thousand in strength. Numerically, therefore, as a loose 'rule of thumb', the divisional number, multiplied by three, gave the centre number of its three brigades, and thus, the two either side! This changed, to a certain extent, once the New Army Divisions got to France. In order to 'stiffen' them with regular troops, some swapped one of their brigades for one from a Regular Army Division, which meant that the brigade number was no longer in sequence. In the event, this swap proved totally unnecessary, as most of

the New Army Divisions performed very professionally without the help of the Regulars! The four-battalion brigade system was to change, with far-reaching effects on the Liverpool Pals, in February 1918, when the three-battalion brigade was

The Orderly Room of the 18th Battalion at Belton Park Camp, Grantham in 1915. Nearest the camera is 26019 Company Sergeant-Major P Lyon.

Some of the 19th Battalion at Grantham in the summer of 1915. Holding the mascot 'Sammy', is 17879 Pte H J Lover, from Birkenhead, who was killed in action during the Battle of Guillemonton 30 July 1916.

21904 Pte J Routledge (left) of the 18th Battalion who is holding a Lee-Metford rifle, in comparison with an unknown private of the same Battalion, who carries a Short Magazine Lee-Enfield rifle. Both these photographs were taken at Knowsley, in early 1915. Both are wearing the 1896 pattern leather equipment, most commonly issued to the New Army.

Signallers of the 17th Battalion on a route march from Belton Park in 1915.

introduced.

Thus, by May 1915, the infantry of 30th Division, was assembled together at Belton Park and could train as one large unit, in the vastness of the surrounding countryside. Brigade training became commonplace, and occasionally, the whole Division trained together. The long-promised modern Short Magazine Lee Enfield rifles finally arrived, not all of them brand-new, and some of them bearing the scars of previous fighting with previous owners, and consequently, the Pals had their first musketry instruction on full-bore ranges, at nearby Great Gonerby – after nearly a year of military service. A Transport Section was also raised, which was quite a difficult task from men who knew more about offices and businesses than horses and mules, and at last training began to reflect the actual war that was being fought across the Channel.

> *At Belton Park we made great strides, were refitted with good clothing and boots, and were issued with modern rifles, the Short Magazine Lee Enfield, a very good weapon, and at that time, the last word in the infantryman's equipment. We fired our Musketry Course, learned something of gas warfare, which had just been used against us, and were introduced to the earliest respirator. It was generally felt that we were ready for battle, but our progress was delayed because our artillery were not ready and in fact were not fully equipped with guns."* Lieutenant E W Willmer, 17th Battalion KLR.

This lack of artillery had, in fact, held back the 30th Division from proceeding overseas, and the Divisional Artillery arrived just before the Pals left Grantham. This gave the infantry very little time to train with the artillery, a situation which was to be rectified later, as it was becoming obvious from what was being learned in France and Belgium, that the static nature of trench warfare necessitated very close co-operation between infantry and artillery. In order that the Division could be kept up to date with the way the war was being waged, it was arranged that parties of officers could go to the front and see for them-selves what conditions were like there. First of all the Divisional Commander, Major-General W Fry and the Brigade Commanders and Majors went, and the Battalion Commanders followed shortly after their

Airing the bedding at Belton Park Camp, in 1915. Note that the soldier nearest the camera is wearing regular 'King's' style shoulder titles, and not the 'Pals' pattern worn by the soldier sitting up behind him.

'A Soldiers' Chorus' – Church Parade and service in the open air at Belton Park in June 1915.

Conducting the service (above), were Canon Linton Smith, Chaplain to the 19th Battalion, and Archdeacon C Wakefield, Dean of Lincoln Cathedral.

return. These visit were labelled 'Cook's Tours' by the officers, and all who went on them agreed that they were very useful. There is no doubt that they helped to prepare the Division for some of the rigours that lay ahead. The final seal of approval was given to the four City Battalions, on 27 August 1915, when the War Office finally and formally took them over as fully trained and equipped units of the British Army.

Once the artillery was up to strength, the Division received orders to proceed to Larkhill Camp on Salisbury Plain for the final part of its training in England. Advance parties from the 89th Brigade left for Larkhill on 31 August 1915, which, coincidentally, was the first anniversary of the raising of the Liverpool City Battalions at St George's Hall. The rest of the Brigade left by rail on 6 and 7 of September, with very mixed feelings. Grantham had been their first taste of soldiering away from home, and almost without exception, they had all enjoyed the experience, the training, the people and the friendships they had made. On the other hand, all also knew that after Larkhill, they would be going overseas to get to grips with the enemy, which was the reason they had volunteered in the first place! As the troop trains drew out of Grantham Station to the strains

One of the last 'Zeppelin Guard' duties at Belton Park before the Pals Brigade left for Larkhill Camp and Salisbury Plain.

of 'Auld Lang Syne', played by the Manchesters' band, the carnival mood was still present, as it had been at Prescot, only this time it was a little more reflective!

Larkhill, Folkestone and France

Larkhill Camp had been a Regular Army Camp before the war, and the difference between it and the other camps, was immediate to all the Pals. For a start, there was less comfort, and more emphasis was placed on training and equipping for the war itself. The camp could accommodate 50,000 men, and altogether, on Salisbury Plain, and the immediate area around it, some 750,000 men were under training. There was also an aerodrome close by, which gave many their first sight of aircraft, which flew constantly overhead. The 89th Brigade was stationed at Canada Lines, which was actually at Amesbury, and very close to the ancient circle of Stonehenge.

> *Then we moved to Salisbury Plain, – Larkhill Camp, in the autumn of 1915 and a couple of big turnip fields away from our quarters, was Stonehenge. Nevertheless, we became more conscious of our present and immediate future there than the past.* **17518 Corporal E G Williams, 19th Battalion KLR.**

One of the first casualties of the move to Larkhill, was Colonel E. F Gosset, Commanding Officer of the 19th Battalion. He was quite simply too old to take an active infantry battalion into trench fighting, and

Men of the 17th Battalion being reviewed on the march at Belton Park, just prior to the move to Larkhill Camp.

either Larkhill finally brought this home to him, or more likely he was gently reminded of the fact by Brigadier-General Stanley. Whichever it was, in early September he relinquished control of the Battalion, although naturally very deeply upset about it, declaring that it was 'like losing a limb'.

There is no doubt he took a wise step in recognising the fact that years were creeping on, and from all that one has seen afterwards, it has brought home to one that age does stand very much in the way. Major Denham, who had been second-in-command to Colonel Gosset, then took over command of the Battalion. **Brigadier General F C Stanley, CMC, DSO.**

Although Major L S Denham, (promoted to the rank of Lieutenant-Colonel), was to lead the Battalion in France during the early Somme fighting, he too, would lose command, before a year had elapsed.

Whilst at Larkhill the Brigade completed another musketry shoot, only their second, the next time that they would fire live ammunition they would be on active service in France! They also practised large-scale manoeuvres over the Plain, and on 23 September 1915, took part in a memorable night exercise which was to descend into total chaos. The idea was, that the whole of the 89th Brigade should advance in the dark against an imaginary fixed position and carry out a mock assault. The night was as dark as pitch, and inevitably, everyone lost their way, and when the apparent objective was reached it was discovered that the 'enemy' there, was a brigade of Manchesters, who were on much the same exercise, from a different direction, and were equally lost! In the resultant confusion, only about a dozen men from the whole Brigade made the objective. The next day the ground they had crossed resembled a real battlefield, full of the debris of discarded kit and equipment.

The two contingents got hopelessly mixed up with each other. It was so dark that they could not sort themselves out and as they were lost in the bargain, the whole thing became a shambles. The story goes that an officer, searching for his own mob,

Salisbury Plain was usually the last training base for infantry troops before they went to France, because it was large enough for divisional training. Some idea of the number of troops quartered there may be obtained by looking at the rows and rows of huts in this photograph, which stretch as far as the horizon.

Officers of the 20th Battalion at Larkhill Camp in October 1915 just before their departure for France. Top row – 2/Lt A Dawson, Lt G H Bradshaw, 2/Lt G A Brighouse, Lt A Macintosh, RAMC, 2/Lt A O Laurie, 2/Lt A H Holden, 2/Lt W H Jowett, 2/Lt H Lancaster. Middle row – 2/Lt S Gooch, Lt & Qr/Mr A C Dawson, 2/Lt G S Sutton, 2/Lt A E Wilson, 2/Lt J G Nixon, Lt R Munro, Lt A B Grant, Lt R E Melly, Lt R D Paterson, 2/Lt J W Musker. Bottom – Capt E C Orford, Capt R H Laird, Capt J D Greenshields, Maj Campbell-Watson, Maj W A Smith, Lt-Col H W Cobham, Lt & Adjt H Bracken, Capt P D Holt, Capt A T Beazley, Capt T Whiting, Capt W L Hicks.

> *met a soldier who happened to be a raw recruit from a pioneer battalion. 'What company are you from?', demanded the officer. 'Wigan Coal and Iron, sir', said the soldier.* **21815 Private C S Hawnaur, 19th Battalion KLR**

Whether or not the officers learned any military lessons from the debacle, it certainly amused the men, who saw more than one pompous officer reduced to humility! Very few Pals were able at the time, to recognise that the lack of coordination and communication which had caused the failure of the exercise might have far more serious implications for the real war!

> *We had a lot of fun, but looking back, the same errors made in training, in the dark, which provided mirth for the troops, and certain disgruntlement for the officers, hoping to gain status at the time, were to come to the same thing in France.* **17518 Corporal E G Williams, 19th Battalion KLR.**

As autumn wore on, the men became fitter and fitter, and as they seemed to be doing the same training over and over again, they became more and more restless. The war didn't seem any nearer to being won, and the Pals were unanimous that once they got 'out there', they would soon sort things out! The first sign that they were actually going overseas was the issuing of all sorts of military kit and stores, most of which they would never use

A meal in the open at Larkhill.

anyway, and finally, on 31st October, they were told that they would be leaving in about a week! King George V normally claimed the right to inspect each New Army Division before it left Great Britain, and this inspection had been fixed for 3 November 1915. However, His Majesty had taken a bad fall from a horse, whilst inspecting his troops in France, on 28 October, and he was unable to get to Larkhill. His place was taken instead by Lord Derby, who was the obvious alternative choice, as he had raised most of the Division anyway. This fact was later recognised when his Eagle and Child badge was adopted as the Divisional insignia for the 30th Division, as well as the cap badge for the four Liverpool City Battalions.

The King did, however, send a personal message to the Division:

> *"Officers, non-commissioned officers and men of the 30th Division.*
> *"On the eve of your departure for active service, I send you my heart-felt good wishes. It is a bitter disappointment to me that, owing to an unfortunate accident, I am unable to see the Division on parade before it leaves England, but I can assure you that my thoughts are with all of you.*
> *Your period of training has been long and arduous, but the time has now come for you to prove on the field of battle the results of your instruction. From the good accounts that I have received of the Division, I am confident that the high traditions of the British Army are safe in your hands, and that, with your comrades now in the field, you will maintain the unceasing efforts necessary to bring this war to a victorious ending. Goodbye and God speed. George R.I.* **His Majesty King George V.**

Lord Derby himself was a little more realistic about the way in which the war might be brought to a victorious ending. He gave his views on the subject in a letter sent to his brother on 4 November 1915.

> *This war is only going to come to an end by killing Germans, and I am perfectly certain that at that game, the 89th Brigade will more than hold their own.* **The Rt. Hon. Earl of Derby**

From 1 November 1915, advance parties had begun to leave each of the Battalions for France, and the main parties were detailed to leave a week later. However, on 6 November, the 17th Battalion and the Brigade suffered its first officer casualty, with the tragic death of Honourary Lieutenant and Quartermaster C E Ryder.[8] In the early hours of the morning he was found dying, shot through the head, not far from Stonehenge, with a service revolver by his side. A later inquest decided that he had committed suicide whilst temporarily insane, and blamed pressure of work for this lapse! The same day, most of the Brigade transport left for Southampton, and there embarked on board the SS *Nirvana,* for Havre, in advance of the rest of the Brigade.

On the following day, 7 November 1915, the rest of the four Liverpool Battalions left Larkhill for the last time, and entrained for Folkestone. The 17th and 18th Battalions were the first to leave, by virtue of their seniority, and they arrived at the harbour at about midday. The 17th embarked on the SS *Princess Victoria,* a former Portpatrick and Wigtownshire Railways steamer, and the 18th on the former South Eastern & Chatham Railways boat, the SS *Invicta.* They left harbour, and England, escorted by two destroyers and an airship, to spot for submarines, at about 2.30 pm.

Some of the 17th Battalion at Larkhill: standing: 16080 Cpl W E Holt, Cpl Rigers, 24924 L/Cpl R D Rodgers, L/Cpl Lodge, 15491 Sgt W Milner, 16098 L/Cpl N J Tunnington, seated: 15904 Cpl R L Spence, 2/L J M Sproat, Sgt Andler, Cpl Colisanan.

The 19th and 20th Battalions left Larkhill after midday and arrived at Folkestone in the early evening, by which time it was already dark. Also, all sailings across the Channel had been cancelled because of a submarine scare, and for a while it was thought that they might have to encamp in the Folkestone area. Eventually the scare passed, only to be replaced by a report that two Zeppelins had been sighted over the Channel. As a consequence, when men and stores were eventually loaded onto the two boats that were to carry them, the whole of the embarkation had to be carried out in the dark, no lights being allowed. Eventually, by about 9.30 pm, they were laden and ready to sail, and with a destroyer escort either side, they too slipped quietly out of the harbour.

Now at last the Pals would meet their destinies, and this time the mood was a mixture of the expectant and the sombre. One soldier from the 19th Battalion had written a short letter from the train on the way to Folkestone to his home in Liverpool, and had asked a railway official to post it for him. His views at the time probably echoed those of many of his comrades.

> *Well, we're away at last, and 'tho one feels that it's a solemn occasion to be in England for perhaps the last time, I think that the predominant feeling in every chap's heart – in mine at any rate – is one of pride and great content at being chosen to fight and endure for our dear ones and the old country.* **25576 PrivateW B Owens, 19th Battalion KLR.**

For Private Owens, it would indeed be the last time he would be in England, for he was killed in action nine months later, on 30 July 1916, during the Battle of Guillemont, without ever seeing his dear ones or his country again.

For the men of the 17th Battalion, who had sailed in daylight, earlier in the day, the farewell to England had been tinged with sardonic humour, and perhaps a hint of grim reality.

No.6 Platoon, B Company, of the 17th Battalion with shaven heads on the day before they left Larkhill for France, 6 November 1915.

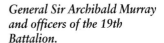

General Sir Archibald Murray and officers of the 19th Battalion.

As the boat left the quay, we all had our own silent thoughts, which were interrupted by a leave boat coming the other way. As we passed each other, someone on the other boat shouted 'Are we downhearted?', to which we all shouted back the ritual, 'No!'. The same voice was then heard from across the water, 'Well you bloody well soon will be!'. **15771 Private W L P Dunn, 17th Battalion KLR.**

NOTES
1. The basic pay of a private soldier throughout the Great War was one shilling (five new pence) per day.
2. The History of the 89th Brigade. 1914 – 1918, by Brigadier General F C Stanley, *Liverpool Daily Post* 1919.
3. This film footage, unfortunately, does not seem to have survived.
4. This tower, Bellmont Tower, (Also spelt Belmont), was built for Viscount Tyrconnel in 1751, and was always a feature of Belton Park, as it is today. Its configuration, when viewed from Belton House, prompted the nickname – 'Lord Brownlow's trousers', amongst the locals. Apart from the 'splendid view' it gave to Private Owens, it was also used each night by a sergeant and twelve armed men who mounted a 'Zeppelin Guard', on the top, presumably to shoot down any airship that might try to bomb the camp!
5. A.S.C. stands for Army Service Corps, whose job it was to fetch and carry all army stores and equipment. Its men were not respected by the infantry, who considered they had a 'cushy' time, and nicknamed them 'Ally Sloper's Cavalry', after a picture comic character of the time.
6. Even today, this 'blue' colour persists, in the names of public houses in the Grantham area, such as 'The Blue Boar' in the town centre itself. Its origins lie in the political in-fighting which existed in the early years of the nineteenth century, and particularly the political ambitions of a local landowner, Sir William Manners. He was a Whig, whose party colour was blue, and to gain votes, he offered the bribe of 'blue beer', or free beer at the pubs he owned. To ensure that would-be voters knew where to go, he renamed his pubs so that 'blue' figured in their titles. It is ironic that today, blue is the party colour of the Whigs' great rivals, the Tories, and that Britain's first lady Prime Minister, was born in Grantham!
7. This change was brought about by Army Council Instruction 96, of April 1915.
8. By this time, six soldiers from the four Battalions had died during training. They were: 21976 Private M F Jackson, 17th Battalion; 24649 Private A A Dodd, 17050 Private E Ellis, 24573 Private W Hayworth and 16578 Private P Hoban, 18th Battalion and 21522 Sergeant G T Hull, 19th Battalion.

Chapter Four

'We were full of beans
– we were going to sort them out!'

France and the real war

As the last infantrymen of the 89th Brigade were crossing the English Channel to Boulogne, the Transport Sections of each Battalion were already in France. They had left Larkhill on 6 November 1915, with their mules, horses and wagons, and embarked from Southampton, for the longer crossing to Le Havre. The crossing was uneventful, apart from the death en route of one of the 17th Battalion's mules. The 18th Battalion was also to lose a mule, which would be the Brigade's first 'casualty' in France.

After disembarkation at Le Havre, the Transport Sections were loaded onto a very slow moving train to take them to Pont Remy, and the newly-established Brigade Headquarters at Ailly, not far from Abbeville, on the River Somme. As such, they became the first of the Pals to suffer the vagaries of the French Railway system, that plagued the British Army throughout the war years.

> *We got to Le Havre all right, and they put us on a train, horses, wagons, mules and their drivers, and on the trucks, it said 'Hommes 40, Chevaux 8'. The horses had their harnesses on, they were head to head, each four horses, heads facing each other, and tied at the top. They had the harness chucked in the middle, and a bit of fodder or something for them to eat, in a nosebag. Although there were eight horses in, there were only two drivers, and there wasn't so much room in between them. Anyhow, we got there, and one of the mules got hung while we were going, he twisted himself so many times around with his chains, that he snuffed it, and they had to chuck him out! We were only going to Abbeville from Le Havre, but it was a slow train, – stop, start, stop, start – and there was plenty of time to go to the engine to get a can scalded![1]* **21915 Lance Corporal W Heyes, 18th Battalion KLR.**

For many of the Pals, now safely landed at Boulogne, it was their first sight of a foreign country, and even those who had travelled the globe pre-war on business, had not visited France. Although horrified by Germany's 'rape' of Belgium and France in 1914, many British people still saw the French as their 'natural enemies', having fought them for the best part of 900 years. The newly-formed Anglo-French friendship fostered by the recent 'Entente Cordiale'[2] was still probably tolerated more than accepted by the majority of the British people, before the German invasion of Belgium and France in August 1914.

15296 Pte L Bateman and 15276 L/Cpl A P Philpot of B Company, 17th Battalion. The original inscription on the back of the photograph, written by Philpot after the war, reads: Two 'Real Stalwarts'. The other gentleman is a Private Bateman. He was eventually discharged with heart trouble. This was taken before we left Larkhill for France. The 'gentleman' on the right is the same as he was 1914–18, domineering, flamboyant, overpowering, unpopular, ex-Lance Corporal unpaid, eventually reduced to common private. Dolly says it must be awful tied to a man with the above mentioned reputation. Pte Bateman was sent back to England in March 1916 with heart trouble, and Pte Philpot (he had lost his Lance Corporal's stripe for 'neglect of duty' on 25 November 1915) was also returned to England in August 1916, after being wounded in the shoulder during the Battle of Guillemont, on 30 July 1916.

The SS Invicta *former South Eastern & Chatham Railways ship, which carried the 18th Battalion to France on 7 November 1915.*

I had never been to France before, so it was a new experience for me in every way, as it was for most of the officers and men of the Battalion. Foreign travel was not so widely practised as it is today. There were not the facilities, and few had the surplus money. It was obvious, however, that it is a very different thing to have war in one's country, a thing we were to learn, at least in part, in the Second War. In Britain in the First War, it seemed remote. There was nothing remote about it in France! **Lieutenant E W Willmer, 17th Battalion KLR.**

To the people of Boulogne, the war was obviously far from remote, for despite the fact that the war was already fifteen months old, they still turned out in force to give yet another marching band of 'Tommies', a tumultuous welcome, as each of the Pals Battalions marched through their town. The Pals responded in like fashion, at least until they found they needed all the breath they had, just to keep up the pace, as they came upon the very steep hill that runs through the town centre itself.

When we arrived at Boulogne, the French people made a heck of a fuss of us, giving us all sorts of things, flowers, sweets and all sorts, and the children as well, made a heck of a fuss. Then we had to climb a huge hill to get to our rest camp, and I didn't know the name of it until a number of years afterwards, and this friend told me it was called Blanket Hill, by the troops. **24644 Private W Stubbs, 18th Battalion KLR.**

When we landed at Boulogne, we had that damned hill to climb, it was a swine of a thing, we had 130-odd pounds on our back, you know, and everyone was cursing. We were exhausted when we got to the top. **16400 Private S R Steele, 18th Battalion KLR.**

When they all got to the top of the hill, they found themselves at Ostrehove Rest Camp, a tented camp built specially by the British Expeditionary Force for newly arrived troops in transit. There is no doubt that an air of expectant optimism filled all ranks of the Brigade, especially now they had actually arrived in France. Most of them could not wait to get to grips with the Germans, when they would surely end the war in spectacular style!

Once we were actually in France, we were all very anxious to get into action before it was all finished. There was a marvellous feeling of adventure, – we were full of beans – we were going to sort them out! **17122 Private W Gregory, 18th Battalion KLR.**

The Pals only stayed one night at Ostrehove Camp, and on the evening of 8 November 1915, and throughout the early hours of the following morning, battalion by battalion, they marched back into the town, to Boulogne Central Station. From there they entrained for Pont Remy, and arrived for the most part, before their Transport Sections, having covered the shorter distance in faster trains. Each battalion was then billeted in a separate village, the 17th Battalion at Bellancourt, the 18th at Monflieres, the 19th at Buigny-l'Abbe and the 20th at Pont Remy itself. They remained in the area for about ten days, by which time they had begun battle training in earnest, and had sorted out most of the kit that had been misplaced in transit. Then they moved nearer still to the front line, by marching to the Vignacourt area, north-east of Amiens, where they were given training for the first time, in up to date weapons and methods of trench warfare. The occasional rumble of guns in the air also reminded them that the war was not very far away!

We entrained at Boulogne for quite a short trip to a little village called Pont Remy, not far from Abbeville, where we went into various phases of infantry training of that time – route marches, bombing practice, (the Mills Bomb had just been invented), respirator drill, which included a visit to an improvised gas chamber, through which we all went to test the masks, foot inspections, kit inspections, night operations and so on. It was winter, and rather wet, and when the wind was in the right direction, there was the continuous, but still distant sound of gunfire, to which we were to get accustomed in the months and years ahead. **Lieutenant E W Willmer, 17th Battalion KLR.**

In early December 1915, Brigadier-General Stanley was given the news that the 89th Brigade was going to have to swap one of its battalions with one from a Regular Army brigade, in order to 'stiffen' it for the forthcoming fighting. This was in line with the general policy of XIII Corps, whose staff, at this stage had underestimated the quality of the New Army units, and it also entailed the swapping of a complete New Army brigade from the 30th Division for one in the 7th Division, which was a Regular Army Division. The brigade chosen was the 91st, which was replaced in the 30th Division by the 21st Brigade. It was a battalion from the 21st Brigade which had to swap with one from the 89th, and Stanley himself was given the task of selecting which battalion he was to lose. Naturally, it was a very difficult choice to make, as he had personally helped to raise all four battalions in his Brigade, but eventually, he decided to transfer the 18th Battalion.

On the 12th December I saw General Kempster, and he told me it had been definitely decided that his Brigade, the 91st, was to go the the 7th Division. He was very sick about it, and doubted whether he would stay on much longer. As it was left to me to select the battalion which was to leave us, after much mental tribulation and many regrets, I decided to send Colonel Trotter's, the 18th. **Brigadier-General F C Stanley, CMG, DSO.**

It is probable that Stanley chose the 18th because it was the most proficient and professional battalion in the brigade, led, as it was, by Lieutenant-Colonel Trotter, a veteran of the Boer War, and a soldier through and through. Stanley probably considered that it could best survive, out of the four battalions, without the immediate close-knit 'family circle' atmosphere of the Pals. In the event, the 21st Brigade was still part of the 30th Division, and the 18th Battalion was never far away from the others, until it rejoined the depleted brigade, in 1918! The Battalion that Stanley received in return, was the 2nd Battalion, The Bedfordshire Regiment, commanded by Lieutenant-Colonel H S Poyntz, whose association with the men from Liverpool over the next two years was both fruitful and highly successful.

By this time, 1915 was drawing to a close, and at last the Brigade was given its first taste of action, and the life that was to lie ahead for the next three years. It was decided to give each battalion first-hand instruction in trench warfare, by attaching it to an experienced Regular Army or Territorial Force battalion actually in the front line. Thus, on 16 December 1915, the brigade began the move forward to what was the northern end of the Somme sector, and the real war, at last!

Each battalion was assigned to a different part of the line, and then went into the trenches a company at a time. The 17th Battalion was sent to Englebelmer on 18 December, where the trenches were in a deplorable state, and were suffering the ravages of a very cold wet winter. They separated into companies, and went into the line at Mesnil, with the 1st Battalion, The Rifle Brigade, and at Auchonvillers with the 2nd Battalion, The Royal Irish Rifles.

Eventually, after a long march partly at dead of night, we reached the village of Englebelmer,

A cross-Channel steamer with British troops on board at Boulogne.

The ruins of the village of Hebuterne.

and each company was allotted to a company, much depleted, of the Royal Irish Rifles, a happy-go-lucky crowd, who were extremely kind to us, and taught us a lot. It was a quiet section of the line and nothing had happened there for some months, we were told. A few odd shells came over and an even fewer number were sent back. Buildings which had been damaged were further wrecked and new shell-holes were created, but very few men were hit, and most of the time was spent in trying to make the trenches fit for occupation by human beings. The communication trenches were in such a mess that most of them were unusable. One waited 'till nightfall and then walked over the top, to the frontline. But the front line had to be kept navigable, and that often meant keeping the water level below that of the fire step, and confining the mud to the bottom of the trench itself. With gumboots, one waded through it, but the firestep was dry land. Dugouts were mostly rough shelters, and were not proof against enemy action. **Lieutenant E W Willmer, 17th Battalion KLR.**

15206 Private E C Ecroyd, of 'A' Company, became the battalion's first casualty when he was shot in the chest by a rifle bullet on 20 December. He was evacuated to Rouen, and eventually to England, where he made a complete recovery.[3] The battalion came out of the line on Christmas Eve.

The 20th Battalion was sent to Bienvillers, also on 18 December, where it was split up into platoons and attached to the 6th and the 8th Battalions of The Leicestershire Regiment, who gave instruction on warfare to those troops who were out of the line as well as those who were in the trenches. Its tour of attachment passed quite peacefully, and it too came out of the line on Christmas Eve.

The 19th Battalion was attached to the 7th Battalion, The Warwickshire Regiment, and also went up the line on 18 December. The Battalion split into two halves, two companies going into the line at Fonquevillers, more commonly known as 'Funky Villas' to the troops, and the other two going to nearby La Haie, to practise attack formations, and to be used as work parties. On 20 December the two companies swapped over duties so that when they left the trenches, also on Christmas Eve, the whole Battalion had experienced front line service, albeit uneventfully.

Typical postcard sent home by the British Army during the war.

At Fonquevillers it was very quiet, the Warwicks were in there, and in December 1915, nobody was annoying anybody unnecessarily. We were about four or five days in the trenches, when we got our first baptism of machine-gun fire, knocking some of the tiles off the roofs in the village, – they couldn't get down far enough to ground level. They had guns firing from Gommecourt which could hit the tops of the tiled roofs of houses which were still standing, and there were still walls, high walls of gardens and orchards, and the bell tents that the soldiers were in would project just a bit, so that there were bullet holes in those, – but that was a quiet front. **17158 Corporal E G Williams, 19th Battalion KLR.**

The 18th Battalion had not yet switched brigades, and they too went up the line on 18 December, at Hebuterne, their different companies attached to battalions of five different infantry regiments. Their time in the front line was not quite so uneventful as the other three Pals battalions, however. No.2 Company[4] had been attached to the Buckinghamshire Battalion of the Oxfordshire and Buckinghamshire Light Infantry, and on the night of 20 December men from both battalions went out on a combined patrol into No Man's Land, with a Lewis Gun team from the 18th. They encountered a German patrol, which opened fire upon them, and they returned fire. Although two or more Germans were seen to fall, one of the Pals, 24620 Lance-Corporal R Rezin was shot through the spine and was killed. His body was later recovered, and he was buried the next day, in Hebuterne Military Cemetery, where his remains still lie. Thus, he has the doubtful distiction of being the first of the Liverpool Pals to be killed in action.

> We went down to Hebuterne, near Gommecourt, and we went in for a day and a night with the Oxford and Bucks Light Infantry, to get an idea of trench warfare. We had to walk right over the top to get to the front line, there were no communication trenches. We had a fellow killed within five minutes. We were sent out the following night and we found his cap. I've forgotten his name now, but it was a terrible feeling to lose a man. **17122 Private W Gregory, 18th Battalion KLR.**

Lance-Corporal Rezin's widowed mother, Ella, at her home in Urmston, near Manchester, no doubt received the dreaded telegram informing her of her son Reginald's death, just in time for Christmas, and no doubt, for the rest of her life, Christmas would never be the same again. Hers was the first of over 2,750 telegrams that would be delivered to the families of Liverpool Pals, before the decade was finished, and for whom also, life, as well as Christmas, would never be the same again!

Like the other three battalions, the 18th Battalion also left the trenches on Christmas Eve. It marched to Louvencourt, in time for Christmas Day. On that day it left the 89th Brigade and became part of Brigadier-General The Honourable C J Sackville-West's 21st Brigade.

The Kaiser's birthday

All four Pals battalions performed extremely well during their time in the front line, and received fulsome praise from the respective brigade commanders of the units to whom they were attached. The following extract from a confidential report on the 20th Battalion, proves the point.

> *20th Battalion The King's (Liverpool Regiment) Officers – Military Knowledge, average; Keenness, very pronounced; Control over their men, good; Discipline, good. NCOs – Class of NCO, a good class; Military knowledge, average; Keenness, very pronounced; Control over their men, good; Discipline, good. Men – Discipline, very good for New Army; Keenness, very pronounced; Standard of Training, rather above the average of the New Army. General Remarks – This is certainly the best Battalion which has been sent to this Brigade for instruction. It is well commanded and has really a good tone all through. The men get through their work quickly, thoroughly and cheerfully, and all ranks are keen to learn. This should be a really good Battalion in a very short time.* **Brigadier-General E G T Bainbridge, Commanding 110th Infantry Brigade.**

The German HQ at Longueval decorated in honour of the Kaiser's birthday, 29 January 1916. The Germans attacked the British and French positions south of the River Somme on this date, and gave the Emperor the village of Frise as a birthday present.

As a result, they were adjudged fit to take up positions in the line, and with the rest of the 30th Division, commanded by Major-General W Fry, were allotted a sector on the southern edge of the Somme battle front, near Carnoy, in early January 1916. Little did they know that this area would be their 'home' for the next ten months!

The British Army had recently taken over this

Left: *Winter conditions in the trenches.*

Right: *17122 PteW Gregory, 18th Battalion.*

part of the line from the French, and in common with all recently vacated French trenches, the Pals found them badly made, badly maintained and with the barest of sanitary arrangements. Whilst they set about consolidation and renovation, the weather could not have been worse, and the four battalions began to experience the realities of life in the front line, and all that meant. All four also began to experience the steady debilitation of casualties, that was the necessary consequence of trench warfare, and a warning of what might be ahead was also given!

> *The weather was of the vilest. It did nothing but rain and everywhere it was a bog. The trenches in our piece were very bad and there were not even any duckboards.[5] We buoyed ourselves up with the hope that the Germans were in a worse plight than ourselves, as they shouted from the German trench: We are Saxons, and after the 29th, you can have our trenches and the ********* Kaiser too.' This looked as if they were pretty fed up, but we did not, at the time, understand their reference to the 29th.[6]* **Brigadier-General F C Stanley, CMC, DSO.**

29 January was, in fact, the birthday of Kaiser Wilhelm II, and in an attempt to celebrate this in style, the Germans determined to capture a sizeable piece of the French line south of the River Somme near Frise. At this point, the French lines jutted out to form a salient, and from about the middle of the month, the German artillery began an intensive bombardment on the area, and on all the British lines north of the river. On 28 January, this bombardment intensified all along the line, and on that day, according to Stanley, on a front of one and a half miles, they fired 35,000 shells at the French and British trenches. Later that night, they took Frise from the French, capturing or killing all the troops there. This left the British flank on the river very exposed, and an infantry assault on the 30th Division's trenches was expected by the minute. Eventually, however, on 31 January, the French counter-attacked, and took back most of the lost ground. Although they were not able to re-take Frise until 2 July, during the Somme battle, the line was straightened and the Germans were no longer able to pose a sudden threat.

For all the Pals the bombardment was a real test of their endurance under fire, a test which they passed with flying colours. For the 18th Battalion, however, now serving with the 21st Brigade, it was a testing ground of a different kind.

At about 1am on 29 January the German artillery opened up a fierce bombardment on their front-line trenches near Carnoy, where No.3 Company under Captain Arthur de Bels Adam, held the line. At 2am, Adam reported that his company and forward saps were under attack by trench mortars and rifle grenades. About half an hour later, his left flank was attacked by a body of about one hundred men, probably from the 3rd Oberschlesisches Infanterie-Regiment Nr.62, and the 4th Oberschlesisches Infanterie-Regiment Nr.63. Some of these had got behind his forward sentries in their saps, and had entered a disused section of trench. These Silesians were engaged by Adam's men, with rifle fire and grenades, supported by local artillery. After a fight lasting fifteen minutes, the enemy withdrew. Although the Battalion lost two killed, (16808 Private R Eaton and 16806 Private J A Eady), and eight wounded, they must have killed or wounded an equal number of Germans who were taken back to their own lines. Two German bodies were later discovered in the disused trench, and four wounded were taken prisoner, including an officer wearing the ribbon of the Iron Cross, Second Class. He was Leutnant der Reserve O Siebert, who died of his wounds the same day.[7]

Later in the morning, the German bombardment continuing, Captain Adam's brother, Lieutenant Charles Adam, was badly hit in both legs, and although evacuated to England, he would never walk properly again. For their action on the Kaiser's birthday, Captain Adam was later awarded the Military Cross, and 16709 Lance-

The 18th Battalion Transport Section.

Corporal A Cohen was awarded the Distinguished Conduct Medal. Captain Adam was to be killed leading his Company, in the assault on the Glatz Redoubt south of Montauban, on 1 July 1916.

Field punishment No. 1
Once the four battalions had taken up their regular places in the line, they too learned the tedium of trench warfare, and the rigours of living out in the open during winter.

> *We were at Carnoy, for a long time, and that's where I got my trench feet. We were in water for eight days, four in the front, and four in the reserve line. It was all water and mud. I got a lift on the Colonel's horse coming out of the line. I couldn't walk, and he lifted me onto the horse. I had the whole of the next day off. They had to cut my boots off, and I jumped from size seven to size nine overnight![9] 17122 Private W Gregory, 18th Battalion KLR.*

Even as early as March 1916, however, the dangers of trench warfare must have begun to make themselves obvious to some of the Pals. Already, Private Heyes of the 18th Battalion Transport Section was able to offer advice on survival to a friend, 26611 Private J Woods, 21st Battalion The King's (Liverpool Regiment), who was still in training with one of the two Pals Reserve Battalions back home in Knowsley Park.

> *You seem to be a bit put out because your eye is not quite up to the mark, but take my advice and stop at Prescot as long as you can. You must not think that you get a bed to lie on and lots of blankets here, you get shoved in anywhere, and nothing to lie on but the bare floor, with rats running over you the whole of the time, keeping you awake. We have been standing to, now, for over a week, expecting being shelled out day and night. Sometimes they will come and get you out about midnight, make you put boots and putties on, and tell you to sleep like that until further*

Left: 26611 Pte J Woods 21st and 18th Battalion. He was later killed in action during the Battle of the Menin Road, near Ypres, Belgium, serving with The Queen's Own (Royal West Kent) Regiment.

Right: 21915 Pte W Heyes, 18th Battalion, who advised Pte Woods not to be too eager to get to the trenches.

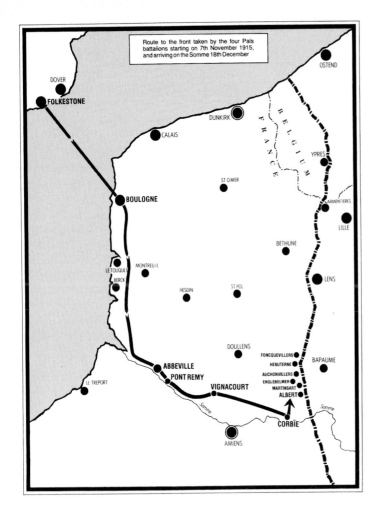

Route to the front taken by the four Pals battalions starting on 7th November 1915, and arriving on the Somme 18th December

notice. *You talk about the excitement being alright, but I can tell you, it is no cop, it is alright seeing them burst around you, but you don't know when one will come and lay someone out, and I can tell you we have had some fellows badly wounded. You will be alright if you can get leave for the Spring and hay time and harvest. I am shoeing now at the Army Service Corps for about a month and it is a better job than the infantry. You ought to try to get a job at cold shoeing, you would soon pick it up. Don't be over-anxious to be a good shot, and pretend you can't throw bombs, as it is a rotten job. Any job on the Head Quarters staff, or brigade, is all right. You always want to keep your eyes open for these jobs, and ask for them. An officer's groom is fine, get the Lieutenant Quartermaster's, if possible, as he never leaves the Head Quarters.* **21915 Lance-Corporal W Heyes, 18th Battalion KLR.**

Evidently, Private Woods did not take his friend's advice, for having completed his training, he was eventually transferred to the Royal West Kent Regiment, and was killed in action on 22 September 1917, during the Third Battle of Ypres. For some, however, as the weather improved, it was possible to imagine that the war was far away!

At that time, it was a very quiet part of the line, and we dropped into some sort of routine, a period in the front line, then in support, then in reserve, which often meant some days in camp or billets at Etineham, a pleasant little village on the Somme, some miles behind. When Spring came, it was a lovely place and life was good. Even in Maricourt Wood, which still had some trees standing, Spring was very beautiful, with bluebells, anemones, primroses and the young green of fresh leaves. When the firing stopped for a few minutes, as it did from time to time, it was hard to imagine that there was a war on at all! **Lieutenant E W Willmer, 17th Battalion KLR.**

Despite the onset of Spring, there was no doubt that there was a war on. In the months of February and March alone, in the 'quiet part of the line', thirty one men lost their lives on active service, serving with the four battalions. For at least one of these, death was almost certainly avoidable!

In the middle of March 1916, the four battalions came out of the line for rest and training, and the 17th Battalion marched to billets at Franvillers, near Corbie, arriving on 18 March. It was normal practice that any battalion coming out of the line was kept busy with fatigues and inspections to stop any hint of 'slack-ness' and, as a consequence, many troops preferred front line service to duties in the rear areas. On 20 March 'C' Company was paraded, for a sudden kit inspection, during which it was discovered that eleven soldiers did not have gas respirators. They were immediately brought in front of the Battalion Commander, Lieutenant-Colonel B C Fairfax, and, their 'guilt' indisputable, were sentenced to one day each of the archaic Field Punishment Number One.

> *We were all given Field Punishment Number One, and as the Colonel passed sentence on us, he said : 'I award you this punishment as an example to the Battalion – I am your father, you are my children.* **15771 Private W L Dunn, 17th Battalion KLR.**

The punishment, sometimes known as 'Crucifixion', was first introduced in the British Army, in 1881, as a substitute for flogging, which had finally been abolished in that year. Field Punishment No.1 was both humiliating and sadistic, in that it demanded that the miscreant be tied to a waggon wheel, or some other such fixed object, for a specified length of time (but no longer than two hours in any day) in such a place as the rest of his unit could witness his torment. When not tied, the victim could be made to perform physical tasks, as if he were undergoing punishment with hard labour. As such, the actual tying up was usually unpleasant rather than brutal, but it was certainly unneccessary as a punishment for citizen soldiers who had volunteered to fight for King and Country in a time of national emergency, and who had volunteered at that, with the best possible motives! Possibly the oldest amongst the eleven,[10] at thirty-six years, was 15590 Private S B Heyes, who came from Wavertree, a middle-class suberb of Liverpool. Sidney Heyes was one of the first to enlist, in August 1914, as part of the original Corn Exchange contingent, and he was also a pre-war international hockey player. As each man was examined by the Battalion Medical Officer, Lieutenant T B Dakin, to see if he was fit to undergo punishment, Heyes was found to be unwell, but Dakin was of the opinion that he was 'quite able to be tied to a wheel"!

However, on this occasion, the punishment did not merely consist of being tied to a wheel. The eleven were first made to dig a large hole to bury some surplus food, and then made to march, at the double, for about ten minutes, around the village square. After this, they were tied to waggon wheels in the square for a further ten minutes, and then made to march at the double again. This routine had continued for about two hours, when it was abruptly halted by a tragic happening.

> *After we had been tied up for the second time, and released, we were once again made to run around the square in double time. It was then that Private Heyes collapsed on the road in front of me, crying 'I can't go on, I can't go on.' I stopped and tried to help him up, and he just died in my arms! Some time later, when I was on leave I had to go and tell his mother what had happened to him.'* **15771 Private W L Dunn, 17th Battalion KLR.**

Heyes' body was taken to No.98 Field Ambulance, where death was confirmed, and where a post mortem examination was performed the following day.[11] His cause of death was found to be 'Cardiac Syncope', or a heart attack, and he was buried in the cemetery at Mericourt-l'Abbe, now called Mericourt-l'Abbe Communal Cemetery Extension. His Death Certificate, however, simply lists his cause of death as 'Died of Sickness'.

Not unnaturally, when the exact circumstances of his death became known to his family, they were very anxious that Field Punishment No.1 should be exposed to the public, and if possible, stopped altogether. Consequently, the press took up the cause, the newspaper 'The Sunday Chronicle', being in the forefront of the protest, with its flamboyant leader writer, Robert Blatchford taking a very personal interest in the case, and all others cases concerning this type of field punishment.

Once the circumstances of Heyes' fate became common knowledge, the public expressed natural outrage, and as a result, it was not long before Parliament became involved. The Quaker MP Philip Morrell especially, became relentless in his pursuit of the truth concerning those cases that had come to light. During a debate in the House of Commons on 21 December 1916, he brought up the case of Private Heyes' death, although he did not actually name him.[12] What he did do, however, was to cite an interview given to 'The Evening Standard', on 14 November 1916, by the Director of Personnel Service, General B E W Childs. General Childs' interview is quoted here in full, to show the contrast between the official version

of the story, and that of Private Dunn, told to the author, in an interview given in 1970, and also later written down in his memoirs, which are now in the Liddle Collection in the Edward Boyle Library, at Leeds University.

The soldier concerned, together with eleven other men, was sentenced to one day's field punishment No.1. On the morning upon which he was sentenced, he was medically inspected by the medical officer of the battalion, who certified him as fit to undergo the punishment. The field punishment consisted of fatigue. From 1.45 to 4.40, the men were employed in digging a hole for the disposal of rubbish. From 4.40 until 6, they were doing nothing. They were then confined under the conditions laid down in the Rules for Field Punishment, for half an hour. Here, it should be noted that the maximum period allowed is two hours. Subsequently to this, the soldier concerned, whilst on the march, asked leave of the provost-sergeant, to fall out, as he did not feel well. The sergeant himself took the man to the medical inspection room, where he was placed on a stretcher and made as comfortable as possible. He complained of acute pains below the right lung and diffi-culty in breathing. He shortly afterwards collapsed and died.

A post-mortem examination was held the next day, at which the Lieutenant-Colonel, a major, two captains and an expert bacteriologist were present. They investigated every organ of the man's body, and apart from some slight trace of fatty disease of the heart, there was no other evidence to the naked eye of the cause of death. In the words of the finding of the post-mortem, the only sugges-tion is that an acute attack of dilation of the heart supervened. **General B E W Childs, Director of Personnel Services.**

The incredible disparity in the details of the two accounts surrounding Heyes' actual death, after the passage of so much time, cannot easily be explained, but it would appear that Childs, at least, was either presented with false information, or was acting on instructions which did not altogether reflect the truth!

In its own way, Heyes' death may well have prevented others suffering a similar fate, because if nothing else, the outcry that accompanied his death probably made commanding officers more circumspect about awarding such a punishment. It was not, however, until five years after the war, in 1923, that Field Punishment No.1 was finally abolished. Heyes' name is one of those inscribed on a roll of honour which remembers all the Great War dead from the Liverpool Corn Exchange, in St Nicholas', the Parish Church of Liverpool, not far from the River Mersey.

The 'Big Push' preparations

As Spring moved towards early Summer, the four battalions became fitter, tougher and more experienced. Everyone, including the Germans, knew that the British Army would attack on the Somme, it was only a question of when! The 'Big Push' had been expected for some time, but following the sudden German assault on the French at Verdun in February 1916, it became vital it should take place as soon as possible to ease the unremitting blood-letting pressure that the French Army was suffering. To the men of the New Armies already on the Somme, waiting to do their bit, it could not come soon enough.

The 30th Division spent nearly all of April out of the trenches, undertaking dreaded carrying parties before taking over the line near Maricourt again, on the last day of the month. On 17 May, Major-General J S M Shea took command of the Division. Originally, he had been a cavalry officer in the Indian Army, but over the next twelve months he proved himself to be a popular, professional and competent infantry leader. It is surprising, therefore, that his eventual demise from command of the 30th Division was somewhat 'under a cloud', although he did later redeem himself in command of the 60th Division, under Allenby, in Palestine.

On 25 May the battalions of the 89th Brigade came out of the line and were sent back towards Abbeville for more specialist training for the 'push'. This training consisted of practising assaults on trench systems which were supposed to be duplicates of those that they would encounter in the actual attack. It was arduous work, and was often hated more than actually being in the line, as it went on relentlessly, day after day, and often into the night as well!

We dug, I suppose, from six to seven thousand yards of trenches; of course not to full depth, but enough to show what it looked like. Here we practised every day, getting every man to know exactly what was required of him and what the ground would look like on the day. They all tumbled to it very fairly well, and certainly our practice improved all of us very much indeed.

We practised all day and every day. First, battalions singly, and then two or more battalions together. All this was a happy life, and we felt that we were in for a good show and one which ought to have a great bearing on the length of the war. **Brigadier-General F C Stanley, CMG, DSO.**

The 18th Battalion, as part of the 21st Brigade, continued to serve in and out of the trenches for the rest of May, until finally released for training at Fourdrinoy and Vaux sur Somme, on 13 June. It did not return to front-line duty until 27 June, when it took over from the 2nd Battalion The Wiltshire Regiment, at Maricourt.

By this time the other three Pals Battalions were already in the trenches, also at Maricourt, the 17th and 19th having moved into the front line on the night of 17 June, and the 20th having followed five nights later, on the 23. Everyone now knew their role in the imminent attack, and if only because of the unbelievable amount of guns and equipment that filled the Somme valley, no one believed that it could possibly fail!

Meantime, all the other branches of the Army were busy preparing for the great day. The signallers were laying miles of telegraph line, the Army Service Corps (ASC – they were not Royal at that time), were establishing their dumps and supply routes, and most of all, the Artillery were amassing the greatest concentration of guns of all time and I doubt if greater has ever been brought together since. They were all shapes and sizes. The really heavy guns were at the rear, but not so very far behind the front line. Even in Maricourt Village there were 6-inch howitzers which nearly deafened one whenever they fired. The 18-pounders (field guns) were literally wheel to wheel, and positioned quite close to the front line so that when the line moved forward, they would still have the Germans within range. **Lieutenant E W Willmer, 17th Battalion KLR.**

This huge artillery concentration included Kitchener Volunteers from Merseyside, who regarded themselves as 'Liverpool Artillery Pals'. They were part of the 30th Divisional Artillery, and they were more properly known as 125th Heavy Battery of the Royal Garrison Artillery, and the CXLVIIIth, CXLIXth, CLth and CLIst Brigades of the Royal Field Artillery. All of them bore the title 'County Palatine'

Major-General J S M Shea, commander of the 30th Division from May 1916 until April 1917.

77

to show that they had been raised in the County Palatine of Lancashire. A County Palatine was the feudal province of an Earl, and it goes without saying that this old title had been resurrected to honour the Earl who had raised these new units in 1915, and helped to provide their horses, – The Right Honourable The Earl of Derby.[13] The 30th Divisional Signals Company, which was also in the line, bore the title 'County Palatine' like the artillery, as it too, had been raised on Merseyside. On 24 June 1916 the fiercest bombardment then known suddenly opened up on the German trenches. Haig had entrusted General Sir Henry Rawlinson, who commanded the newly-formed Fourth Army, with the task of making the main assault on the Somme. Rawlinson had studied very closely, the devastating German artillery barrage which had opened the Battle of Verdun, in February 1916, and he realised that it came very close to winning the battle for the Germans in the first few days of the offensive. Thus, he sought to employ similar tactics on the Somme. As a result, he managed to muster 1,437 guns (or one gun every fifteen yards) along the front, to bombard the German lines and pulverise any opposition. This bombardment was to last five days and five nights, after which he expected his infantry would be able to walk into the German lines without any opposition.

The great and eagerly expected 'U' day[14] commenced with the simultaneous fall of a row of thirty tall elms, which lined the route nationale. Their destruction was necessary to clear the field of fire for certain 75mm batteries. By 10am, the bombardment was in full swing, and the unbelievable din which we were to have night and day for the next week, commenced. The 4.5-inch Howitzers and many of the siege guns were not shooting on this, the first day, and so, after a morning spent in final inspections, I went down to the front line in the afternoon to study our zone. . . . . .

Going back to the battery, I passed through the grounds of Maricourt chateau: the noise there baffled all description. About thirty light, medium and heavy trench mortar batteries were in action near the chateau, and being in the apex of the salient, they were all firing at once, in three different directions. Certainly the most wonderful thing in a big show is the stupendous noise thereof. **Lieutenant-Colonel N Fraser-Tytler, DSO, TD, CLIst (County Palatine) Brigade, Royal Field Artillery.**

The Battle of the Somme had begun!

NOTES

1   Lance-Corporal Heyes' home was in Bickerstaffe, just outside Liverpool, where he still lived, at the time of writing, in his mid nineties. Although not far from Liverpool, Bickerstaffe is still very rural, and as such the people there talk with a Lancashire dialect, rather than the city 'Scouse' accent. Thus, when Lance-Corporal Heyes talks of going to the engine 'to get a can scalded', he means opening a stop cock on the steam locomotive, to get a mess tin full of boiling water to brew up some tea.

2   The 'entente cordiale', was the name given by the press to the new movement of agreement between France and Great Britain, following the Anglo-French Entente of 1904. This entente, or loose agreement, which was not an alliance, and which was not binding on either country, dealt largely with Anglo-French colonial rivalry in North Africa before the turn of the century. Any misgivings that the French people might have had about their natural enemy were apparently dispelled by the visit to Paris of King Edward VII, a year earlier.

3   Ecroyd rejoined the Battalion, as a corporal, in May 1916, only to be wounded again in July, during the fighting for Trones Wood, this time in the head and the left foot. He was once more evacuated to England, and once more recovered, to survive the war.

4   Most infantry battalions at this time, designated their four companies by letters, ie 'A' Company, 'B' Company, 'C' Company and 'D' Company, but Lieutenant-Colonel Trotter preferred a numerical system. Thus, the 18th Battalion's companies were No.l Company, No.2 Company, No.3 Company and No.4 Company. The 19th Battalion also followed this pattern, until late October or early November 1916, just after Lieutenant-Colonel G.Rollo (former second in command of the 17th Battalion), had taken command of the Battalion, when it changed to the lettered system. The 18th Battalion continued with the numerical system until its absorption of the 14th Battalion The King's (Liverpool Regiment), in August 1918, when it too, finally adopted the lettered system.

5 Duckboards were planks of wood laid across joists of timber in the trench bottom in the hope that they would provide a firm and dry footing.

6 Saxony was one of the minor states of the German Empire, whose troops made up approximately seven per cent of the German Army. They were known to be less than enthusiastic about their Prussian neighbours, and often professed an affinity with British troops, whom they considered, had the same ethnic origins. In fact, in one section of trenches, before the Somme battle, one Saxon unit is reputed to have erected a sign-post, which stated 'We are Saxons, You are Anglo-Saxons, Don't fire, and We won't!' This attitude is also mentioned by Patrick MacGill in his book 'The Red Horizon', first published in 1916, when he describes his arrival in the front line, opposite a Saxon unit, to be told 'They're quiet fellows, the Saxons . . . there's a kind of understanding between us. Don't fire at us, and we'll not fire at you.

7 Leutnant der Reserve Siebert, of Infanterie-Regiment Nr.62, was presumably buried behind the lines by the 18th Battalion, but his grave was subsequently lost. The Volksbund Deutsche Kriegsgrberfrsorge e.V., the German equivalent of the Commonwealth War Graves Commission, gives his date of death as 28 January 1916. This is probably because the raiding party would have left the German trenches on the late evening of that date, and no men from his regiment returned to their trenches, who actually knew his fate.

8 This was Lieutenant-Colonel Edward Henry Trotter, DSO, Commanding Officer of the 18th Battalion, later killed by a shell on 8 July 1916, near Montauban.

9 At the time he was interviewed, in May 1985, Bill Gregory still wore a size nine shoe!

10 The other nine were:- 15858 Private G H Bargery, 15677 Private C Chapman, 15680 Private J Drummond, 15712 Private H E Foulkes, (died at Abbeville, probably of influenza, 08/11/1918, as a Corporal in The Labour Corps), 24938 Corporal P M Greany, (died of wounds in Russia 07/12/1918), 15592 Private S O Henry, 15775 Private C A Holland, (killed in action near Maricourt 11/05/1916), 26117 Private L B Mills, 15757 Private T H Molyneux and 24835 Private E L Phillips, (killed in action at Flers, 12/10/1916).

11 A Field Ambulance was a Royal Army Medical Corps unit, which provided the personnel for dealing with immediate battle casualties from the front line. Normally, three Field Ambulances served a division. Numbers 96, 97 and 98 Field Ambulances served the 30th Division, throughout the war in France and Flanders.

12 Taken from Hansard, 21 December 1916, during a debate on the Consolidated Fund (Appropriation) Bill.

13 See Chapter 3.

14 Originally, the bombardment had been planned to last only five days, with the attack due to commence on 29 June. However, because of foul weather, this had to be postponed for two days, and, necessarily, the bombardment extended by a further two days. Thus the countdown, which had made the first day of the bombardment on 24 June, 'U' Day, would have made 29 June 'Z' Day.

Chapter Five

'One mass of dead men, as far as you could see'

The plan

By the end of 1915 the German Army was in a very strong position on the Western Front. It was in control of vast areas of French and Belgian soil, and it occupied virtually all the high ground, which gave it an enormous artillery advantage, in a war dominated by artillery. Thus, it could afford to sit tight and wait for the inevitable attack that the Allies must make if they were to regain the initiative. As a result, it considered its fortifications to be semi-permanent, and set about making them as comfortable and as easy to defend as possible. In contrast, as the Allies were forced by political necessity to try to regain all their lost territories, their front-line soldiers were not encouraged to consider their trenches as anything other than temporary, and were thus not allowed to make them even comfortable, let alone permanent! As their positions facing the Germans were thus invariably on low ground, they also suffered the worst possible drainage conditions.

In December 1915, at a War Council meeting held at Chantilly, four of the main protagonists of the Allied cause, Britain, France, Russia and Italy, determined that they must regain the military initiative by the late summer of the following year. They planned to do this by making three simultaneous attacks on the forces of the Central Powers, on three different fronts. The Anglo/French plan was for a 'Big Push', or joint attack of equal strength, on the German Army in the region of the River Somme in France, at the junction where their own Armies met. However, the Germans pre-empted any such possibility in February 1916, when they suddenly and with devastating ferocity, launched a mass attack on the French lines at Verdun. The subsequent blood-letting there, inevitably called for a dramatic rethink of the whole situation.

The British Commander in the field, General Sir Douglas Haig, was forced to alter his plans for 'The Big Push', because of Verdun. The French had soon realised that they could not hold out against the relent-

A typical German fire trench with carefully prepared wooden supports and fire step.

less German assault on their eastern border much beyond midsummer without the help of a large-scale attack elsewhere on the Western Front. Moreover, because of their losses at Verdun, they also estimated that any joint offensive with the British, could only be mounted along eight miles of their front, as opposed to the twenty-five miles that was originally intended.

Haig was reluctant to commence the British offensive early, as the attack was to be made largely by men of the New Army, whom he considered needed more front-line experience, but the very real possibility of a sudden collapse of the French Army made his mind up for him. Thus, he eventually, decided to attack the German Army on the Somme, in late June of 1916.

The main force of the attack was to be carried out by the newly formed Fourth Army, under the command of General Sir Henry Rawlinson, an infantryman and a veteran soldier with much experience of armies worldwide – a rarity in those days. Haig's plan was that the Fourth Army would capture the high ground of the Pozieres Ridge and the land either side of it, on a front of about fifteen miles, which would allow cavalry and infantry to push through the resultant gap, and then break out into open country. The actual planning of the attack he left to Rawlinson, as he considered that his own experience as a cavalry officer, was not sufficient for a battle which would largely depend on infantry tactics.

General Sir Henry Rawlinson

Rawlinson's plan, however, was fairly simple. He realised that the massive bombardment that the German Army had mounted on the French front line trenches at the beginning of the Verdun assault had only failed by the merest margin to win the battle in its early phases. He reasoned that if he could use the same tactics on the Somme, but with a longer and larger concentration of artillery, then it ought to be possible to destroy the German Army in its trenches, sytematically, step by step, and then use his own infantry merely to take possession of the decimated German positions.

He placed great faith in the effectiveness of his artillery, and the efficiency of British industry to deliver sufficient guns in sufficient time to mount the largest bombardment in the history of warfare! It was probably because of this that he felt confident enough to ignore the lessons that both the French and German Armies had learned about infantry attack, at Verdun. Experience had taught them both that the best chance of success for any infantry unit lay in making rapid, sudden forward movements, from shell hole to shell hole, to confuse and surprise the enemy defenders. Instead, believing that no defenders would survive the ferocity of his bombardment, Rawlinson was to order his infantry to advance across No Man's Land at a slow, almost leisurely pace, carrying enough equipment on their backs to keep them going through several further days of expected breakthrough.

It was also Rawlinson's intention that while the British attack was moving forward, the French Army would mount their promised offensive, on the now limited scale, on its own front. This was to achieve a success which might divert the Germans from the main British assault, and to contain any attempt that the enemy might make to break through in force from the south to rescue their beleaguered comrades.

In all, thirteen British divisions were to make the assault, and the 30th Division, with its four battalions of Liverpool Pals, was given the position of honour, on the extreme right of the British Army, next to the French. The 18th Battalion, as part of the 21st Brigade, was to attack from assembly trenches to the east of Talus Bois, a jutting finger of woodland which pointed towards the German trenches below the southeastern corner of the fortified village of Montauban. Its objectives were the German trenches and fortifications of the Glatz Redoubt, to a point where a railway track cut the road running from the southwest corner of the village. The Glatz Redoubt, like all the fortified German positions was a warren of trenches, saps and hedgerows, which could hide machine gunners, riflemen, and snipers. At this point in the British line, No Man's Land was quite wide, at about 500 yards. Once the 21st Brigade's objectives had been taken, the 90th Brigade, itself part of the 30th Division, was to follow through the newly captured positions and take Montauban itself.

The 20th Battalion was to attack from assembly trenches just north of the village of Maricourt, and its objectives were the German reserve – in trenches known as Casement and Dublin Trenches. Dublin Trench connected the Glatz Redoubt with another redoubt, known as Dublin Redoubt, which was in the French sector. Once Montauban had fallen, the 20th Battalion was to send a company forward to capture the Briqueterie, which was in reality, only the remains of the old Montauban brickworks, to the west of the road that connected Bernafay Wood with Maricourt. The Briqueterie was known to be heavily fortified by a small garrison of Germans.

Field-Marshal Sir Douglas Haig

The 17th Battalion was to be on the right of the 20th, and was also to attack from north of Maricourt. Its objectives were similar to those of the 20th Battalion, except that it was to keep in contact with the French Army on its right, as it was the most easterly unit of the British Army on the Western Front. The French next to it were the 3me Bataillon of the 153me Regiment d'Infanterie, who were part of the 39me Division of the French 20me Corps, better known as the Corps de Fer, or Iron Corps. The Corps de Fer had distinguished itself since the beginning of the war, and had an excellent reputation under fire. Relations between the British and the French were extremely cordial, and could not have been better.

The 19th Battalion was to be held in reserve at the Chateau in Maricourt Village, to be used in case of a German counter-attack, but otherwise as a carrying battalion for the rest of the 89th Brigade.

Waiting

The bombardment which opened up on 24 June, left the Germans in no doubt that the start of the long awaited offensive was imminent. Curiously silent at first, eventually, on 26 June, the German artillery began its inevitable counter barrage, and the 17th and 20th Battalions in their front line trenches, suffered quite serious losses. On the late evening of the 26th, whilst relieving the 2nd Bedfords, in the front line near Maricourt, the 20th Battalion endured a fierce bombardment, which killed nine other ranks, fatally wounded Second Lieutenant W H Jowett, and wounded three other officers, and forty-seven men. Second-Lieutenant Jowett, who died from his wounds two days later, was the first of twenty-three officers from the Battalion who would lose their lives over the next two years. During the course of the same night, the 17th Battalion lost seventeen other ranks killed and fifty-seven wounded.

By this time the weather had broken, and rain was lashing down along most of the Somme front. It would obviously have made conditions for an assault impossible, and on the 28th Haig took the courageous decision to postpone the attack for 48 hours in the hope that the weather would clear. This now meant that the greatest attack in the history of warfare was put off until Saturday 1 July, and was timed to commence at 7.30 in the morning.[1] Although this was some three hours after dawn, the traditional time for infantry assaults, it was felt that the daylight at 7.30 would make for better artillery observation and accuracy, upon which, after all, Rawlinson had based most of his expectations. In any case, the French, who were supposed to be an integral part of the plan, also favoured an attack in daylight. In fact, General Foch, in command of the French Armies on the right of the British, and General Fayolle, commander of the French Sixth Army, which would operate on the immediate right of the Pals, virtually insisted on this time.

General Staff in the Great War are often criticised for lack of foresight and planning ability, but despite the fact that the postponement of the attack meant altering the plans of battle, extending the five day bombardment for another two days and resupplying upwards of half a million men in the forward areas, it seems to have gone remarkably smoothly. The extra strain of waiting another two days before going 'over the top', however, must have been unbearable for many of the British infantry. For the German Army, the extra two days wait must have been infinitely worse!

The majority of the German troops suffering the incessant British bombardment, were sheltering deep in the ground, in carefully prepared dugouts often fifty or sixty feet deep. Although at least one dugout of this nature had been captured by the Allies earlier in the year, it does not seem to have occurred to the planners of the 'Big Push', that the same type

A rare example of 'an appreciation of good services' citation issued by the 89th Brigade to Pals who had distinguished themselves in the field. Lieutenant G S Sutton, who witnessed Pte Hamilton's conduct was killed in action in March 1918, whilst serving with the 19th Battalion.

might have been in regular use on the Somme. This depth meant that most of the defenders were virtually immune from all but the heaviest calibre artillery shells, as they cowered deep in the earth. However, the ferocity of the bombardment did mean that no relief troops could get through to the front line, and no hot or fresh food could be delivered either. Also, all the defenders were only too well aware that the attack was coming, and they knew only too well that as soon as the gunfire lifted from their forward lines, they would have to race to the surface and try to repel the largest assault in the history of warfare. The certainty of this knowledge, and the shock to the nervous system of the 24-hour-a-day constant barrage can have done little for German morale!

Despite the fact that the deep German dugouts were largely left intact, the whole of the surface had been so pulverised, that it resembled a vast moonscape. Watching this barrage fall on the enemy's lines, very few of the Liverpool Pals could ever have believed that any Germans could have remained alive underneath its fury!

The stage was set and finally the date of the attack was given to us as 1 July at 7.30 am, a date and time one is never likely to forget. The artillery bombardment was to begin seven days before the attack and each battery was allotted targets to prepare the way for the infantry. The wire had to be cut so that infantry could move through it, the trenches and dugouts had to be blown to pieces so that there would be few if any living Germans left. Communication trenches and roads had to be battered so that fresh troops and supplies could not be brought into action. Nothing was to be left to chance. The noise of that week, day and night without respite, and increasing as zero hour approached, can never be forgotten by those who experienced it. It was terrible, but what it must have been like at the receiving end, I can only imagine. One could almost be sorry for the enemy. **Lieutenant E W Willmer, 17th Battalion KLR.**

We had watched the preparations for advance for several weeks. Less than a month ago, the valleys were all ablaze with the colour of an endless variety of wild flowers. Gradually, the smiling valley became dissected, and now, in place of the wild flowers, one noticed gun muzzles glinting mischievously and menacingly out of their excavated emplacements. The valleys simply bristled with them – guns of all calibres. There must have been thousands of them, and each had a huge stack of 'Kultur' – 'Kultur' of the German kind, waiting to be pumped into the enemy's lines.

The earth trembled and quaked when the bombardment commenced. Shells whistled and screeched through the air in a perpetual groan; the valleys reeked with the smell of smoke and powder, and at night the darkness was rent by myriads of lightning-like flashes which seemed to dance joyously along the surrounding hilltops. The din on one side of the line was deafening, on the other side it must have been nerve-shattering. For days the cannonade lasted with indescribable intensity. The German trenches were severely battered and their earthworks swept completely away.

At last the artillery had done its preliminary work, and done it well, and now it was the turn of the infantry! **17993 Private F Winn, 19th Battalion KLR.**

A 6-inch BL Mark VII field gun firing near Reningelst near Ypres) in June 1916. Similar guns took part in the bombardment of German positions on the Somme.

It must have seemed to the watching British infantry, that nothing could survive the torrent of shells which had obliterated the German lines, but the fact remains that the vast majority of Germans were still alive, although badly shaken. What the Royal Field Artillery needed, to destroy deep dugouts, was many more heavy howitzers. A howitzer was a field piece designed to fire a shell high into the air at a steep angle, which would then descend and explode, a short distance away. Howitzers were therefore ideal for trench warfare, especially if fitted with time fuzes, which would allow the shell to burrow deep into the earth before exploding.

The majority of light artillery weapons laying down the barrage before the attack on the Somme, however, were 18 Pounder field guns. These were ideal for harassing enemy troops in the open, and if fired by experienced crews, were very accurate and very reliable, but they were not designed for destroying deep fortifications. In addition to this, a great number of them were firing shrapnel shells[2], which did not harm the German defenders in their dugouts or destroy their barbed wire: If the infantry assault was to succeed, it was vital that the barbed wire, often in belts twenty feet thick, was destroyed. It is also estimated that out of all the British shells fired at this stage of the war, some thirty per cent failed to explode.

However, as far as artillery support was concerned, the Pals were perhaps luckier than most infantry, in that they held the right of the line next to the French. The French knew full well from their experiences at Verdun that a greater concentration of howitzers was needed to dislodge a well-entrenched enemy, and as their objectives north of Maricourt also included the trench system in front of Montauban, they employed their own artillery to bombard the village, with subsequent devastating effect. This was to be of vital significance for the success of the attack.

Even though they knew they were facing the possibility of death or mutilating wounds, the mood of the men the night before the attack was fairly buoyant. In the first place, they were keen to get to grips with the enemy, and 'put him in his place', and in the second, given all the evidence of their own eyes and ears over the last three months, they did not see how they could possibly fail.

> On the night before the 1st of July my Battalion stood in a field, and sang that song 'They'll Never Believe Us', but at the end, they changed it to 'They'll never believe us, that out of all this world they've chosen us'. This was because we thought that we were very lucky to be given the chance to go down in history. **24644 Private W Stubbs, 18th Battalion KLR.**

Certainly, Brigadier General Stanley was in no doubt about the outcome of the attack. On 30 June, he sent the following message to all the four battalions of the 89th Brigade:

> The day has at last come when we are to take the Offensive on a large scale, and the result of this will have a great effect on the course and duration of the war.
>
> It is with the utmost confidence that we go forward, the Battalions of which the City of Liverpool is so justly proud, determined to make a name for themselves in their first attack, and the 2nd Battalion Bedfordshire Regiment to add still more to their glorious record.
>
> The 89th Brigade occupies the most honourable position in the whole of the British Army, because not only are we on the extreme right, but we are fighting side by side with the celebrated French Corps de Fer.
>
> One and all will strive to prove ourselves worthy of this distinction, and to show our French neighbours what the British can do. That being so, success is assured.
>
> With these words I wish you the best of luck and a glorious victory. **Brigadier-General F C Stanley, CMG, DSO.**

Once the troops had moved up into the line to wait for morning, however, very few of them got any sleep. Apart from the constant drudge of carrying trench stores up the line for the attack, and the incessant thunder of the guns, the certainty of what the morning would bring, and the uncertainties of the day that would follow, made sleep an impossibility.

Each man would carry, in addition to his usual

The congested roads of the Somme show the strength of the British presence.

A German dug-out position to the rear of the village of Montauban in early 1916; probably the trench known to the British as Montauban Alley.

steel helmet, rifle, bayonet and equipment, a waterproof sheet, iron rations, one hundred and seventy rounds of rifle ammunition, two Mills grenades, and either a pick or a shovel. Those who were in the first wave, runners, bombers and stretcher bearers, all needed a high degree of mobility. As a result, they would not carry either a pick or a shovel, and only fifty rounds of ammunition. Bombers, whose job it would be to clear trenches and dugouts, would carry ten Mills grenades instead.

The bombardment continued unabated throughout the night, and then, with a little more than an hour to go before 'zero', all the guns on the front opened up with a barrage that made it almost impossible to speak. Thousands upon thousands of shells hurtled onto the German lines, with an intensity that made the previous week's effort seem almost minimal. The Germans now knew that the attack was coming soon, and tensed themselves for the inevitable moment when the barrage would shift from their front line, and the enemy infantry would come.

Along the trenches manned by the Pals, Russian saps had been dug by Tunnelling Companies of the Royal Engineers, out into No Man's Land. Russian saps were tunnels dug just beneath the surface, and at about the same time that the final bombardment began, the ends of these tunnels were opened to the air, and men of the 89th Trench Mortar Battery[3], positioned Stokes Mortars[4] to fire on the German trenches. It was at the end of one of these Russian saps, that what was probably the first fatality of 1 July amongst the Pals in the field, occurred.

They put me as a shell carrier on the Trench Mortar Battery. We had four guns, and a chap called Pringle was the gunner on our gun, and he and his mate had laid out about a dozen shells on a sap head into No Mans Land. There was a hole at the end where the shells would be fired through. In their excitement, they knocked one of them over, and it started to go off. His mate went to one end of the sap, but Pringle tried to get out of the hole, and when the shell went off it killed him.[5] **25561 Private H Redhead, 20th Battalion KLR attached to The 89th Trench Mortar Battery.**

At exactly 07.22 the Stokes shells of the 89th Trench Mortar Battery began to register hits on what was left of the German front line.

All night long the bombardment had continued,

Two views of a German Officers' Mess in the village of Montauban during the winter of 1915/1916.

but at 6.25am the final intense bombardment started. Until 7.15am, observation was practically impossible owing to the eddies of mist, rising smoke, flashes of bursting shells, and all one could see was the blurred outline of some miles of what appeared to be volcanoes in eruption. At 7.20 am. rows of steel helmets and the glitter of bayonets were to be seen all along the front line. At 7.25 am, the scaling ladders having been placed in position, a steady stream of men flowed over the parapet, and waited in the tall grass till all were there, and then formed up; at 7.30 am the flag fell and they were off, the mist lifting just enough to show the long line of divisions attacking: on our right the 39th French division; in front of us our 30th division; on our left the 18th.' **Lieutenant Colonel N Fraser-Tytler, DSO, TD,CLIst (County Palatine) Brigade, Royal Field Artillery.**

The waiting was over!

1 July: The 18th Battalion

The 18th Battalion attack began from assembly trenches east of Talus Boise which gave the men 500 yards of No Man's Land to cross, before they reached the German front line, named Silesia Trench. Four waves, each of roughly company-strength were to make the assault, and the leading wave at least was out in No Man's Land and over the parapet by the time that the whistle blew at 7.30 am. Perhaps because of the efficiency of the French heavy and medium mortars, the barbed wire in front of Silesia Trench was obliterated, and the first wave had little difficulty in reaching what was left of the German front line before most of the defenders had got to the surface.

Along all of the 30th Division front, the German front line was defended by the 6 Bayerisches Reserve Infanterie-Regiment, (6th Bavarian Reserve Infantry Regiment). Its troops came from Amberg, in the Bavarian Palatinate, and it had already suffered grievous loss, when a heavy shell had penetrated a command post in the Glatz Redoubt and killed nearly all the regimental staff. The 3 Oberschlesisches Infanterie-Regiment Nr. 62, (The 3rd Upper Silesian Infantry Regiment No.62) held the lines to the rear of the Bavarians. Its troops came from Cosel and Ratibor, in Upper Silesia.

Second-Lieutenant E Fitzbrown was the first officer from the 18th Battalion to reach Silesia Trench and as he led his men forward over the obliterated position, the Germans were already retreating towards their support line and beyond. Fitzbrown emptied his revolver at them as they ran, and accompanied by the rest of the first wave, pressed on towards this support line.

This photograph of a German trench at Montauban in early 1916 shows a portable notice board which displays maps of German successes on the Eastern front.

There, the British came under machine-gun fire from the high ground on the right, until the gun was finally silenced. Its crew, who fired until the last moment, attempted to surrender but was killed. The German support line was then sytematically cleared by bombing parties, who killed or captured what remained of the defenders.

Thirty Bavarian prisoners were captured. Inevitably, however, the Battalion had taken losses as well.

By this time the Germans were beginning to recover from the effects of the shelling, especially in the reserve and rear line areas, and before the British barrage moved forward onto the high ground just in front of Montauban, they began to bring machine guns out onto the trench parapets, which were well concealed by hedges. From there, they were able to pour a devastating fire onto the third and fourth waves of attacking Pals who were either still leaving the British front line or just crossing No Man's Land.

We had to get through all the barbed wire on the way. First of all you had to get up the firestep, – the ladders were on the firestep, and our Platoon Commander, he'd just got half way up, and he was shot through the head, he'd gone before we got out of the trench! His name was Golds. I was one of four that eventually got to the German lines out of sixteen. It was machine-gun fire that was hitting us, you could see the flash as it passed, and you knew that your friends had been hit. They were lying all over the place and some were screaming. It was awful, and I was only a kid of nineteen.

We had to keep going. I must have got a clout in the leg. We had an entrenching tool that was fastened at the side of the leg, but we always fastened ours at the front, to give you a bit of protection. I was carrying eight bombs in my bag,[6] and I must have been hit with some machine gun fire, because I had quite a depression in the entrenching tool. It must have hit the metal and made the depression, which made my leg quite stiff. I lost all my bombs when I made a dive into a shell hole.
17122 Private W Gregory, 18th Battalion KLR.

Meanwhile, as the first three waves began to move forward towards the German reserve line, known as Alt Trench and then on to the Glatz Redoubt itself, they suddenly came under enfilading fire from the left. This was from a machine gun which the Germans had sited at a strongpoint in Alt Trench. The gun was itself protected by a party of snipers and bombers, who, hidden by a rough hedge, were dug into a position in Alt Trench, at its junction with a communication trench known as Alt Alley. These bombers and snipers were themselves protected by rifle fire from another communication trench, Train Alley, which snaked back up the high ground and ran into Montauban itself. The machine-gun fire was devastating, and it is certain that nearly all of the Battalion's 500 plus casualties that day, were caused by that one gun! The situation was made worse by the fact that the men, out in the open, could not move forward, because the British barrage was still pulverising the area around the Glatz Redoubt.

Captain A de B Adam, who had played such a prominent part in repulsing the German attack on the Kaiser's birthday, now tried to lead his men forward to take the enemy position, but was hit by sniper fire almost before he had left Silesia Support line. Nevertheless, he led his men forward again toward the hedge, until he was hit once more, his servant 16818 Private F S Haslam dashing forward to dress his wounds. More bombing parties then moved up, but they were also hit by sniper fire or stick grenades. Second

Above: Something to 'crow about' after the first day of the British offensive on the Somme; a fanciful and dramatic picture of the successful capture of Montauban, which appeared in a popular British magazine in July 1916.

Lieutenant G A Herdman had his head blown completely apart by a German stick grenade at this stage, whilst the tenacious Fitzbrown was shot through the head and also killed.

Despite being seriously wounded, Adam now called up a party of bombers to reconnoitre the enemy's position, and shortly afterwards he was hit and killed by a German grenade, which also wounded Private Haslam. The bombing party, led by Lieutenant H C Watkins, then pushed slowly forward towards the hedge, and the Lieutenant ordered his best throwers to hurl Mills grenades at the German stronghold. With either great luck or superb skill, one of these grenades landed amongst the German bombers, killing two and scattering the rest. This allowed the bombing party to force their way up Alt Trench and Train Alley, shooting Germans in the hedges, and bombing them in their dugouts! Once more, about thirty Germans were captured, including one medical officer.

With the machine gun crew killed or captured, the survivors of the Battalion were able to push forward towards the right flank, and take the original objective, the Glatz Redoubt, which fell at 8.35 am. For their actions that day, Lieutenant Watkins was awarded the Military Cross and Private Haslam the Military Medal. 16993 Corporal T T Richards was also awarded the Military Medal for charging and taking a German machine gun, and killing its crew. This was probably the same gun crew that tried in vain to surrender, after firing until the last possible moment, from the high ground on the right flank!

Despite the relatively close proximity of the events taking place that morning, the commanding officer of the 18th Battalion, Lieutenant-Colonel E H Trotter, DSO in his dugout behind the British front line, did not know what was happening in front of him. He had missed seeing the Battalion go over the top, by a minute, and because of the carnage below the Alt Trench, no one had returned to report back to him. As a result, he sent a runner to find out what was happening.

A field medical card issued on the evening of 1 July l916 by 96 Field Ambulance RAMC at Maricourt, to 17122 Pte W Gregory, of the 18th Battalion. This type of card was always contained in a see-through envelope, and tied to the wounded man's epaulette. The letters 'ATS' refer to anti-tetanus serum, and 'GSW' stands for Gun Shot Wounds. Pte Gregory was wounded, as the card suggests, in his right arm and right leg, but by the explosion of a shell, not gun shot.

I was with the Colonel, the Adjutant and the Signal Officer, I was not only a linesman, I was the Signal Officer's batman. The Colonel called me over and he said "Go and see what's happened". So I went along the trench, and I got up on top, and the first thing that I saw, was all the dead, fellows just lying there, higgledy-piggledy all over the place, some two, three and four high, – one mass of dead men as far as you could see, right and left! The first one I came across was a captain. Captain Brockbank, he'd been hit twice. One of the fellows told me afterwards that he'd been hit, and got up, and then been hit again in the throat. Further on, I found Lieutenant Withy and Lieutenant Herdman, he'd had his head blown off by a German bomb, one of those potato mashers . . .

Then I bumped into my pal George Kirkpatrick, he lived in Rutland Street, Bootle. He had a bandage round him and I said 'What have you got', and he said 'I've got a bullet in the head'. I said 'Ah – get away, you'd be dead if you had a bullet in the head! How did you get that with a helmet?' He said 'It went right through the helmet!' The bullet must have hit him square on, – they normally glanced off and left a groove, but his went in, – into his head. So I helped him down, and lowered him down into the trench, and told him where to go, and howled at one of the stretcher bearers, but he said 'We're dead beat, if he can walk, he'll have to walk', and he had to walk down winding communication trenches, three miles to Bray, to the main dressing station, with a bullet in his head. He died twenty-one days later in hospital in England.[7]

When I went back and reported to the Colonel – I could see he was upset and he said 'Are you sure?' and I said 'Yes sir, I wouldn't make a story up like that, now would I?' So he turned to the Signal Officer and he said 'You'd better go and check'. So we got over the top, and I took him to where the Captain was lying, and he looked at me, and he says 'You wouldn't think he was dead, would you?' And you wouldn't have done, only for the white bandage around his throat. So we walked further along, and we found Lieutenant Withy, and you couldn't see where he'd been hit, but he was dead, I think he'd been killed by shellfire. There was a little tin box by his side, full of capsules, little tubes, and one of them was missing, – it was lying by his side. It looked as if he'd poisoned himself.[8] We came back then and we reported to the Colonel, and the Signal Officer told him 'It's true enough, there's very few of the men left, we'll be lucky if there's 150 left.' **16400 Private S R Steele, 18th Battalion KLR.**

By this time the Glatz redoubt had been occupied by men of the 18th Battalion, who held it, whilst the 16th and 17th Battalions of the The Manchester Regiment (1st and 2nd Manchester Pals) and the 2nd Battalion of the Royal Scots Fusiliers, all of the 90th Brigade, followed through. These troops eventually captured the village of Montauban itself, at approximately 10.05 am, after themselves coming under heavy machine-gun fire, which they were eventually able to silence with a Lewis gun. To the men of the 2nd Liverpool City Battalion, the Glatz Redoubt was a strange contradiction of obliterated trenches, and deep, safe, secure dugouts.

When we got to the Glatz Redoubt it was in a right mess. There were bodies everywhere, in all kinds of attitudes, some on fire and

The effects of a British bombardment on German positions on the Somme.

burning from the British bombardment. Debris and deserted equipment littered the area, and papers were fluttering round in the breeze. There were only two of us at the start, and then we met with others coming in. Then a German came up, we thought, out of the ground, but he was coming up some steps, – proper wooden steps, about twelve or thirteen. They led down to a huge dugout, with wire beds in three tiers, with enough beds to take about one hundred people. This little German came out and put his hands in the air, and I couldn't shoot him. I just indicated for him to go over the top to our lines, and he just scampered off, I'll always remember that little lad, he had a little round pork pie hat. I couldn't face the shooting.

Then we went down into the dugout to do a bit of 'souveniring'. The Germans must have left pretty quickly because they hadn't even taken their coats, so we went through the pockets. I got a German soldiers pay book, some buttons and a spiked helmet. One other souvenir I collected from a German officer's tunic was a diary, which I later handed over to an officer at a dressing station. The remainder of the 'loot', like tunic buttons, I found a ready market for, at the base at Rouen, where the Army Service Corps blokes who were on 7/- a day, pay, (ours was only 1/-), would pay anything for these small souvenirs. I hung onto the pay book and the officer's helmet, which my mother later threw out, because it had bloodstains on the inside![9]

Another of our lads also got a spiked helmet which he put on his head. He was still wearing it when we went back up to the trench, and he was larking around with it. Just then, one of the Manchesters who were coming up behind us, came round the traverse, and seeing him with this helmet on, must have thought he was a German, and shot him dead! It couldn't be helped, it was just one of those things that happens in war! **17122 Private W Gregory, 18th Battalion KLR.**

By nightfall, the full cost of the Battalion's success of the day was beginning to be felt, as roll calls were made, and the process of collecting the wounded and identifying and burying the dead began.

The Stretcher Bearers, whose ranks were thinned by casualties, worked incessantly on the wounded of every Battalion they came across. Many of our own and the wounded of the 90th Brigade remained on the field for forty hours. The Medical Officer and every available officer went out to deal with them and marked their positions and the 2nd Wiltshire Regiment, overworked as they were, gave valuable help in clearing the race course[10] into our front trenches. I cannot help mentioning that I never heard a man of my Battalion make a single complaint or request that he should be moved, but seemed to look upon it as a kindness that officers and the Medical Officer should come out and do their best for them when they were helpless.

Every fighting officer was hit by enemy bullets or shells except one, and he was accidentally bayoneted as he crossed a trench.

The Officers of the Battalion Staff alone were untouched.

I estimate our casualties at about five hundred.

I told both Officers and men that there were to be no SOS messages and the REDOUBT was to be carried by themselves without causing the Brigadier to use his reserves, so during the battle the estimates were put at a lower figure than the actual number.

I cannot speak too highly of the gallantry of the Officers and men. The men amply repaid the care and kindness of their Company Officers, who have always tried to lead and not to drive. As laid down in my first lecture to the Battalion when formed, in the words of Prince Kraft, "Men follow their Officers not from fear, but from love of their Regiment where everything had always and at all times gone well with them. **Lieutenant-Colonel E H Trotter, DSO, Officer Commanding, 18th Battalion KLR.**

It is impossible after so many years to make an accurate assessment of the Battalion's casualties on that day. The final figure would include dead, wounded, missing and prisoners, and Colonel Trotter's estimated figure of five hundred is probably reasonably close to the final figure. Discounting those who were hit on 1 July, and died subsequently of their wounds, the actual number from the Battalion who perished on the day of the attack, was seven officers and one hundred and sixty-four men! As a rough guide, Great War casualties are often quantified in a ratio of dead to wounded and other losses, of approximately one to three. If this rough guide is followed for the 18th Battalion on 1 July, then Trotter's figure would be remarkably close, at least mathematically, at five hundred and thirteen overall!

1 July: The 20th Battalion

The 20th Battalion attack from assembly trenches just north of the village of Maricourt, was made on the left flank of the 89th Brigade. No Man's Land at this point was between four hundred to five hundred yards wide, and at 7.30 am, the Battalion went over the top in four separate columns. No.1 Company was on the right, and No.2 Company was on the left, and they moved forward together, with the other two Companies forming third and fourth waves behind them, at intervals of about one hundred yards each. So rapid was their advance through German artillery fire and on across No Man's Land, (the Battalion War Diary describes it as "as tho' on parade in quick time") that they soon crossed Silesia Trench, the German front line, and without pause moved across the Silesia Support Line. They then continued to push forward, crossed Alt Trench, and captured their first objective, Casement Trench. At this stage they had to pause for the British barrage to lift from their secondary objective, Dublin Trench, but once this happened, Dublin Trench too, was occupied without opposition at 8.30 am. This allowed some men to move down the trench to the left, to join the 18th Battalion in the Glatz Redoubt, and others to move up the trench to the right, towards the Dublin Redoubt, which had fallen to the French at about the same time.

One of these men found a number of Germans in Dublin Trench, who were either unconscious, or badly stunned, from the British bombardment, - but not for long!

> *Some of them were just stirring and coming round. When I reached the next trench parapet, I fell into a barbed wire trap which cut my legs and arms, ripped my uniform and pinned me fast.*
>
> *When I got disentangled, I thought the best thing to do was to fix the trap so that the next man would not fall into it as I had done, so I put my rifle against the fence and tugged at the wire. When I turned round to pick up my rifle, I got a shock. A German had it, and he was pointing it right at me.*
>
> *I thought my last day had come, but for some unknown reason, he didn't fire. Whether I had left the safety catch on or what, I cannot imagine. The next thing I saw was the Jerry running towards Trones Wood, and taking my rifle with him, so I took a rifle from a dead German, but found I didn't know how to use it!* **22080 Private R J Fleetwood, 20th Battalion KLR.**

As the newly captured enemy trenches were all but obliterated by the British shelling, men of No.3 Company, led by Captain T Whiting then began to dig a new trench one hundred and fifty yards to the rear of the old German reserve lines. This was to provide shelter from the inevitable German retaliatory barrage, which would eventually rain fire on the newly captured lines. The positions of these former German trenches would naturally be known to the inch by the German gunners. These, and other events of the attack were described by one NCO of the Battalion in a letter written home to a friend in England three weeks after the attack.

> *Of course, you have had your usual rum ration about half-an-hour before, but I can assure you, you don't get enough to make you feel like telling anyone the history of your past life, let alone getting to the stage of being drunk. At any rate, you push forward until*

This German photograph taken before the Somme battle began, shows the extensive nature of a deep German dugout in the northern corner of Bernafay Wood, just north-east of the village of Montauban.

The exterior of the German staff headquarters building in Montauban before the attack on 1 July.

you arrive at the remnants of the barbed wire. There are always portions of it that have escaped being cut, and the thing that struck me most just here was the amount of lurid language used by fellows who happened to get stuck. It was such glowing hot language that the wire should have melted away by itself under it. Through one fellow becoming entangled, he was hit with an explosive bullet which made an awful mess of his arm.

On you go amidst the shells (some of which burst dangerously near you) and having advanced some fifty yards or so you become strangely cool and confident. We halted at the first German trench which had practically been evacuated. The line in front of us had by this time reached the German second line. Whilst at this halt many of us lit cigarettes and viewed the situation generally. We were now in possession of three lines of German trenches, and it all seemed so easy – much easier than when we had practised it behind the line. Of course shells were dropping all about us, but we took them philosophically. By this time the . . .[11] who were reserve carriers for us, had now come over too, and we watched them from the battered German trench as they came on with coils of barbed wire, ammunition etc., over their shoulders.

Sometimes you would see one of them coming forward laboriously with a big load. Suddenly a 'Jack Johnson'[12] would scream over our heads and appear to burst within twenty yards of the carrier. At any rate, the smoke from it would clear away and you would again see him like the man

Below left: *A German artillery control centre during the Battle of the Somme*
Below right: *A German machine-gun crew probably giving covering fire to attacking troops earlier in the war. By July 1916, all frontline troops on the Western Front had been issued with the 'stahlhelm' or steel helmet for front line duty which gave better protection against shrapnel bursts than the 'pickelhaube' or spiked helmet shown being worn here.*

'off to Philadelphia', striding forward with his 'bundle on his shoulder', as though nothing had happened. Of course what made it all seem so easy was that Fritz had made 'a strategic retreat' – that is to say, he had run like the devil a few miles back.

There were great numbers gave themselves up as prisoners, and they did it in anything but a manly sort of way. When our line pushed on further, we found that the Germans' 4th line was too congested, so had to dig ourselves in. The shelling was very bad just here, but not many of our fellows got hit. It was later on, when we were holding the trench we had dug, that we had a hot time, for the German artillery picked up the range very quickly, and were soon dropping shells right over the parapet. Most of us out here have a great respect for their artillery, but so far a contempt for their infantry. **26540 Lance Corporal J Quinn, 20th Battalion KLR.**[13]

Although the 20th Battalion had achieved all its objectives so far, it had naturally lost men too. By the time that the German reserve lines had fallen, and Lance-Corporal Quinn was in the process of digging in, two Second-Lieutenants, F Barnes and J C Laughlin, had been killed, a Captain H H Robinson, had been wounded, and forty-nine other ranks had been killed or wounded. At about 11.50 am, orders were received from Brigade, to proceed with the second phase of the attack, – the capture of the Briqueterie, to the west of the Maricourt to Bernafay Wood road. A briefing for this operation was given at the north-west corner of Germans Wood, which had recently fallen to the 17th Battalion, and then, at 12.20 pm, No.4 Company, led by Captain E C Orford, began the assault, following a thirty minute hurricane bombardment on the position, by the Royal Field Artillery. To help pin down the German defenders, and to secure the left flank of the attack line, a section of bombers had already moved up the twin communication trenches known as Nord Alley and Chimney Trench. They were able to prevent any attempts to rescue the Germans in the Briqueterie, or cut off any attempt by the garrison to escape. As a consequence, Orford and his men, after a brief fight, were able to overcome the garrison there and capture a colonel and adjutant from the head-quarters staff of the 3 Oberschlesische Infanterie-Regiment Nr. 62, the commander and observer of Feldartillerie Regiment von Clausewitz, (l. Oberschlesisches) Nr.21, one other officer, and forty men. Two machine guns and many documents and maps, were also captured. In the early afternoon, the Germans commenced their long-expected bombardment of their old positions, and this continued almost unabated, throughout most of the night, with its usual professional accuracy. More casualties were caused by this bombardment, and by the end of the day total casualties amongst the Battalion amounted to twenty-three of all ranks killed, and seventy-seven wounded. In view of what had been accomplished, and the numbers engaged in the assault, these figures must be seen as being remarkably small.

1 July: The 17th Battalion

The objectives of the 17th Battalion were very similar to those of the 20th, who were on their left, but No Man's Land on the frontage of the 17th Battalion narrowed to between four hundred and two hundred yards at the point where the German front line trench, Silesia Trench became Faviere Trench. At 7.30 am the first of the four companies in the attack, a mixture of 'A' and 'B' Companies, left the British front line, and with the French on their right, advanced in quick time towards the remains of the German wire. One hundred yards behind them, in the second wave, which consisted of the rest of 'A' and 'B' Companies, the commander of the 17th Battalion, Lieutenant-Colonel B C Fairfax stepped over the parapet at the extreme right of his troops. At exactly the same time, Commandant Le Petit, the commander of the French 153me RI,

French soldiers probably at Verdun attacking at the rush, in marked contrast to the way in which the British were ordered to attack on 1st July.

stepped over the parapet at the extreme left of his troops. Pushing forward to the front of their respective lines of men, the two commanders then led the second wave forward, together, arm in arm, across No Man's Land. 'C' and 'D' Companies, the third and fourth waves, left the British front line slightly earlier than scheduled in order to avoid a sudden German barrage which began to fall on the British assembly trenches, and then they too moved in quick time towards the enemy positions.

They were engaged by weak artillery and small arms fire, but nevertheless crossed over Faviere Trench and Faviere Support Trench, with little difficulty, as the German defenders had already fled, or had stayed in their dugouts. (After the advance some 300 prisoners, mostly from Infanterie-Regiment Nr.62, were captured by The 2nd Battalion The Bedfordshire Regiment, who were the 89th Brigade 'mopper uppers'). As the advance passed through Germans Wood, thirty Germans surrendered to 'A' Company and were sent to the rear, as the Pals pushed on to take Casement Trench. Once the British barrage on Dublin Trench had lifted, it too, was entered, at its western end, and fell without resistance at 8.30 am. At the same time, its eastern end was entered and occupied by the French, as they took the Dublin Redoubt, and once more Fairfax and Le Petit met, somewhere in the middle, and embraced in the spirit of unity, comradeship and victory.

All that remained was to consolidate the position against the inevitable German counter-barrage, and as with the 20th Battalion on the left, carrying parties from the 19th Battalion brought up picks, shovels and trench materials, and new lines were dug, clear of the former German positions, which saved many lives when the German barrage eventually fell.

The action was witnessed first hand, by Lieutenant E.W.Willmer.

On the first day, I was in the support line and got a marvellous view of all that happened. It was a lovely sunny morning and promptly at 7.30, our barrage lifted from the German front line to their support line, and waves of British troops left the trenches and walked out into No Man's Land, in extended line, with bayonets fixed and rifles at the carry. There was no hurry, and so far as our battalion was concerned, very little resistance. Our casualties were small and we gained our objec-

This famous photograph shows men of the 34th Division at La Boisselle walking towards the German lines on 1 July 1916.

tives without trouble, and dug in at our new position. Further north of course, things were very different, and the casualties were appalling! **Lieutenant E W Willmer, 17th Battalion KLR.**

In a letter written after 1 July, very much for home consumption, another soldier from the 17th Battalion described the attack, perhaps with a little more embellishment!

It will please you to know that our Brigade led the attack and also held what we consider a great honour, the extreme right position of our line. It is a strange feeling you get just in the couple of minutes before going over, but once over, everything is all right. Never in my life, and especially considering what we were going through, have I seen our boys so calm, there was no mad rushing about, we simply walked over as if on parade. Shells were dropping like rain, (poet's license) and men were being blown to bits all around, but there was never a halt or falter till we reached our objective.

All that I can say is that I'm lucky to be alive and I have often thought since, that someone must have been pray ing for me during these trying days. We have been dubbed the 'Black Kings', owing to our black ribbon for identification purposes, by the regiments near to us, and they think the world of us and say that the way we advanced was simply great. In fact everyone in Authority is proud of us and it is said General Stanley was absolutely delighted with us, and clapped his hands as he watched the Germans flying for their lives.

The French were marvellous and went into action singing the 'Marseillaise', so it could not be expected of us to go wrong with such gallant comrades beside us. We have spoken to several of them since, and they are very helpful; those who cannot understand our lingo keep saying as we pass 'Bon camarade, Bon Avance'.

We are going into action again shortly, and many will fall. If it should be my turn, and one can never tell, my wish is that you will not grieve, but think only that I have done my bit, 'For how can man die better'.

Of what took place, there is lots to tell, but it takes so long to put it down in writing. One thing I would like to say, is that our Officers were great and fearless, and led us with fine pluck and courage. **15406 Private I Smith, 17th Battalion KLR.**

Thus all the objectives were taken and the victory was complete. Perhaps, however, it was all the more illustrious because it had been achieved with an almost unbelievable lack of casualties. According to the official records *Soldiers Died,* and *Officers Died in the Great War* published by HMSO after the war, not one soldier from the 17th Battalion was killed in action on 1 July. However, if one studies the cemetery records of the Commonwealth War Graves Commission, and a 17th Battalion casualty ledger in the archives of the Imperial War Museum, a further seventeen soldiers are listed as having been killed in action on l/2 July. However, the 17th Battalion positions definitely came under shellfire on 2 July, so it is by no means certain that all seventeen casualties would have occurred on 1 July. Because of the obvious confusion surrounding 1 July, however, it would seem more likely that some men were killed on that day, but their exact day of death could not be irrefutably verified.

Certainly by noon on the 1st, three officers, Captain E C Torrey, Lieutenant D H Scott and Second-Lieutenant P L Wright, and about a hundred other ranks, had been wounded. Scott had led 'A' Company into action, on the right of the line, and had been in touch with his French opposite number, Capitaine Mirascou, who had led the left flank of the French first wave. Mirascou, also, had been wounded, and was evacuated to the same French military hospital as Scott, at Cerisy, where, unfortunately, Scott died of his wounds the following day, on 2 July 1916.

Two soldiers from the Battalion definitely died on 1 July, 15765 Private J R Pearson from Northallerton in Yorkshire, and 15470 Private J S Riding, who came from Liverpool. However, they both died from wounds they received during the German counter-bombardment on their front line trenches at Maricourt on 27 June, and were thus not involved in the fighting of 1 July. Furthermore, apart from Second-Lieutenant Scott, only two of the one hundred men recorded as being wounded on 1 July actually died of their wounds in the seven days following the attack. These were 16119 Private H Crawford, shot through the right side and right thigh, who died on 2 July, and 15309 Private G Cochrane, shot through the right

German prisoners captured early in the Somme battle being looked after by British soldiers who clearly display the complexity of their battle order.

thigh and arm, who died on 6 July. It is tempting to hope that the vast majority of the others might have survived their wounds!

An evaluation

It is not always easy to decide exactly what constitutes success in military terms, but as a general rule, during the Great War, a unit which captured its objectives with a fairly low incidence of casualties could be reasoned to be successful. If one then applies this rule, there can be little doubt that the 30th Division was the most successful of the Divisions that attacked the German lines on 1 July. It achieved all its objectives on time, and without incurring the terrible losses suffered by other divisions further north. If this is so, then there is no doubt that the Pals Battalions from Liverpool and Manchester were the most successful battalions on the day. If one then takes into consideration the casualties suffered by each battalion from the Division, and even allows for those suffered by the 18th King's below the Glatz Redoubt, the three Liverpool Pals Battalions in action that day must have been the most successful of all who went over the top with the first wave on the morning of 1 July. So why were they so successful, when many others failed?

The overwhelming feature in their success must lie with their geographical position on the extreme right of the British line, next to the French. The French Sixth Army, under General Fayolle, was successful in taking all its objectives on 1 July, almost certainly because it had learned the hard lessons of warfare from Verdun. Thus, at least, the right flank of the Pals' attack was secure. Furthermore, the French Army was equipped with many more medium and heavy howitzers, whose shells actually penetrated dugouts, and blew away barbed wire. It can surely be no coincidence that the further away from the French that an attack was made, the less successful it was in quickly achieving its objectives, and consequently, the higher was its casualty rate. This is amply demonstrated by the fact that even from the Pals Battalions, the 17th Battalion on the right, suffered the least casualties, whilst the 18th, on the left, suffered the most.

The width of No Man's Land is also significant. Where it was widest, heavier casualties were suffered, presumably because the Germans, stunned though they were, had more time to recover and race to the surface to repel the attack. Also, rather than walking slowly across No Man's Land with rifles at the port, as instructed, it would seem from the War Diaries and other accounts, that the Pals went 'at quick time', some even starting from over the parapet, when the whistle blew, and as a result, they had crossed the German front and support line trenches before the enemy's opposition had stiffened. It is significant also that the machine gun that caused so many casualties to the 18th Battalion was not firing from the German front line, but from a fortified position further back, probably out of the firing area covered by the French howitzers. Many other British attacks that day failed because the troops could not get past the devastating German machine gun fire that hit them from the German front line whilst they were still exposed and struggling in the middle of No Man's Land.

It is probable, also, that on average all ranks of the Pals were better educated than their Regular Territorial or New Army comrades. This may have allowed them to think out the situation on 1st July to their better advantage. Furthermore, very few, if any, of the officers leading the attacks on that day would ever have thought of making a career in the Army after the war. Thus, they would not hesitate to vary an order slightly, if they thought it might make it work better. This would have been unthinkable for an officer who had ambitions of seeing his career prosper in the future! Certainly the ability to communicate with the French seems to have been a very important factor in the cooperation between the 17th Battalion and the French 153me RI, and it is probable that by reason of education, there would have been a much higher proportion of French speakers in the Pals Battalions than in many of the other battalions engaged on the Somme! This cooperation also extended to the artillery, where there was a particularly close liaison between the French Artillery and the 30th Divisional County Palatine Artillery.[14]

Making comparisons with other units who fought on that day is certainly an odious task, because it is not possible to compare like with like over a front as long as the Somme. However, some facts are inescapable, if one continues to measure success by objectives taken and casualties sustained. Perhaps it is something peculiar to the British character that splendid failures are remembered and revered, yet over-whelming successes are totally ignored. It is nevertheless a puzzling fact that other Pals battalions that were virtually wiped out and gained nothing on 1 July 1916, are remembered today with awed respect, yet few people have ever even heard of the part played by the men from Liverpool:

Notes

1. Many historians neatly label 1 July 1916 as 'The middle day of the middle year of the war', but in fact, as 1916 was a leap year, there was no middle day of that year.

2. Although the term 'shrapnel' is often used to describe any piece of metal shell fragment, technically, a shrapnel shell is one with a hollow centre filled with hundreds of lead or steel balls. It is designed to burst in the air, on a timed fuze, to shower the balls over troops on the ground. It was invented by a Lieutenant, (later General) Shrapnel of the Royal Artillery at the end of the eighteenth century.

3. Trench Mortar Battery troops were infantrymen trained in the use of light trench mortars, and found from within the Brigade, whose numbers they took.

4. The Stokes Mortar was a light mortar developed specifically for trench use by Sir Wilfred Stokes, a celebrated civil engineer. It was first used in action in March 1916. Its shell resembled a silencer box from a car exhaust, and although it did not look particularly martial, it could be deadly, when fired by an experienced crew.

5. 17928 Lance Corporal L. Pringle, 19th Battalion KLR, attached to The 89th Trench Mortar Battery, came from Wallasey in Cheshire, and is buried in Peronne Road Cemetery, Maricourt. Although the post-war HMSO publication *Soldiers Died in the Great War* lists his death as

occurring on 3 July 1916, this is clearly an error. The more reliable source, the Commonwealth War Graves Commission, gives his correct date of death as 1 July 1916, and this date is inscribed on his headstone.

6. The fourth wave of troops were 'mopper-uppers', whose job it was to consolidate trenches won, and clear out dugouts. Private Gregory was probably carrying ten grenades in his bag, not eight, although as a bomber, he shouldn't have had to carry the entrenching tool that probably saved him from a very serious leg wound.

7. 16376 Private G H A Kirkpatrick was evacuated to England, where he eventually died from his wounds nearly three weeks later, on 20 July 1916. He is buried in Bootle Cemetery, Merseyside, not far from where he lived.

8. Lieutenant Basil Withy was not in fact dead when Private Steele saw him. Presumably he must have been wounded, and had taken morphine to ease the pain, before lapsing into unconsciousness. Officers often carried morphine capsules into action with them, and the tin box which Private Steele thought was poison was obviously one of these. Withy was subsequently picked up by stretcher bearers, and taken to No.21 Casualty Clearing Station at Corbie, where he died of his wounds the following day.

9. Private Gregory was wounded in the evening of 1 July, by shrapnel from a German 5.9" shell, which burst as he was searching amongst the dead for a full water bottle. He was eventually evacuated to Rouen, and after recovery, was to be wounded twice more, before finally being transferred to the Royal Flying Corps, in England in 1918. After his demobilisation, he volunteered to fight the Bolsheviks in Russia, and served as a trench mortar battery sergeant with the Royal Fusiliers. In the 1930s, he rediscovered some of his souvenirs, and corresponded with the original owner of the pay book he had found in the dugout, 472 Reservist Michael Theurlein, who came from the village of Langensteinach, not far from Nuernberg.

10. This extract is taken from the 18th Battalion War Diary, and contains Colonel Trotter's second reference to the 'racecourse', which as a specific location, is not identifiable on any of the Army maps of the time. It would appear, however, that he is referring to the land immediately in front of the original British front line. It may be that No Man's Land, which could often be relatively untouched by shelling, was commonly referred to as a 'racecourse', in contrast to the rest of a battlefield. Certainly the Canadian VC flying 'ace' Billy Bishop in his book *Winged Warfare* likened No Man's Land to one, in an early impression of the Western Front: 'At six hundred feet, we were free of most earthly noises and again I looked down. For the first time I saw the front line as it really was, mile upon mile of it. Now running straight, now turning this way or that in an apparently haphazard and unnecessary curve. The depth and complexity of the German trench system surprised me. No Man's land, much wider in places that I had realised from any map, looked like a long neglected racecourse by reason of the distinctive greenness of its bare but relatively undisturbed turf.'

11. In the original letter, the censor had obliterated the name of the carrier unit mentioned by Quinn. It was, of course, the 19th Battalion KLR, who were being held in reserve in Maricourt Chateau.

12. A 'Jack Johnson' was the nickname for a heavy calibre German shell, with a gunpowder filling, which exploded with a great amount of black smoke. It derived its nickname from Jack Johnson an American negro heavyweight boxer, who was very popular at the time. This type of shell was also sometimes called a 'coal box'. British shells were more usually filled with lyddite, which exploded with a burst of yellow smoke.

13. Lance-Corporal Quinn was killed in action nine days later, during the action at Guillemont. His body was not found and identified after the war, and he is commemorated on the Memorial to the Missing of the Somme, at Thiepval.

14. The CXLIXth Brigade, Royal Field Artillery, was, at this time, commanded by Lieutenant-Colonel The Hon G F Stanley, another brother of Lord Derby.

Chapter Six

'His name will never be forgotten in Liverpool'

The aftermath

The night of 1/2 July did not pass entirely well for the victorious Pals, now dug in in the former German lines. The German Army had not yet accepted that Montauban was irretrievably lost to them, as they knew that they still held nearly all of the original front line further north. At this stage, although the village of Mametz had fallen, the Germans had not yet evacuated what had become a salient around Fricourt. Consequently, they shelled their old positions with devastating accuracy from the high ground on the other side of Caterpillar Valley. This caused inevitable casualties amongst the new defenders, and for most it meant a second night in succession without sleep – and after all they had been through! At about 3.30am a patrol of about one hundred Germans advancing in rushes from the corner of Bernafay Wood, tried to recapture the Briqueterie from its 20th Battalion defenders, but it was spotted and repulsed without loss to the Pals. Many German dead were found in front of the position at first light.

2 July found the 18th Battalion in the Glatz Redoubt, the 17th Battalion consolidating Casement and Dublin Trenches and the 20th Battalion also in Dublin Trench, and the Briqueterie. The 19th was still acting as Brigade carriers. Mid-morning, two patrols from the 20th Battalion, led by 23901 Sergeant C B Ambler and 22905 Lance Corporal R Abbott were sent into Bernafay Wood, to see if it was still occupied. They found a scene of total devastation consisting of broken and twisted trees, wrecked and smashed dugouts and many dead bodies. The only live Germans they found were hiding amidst the ruins, and altogether seventeen were brought back to the British lines as prisoners. That evening Ambler returned to the wood, and found it still deserted apart from three more Germans, whom he captured. When the wood was finally occupied, on the night of 3/4 July, by the 27th Brigade of the 9th Division, the Germans shelled it very heavily, causing many casualties, including over thirty of the 20th Battalion, which was still occupying the remains of the Briqueterie.

Lieutenant-Colonel E H Trotter, DSO, Commanding Officer of the 18th Battalion, until his death in action on 8 July 1916.

British troops in the ruins of Montauban after its capture.

The village of Montauban itself was just a heap of rubble, which afforded little cover to its captors.

Our infantry had very little house-to-house fighting in the ruins of Montauban, and the prisoners captured there were half-starved, as during the bombardment hardly any rations had reached them.

One Hun, however, did put up a great fight. From a hidden hole beneath a pile of masonry, he sniped away for thirty-six hours, killing at least seven of our infantry. Everything possible was tried to shift him, but fire, water, smoke candles and packets of bombs with gun-cotton all failed. Finally the Engineers removed the whole mound with gun-cotton, so one never knew the secret of his dug-out or how he kept alive so long. A stout fellow!

In the pleasant wars of by-gone days possibly lace capped old ladies would have been sitting at the porch of their homes, waving their knitting needles to each passing "liberator" while the children gambolled on the lintels or wherever children ought to gambol. Today this village at all events is not a pleasant place; it is almost impossible to trace the line of the streets, hardly a wall higher than a few feet still stands, and all around reeks with the indescribable stench of stale high explosive and unburied remains: the distilled spirit of death brooding over the whole hamlet.

During an extra heavy rainstorm we sheltered in a dugout formerly Battalion HQ of a Bavarian unit. The room was filled with papers and books, and it was interesting to read their typed reports of the effect of our bombardment up to the 26 June: after that, words (on paper) seemed to have failed them. **Lieutenant Colonel N Fraser-Tytler, DSO, TD, CLIst, (County Palatine) Brigade, Royal Field Artillery.**

The ruins of Montauban after its capture, showing the result of the British and French bombardments.

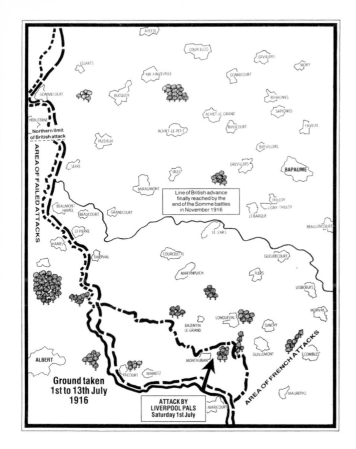

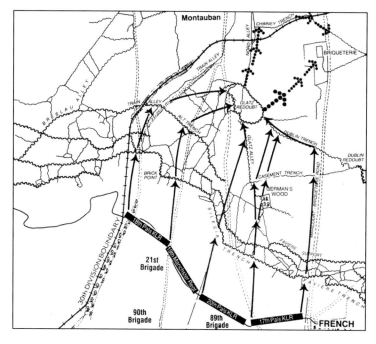

Assaults on 1 July by the Pals battalions: After the Glatz Redoubt had fallen to the 18th Battalion, and both the 17th and 20th had taken their objectives, a successful two pronged attack was launched against the Briqueterie by the 20th Battalion.

The task of recovering the wounded and dead had begun on 1 July, and continued throughout the six hours of darkness to the 2nd. At least the victory at Montauban gave the wounded a better chance of being picked up and treated than elsewhere in the line, as the positions where the Pals had dropped were now in British or French hands. Elsewhere on the Somme front, many men would die in No Man's Land in shell holes or even in front of the enemy's wire, because stretcher bearers just could not get near them without being spotted by German machine gunners or riflemen, who were often in no mood to compromise! Throughout the day of 2 July, stretcher bearers south of Montauban worked ceaselessly to bring in anyone who had any chance of survival. Although they were hampered by the inevitable German shelling, they still managed to carry out their tasks of mercy reasonably well.

Inevitably, it was the bearers of the 18th Battalion, whose infantry had suffered the most casualties, and whose own ranks had been depleted by death or wounds, who were the most overworked. Their biggest problem, apart from combing the battlefield and searching the shell holes, was finding enough unwounded men from the Battalion to help get the casualties back to the Regimental Aid Post in the British front line. From there they would be taken to a Field Dressing Station, or the Advanced Dressing Station at Maricourt, and ultimately, if they lived that long, to No.21 Casualty Clearing Station at Corbie. Lieutenant-Colonel Trotter noted the problem, on the morning of 2 July.

I was out personally this morning with the Doctor and visited the field. The wounded of my Battalion were very cheerful and I had not a single complaint. It was not so with the men of the 90th Brigade who complained bitterly that they had been left since 8am yesterday morning. Our Doctor treated all alike, got food and water off the dead and made them as comfortable on the field as we could, but owing to casualties amongst our stretcher bearers, the work of moving them by this means was of course slow. But I am glad to say that sections of the RAMC Cyclists were beginning to work, and I hope clearing will be satisfactory. The men are mostly in shell holes, so we placed rifles in the ground to mark the spots. **Lieutenant Colonel E H Trotter, DSO, Officer Commanding, 18th Battalion KLR.**

Even if a soldier got to a dressing station fairly soon after being wounded, however, there was no guarantee that his problems were over.

We had our waggon lines at Becordel, and just opposite was a big Field Dressing Station. There were about twelve or eighteen big marquees with doctors in, attending to the wounded. When a chap was wounded, he came down to Field Dressing Station, and they just patched him up, and

British wounded receiving medical attention. The capture of Montauban at least meant that the wounded could be collected from the battlefield and treated.

from there he was moved to a Casualty Clearing Station and they moved the wounded from there probably down to a rail head, to Rouen or one of the big towns where there was a big hospital. Well, there was so many wounded that all these eighteen marquees were full of the wounded, and there was a line about several hundred yards long, with eight or nine rows of chaps lying on stretchers, waiting to be attended to, and some of the chaps were out three, four or five days before they were attended to! Consequently, a lot of them died. Had they been able to attend to them right away, a lot of them would have survived, but they didn't think for one minute that there would be the casualties there were! Many chaps would have survived had they had the hospital accommodation. **292261 Gunner T P Brennan 125th (County Palatine) Heavy Battery, Royal Garrison Artillery.**

One of those who was wounded on 1 July, and not evacuated until 2 July, was 16744 Private John Redhead of No.3 Company of the 18th Battalion. Despite eventually getting back as far as No.13 Stationary Hospital at Boulogne, he still succumbed to his wounds, and died nearly three weeks later, on 20 July. We have already read an account of his brother's experiences on 1 July. He was 25561 Private Harold Redhead, serving with the 20th Battalion, but attached, at the time, to the 89th Trench Mortar Battery. After John's death, another brother, at home in England, wrote to the 18th Battalion to try to get further details of John's fate. The eventual reply he received from the officer commanding his brother's Company, Captain G Ravenscroft, illustrates quite clearly the chaos and confusion which reigned on the battlefield at the time.

'8th August 1916 Dear Sir,
Your letter of the 30th ult. has been handed to me by the Commanding Officer to reply to. Your brother who was respected and liked by both officers and men of the Company, was shot with a rifle or machine gun bullet, in the head on the 1st July. His friends made him as comfortable as possible in the German lines, (where he was hit) getting blankets from a German dugout, and bringing him coffee. He suffered very little pain, and was all along very cheerful, refusing to be moved, until what he thought were worse cases had been attended to. He was taken in by the stretcher bearers on the morning of 2nd July, after which we have no trace of him. You must understand that with the vast numbers of Casualties in the big battle, it was not possible to collect all the wounded at once. All through he was patient and cheerful, and his loss to the Company is great. On behalf of the Company and myself, I beg to offer his family through you, my sincerest sympathies.
I am, Yours truly, Guy Ravenscroft, Captain, No 3 Company.'
Captain G Ravenscroft, Officer Commanding No.3 Company, 18th Battalion KLR.

A shallow British grave showing corpses wrapped in either their issue blankets, or waterproof sheets, and no doubt identified by their green identity tags. After the war, such burials were sometimes disinterred and relocated, or formed the nucleus of a permanent war cemetery. Often field burials were no deeper than this, unless a disused trench was utilised.

Captain Guy Ravenscroft from Birkenhead, Cheshire, would himself die just over two months after writing the letter. He would be killed in action, during the Battalion's assault at Flers on 17 October.

After Private John Redhead's death was notified to his family in Seaforth, Liverpool, his mother sought details of his burial. She received the following reply:

No.13 Stationary Hospital, Boulogne. 2nd September, 1916.

Dear Mrs Redhead,
The Chaplain who wrote to you about your son, Pte J Redhead 16744, has gone up to the front, hence the delay. I have looked up the records and find that your son was buried on 22nd July at 8 am in the Boulogne Cemetery, in the position reserved for the graves of British Soldiers; the funeral was attended by a guard of honour, and everything possible was done as always, to show honour and respect to the memory of our men. The body was of course buried in a coffin.[1] The grave is marked by a simple white cross and will be kept in order carefully by the military authorities, who do this work very well. The number of the grave is 3559 the funeral was taken by one of the Church of England Chaplains. Please accept my deepest sympathy with you in your great trouble. May God comfort – give you his comfort.
Yours very sincerely A H Balleine CF. **Reverend A H Balleine, Chaplain to the Forces, Royal Army Chaplains' Department.**

Once all the wounded had been dealt with, the more macabre task of collecting and burying the dead began. Even before the battle, one soldier remembers plans being made to receive them.

Before the 1st of July, our fellows dug two big trenches, and they said they were for water, but I believe, I was told afterwards, that that was where they piled the dead, for the time being, like sardines. There was so many of them you see, you couldn't bury them at the time, it was utterly impossible. So when they really buried them later – how they got them out I don't know – they must have used blankets . . . there must have been one terrible smell! It was on the left hand side of the road to Carnoy – they told us they were reserve water tanks. **16400 Private S R Steele, 18th Battalion KLR.**

It was probably the same burial ground which a famous war correspondent wrote about on 4 July. Incredibly, his account was printed in the Merseyside newspaper *The Liverpool Daily Post and Mercury*, only two days later!

Closer to the lines there was a scene which would make one weep if one had the weakness of tears after two years of war. Our dead were being buried in a newly made cemetery, and some of their comrades were standing by the open graves and sorting out the crosses – the little wooden crosses which grow in such a harvest across these fields of France.

They were white above the brown earth, and put into neat rows, and labelled with strips of tin[2] bearing the names of those who now have peace. **Philip Gibbs, Official War Correspondent.**

Although a lot of corpses were removed to organised burial grounds in the rear, the majority of the dead were buried more or less where they fell, in shell holes, or portions of captured German trenches.

Everywhere there was bodies . . . I saw them burying hundreds of men. They'd dug a hole as big as a room, and thrown the bodies in, crossing bodies. They had to have masks on because the smell was terrible! The first day of the Somme was terrible, terrible. **292261 Gunner T P Brennan 125th (County Palatine) Heavy Battery, Royal Garrison Artillery.**

The heat of July made it necessary to get the bodies underground as quickly as possible, and although each grave location was marked at the time, the majority of them were subsequently destroyed or lost through shellfire. Consequently, most of the Pals dead from 1 July have no known grave, and are now commemorated on the Memorial to the Missing of the Somme, at Thiepval. The majority of those whose remains were subsequently discovered by the Imperial War Graves Commission are buried today in Bernafay Wood Cemetery, Danzig Alley Cemetery at Mametz, and Peronne Road Cemetery at Maricourt. It was at this latter that Private Steele saw 'water tanks' being dug, and Gibbs saw burials taking place on 4 July. Gibbs' account continued with a very graphic description of the battlefield area which the Pals had recently captured.

As I went over the battlefield of Montauban the enemy's shells and our own were falling over Bernafay Wood, where each side held part of the ground. A little to my left Mametz was being pounded heavily by the German gunners, and they were flinging shrapnel and 'crumps' into the ragged fringe of trees just in front of me, which marks the place where the village of Montauban once stood. They were also barraging a line of trench just below the trees, and keeping a steady flow of five-point-nines into one end of the wood to the right of Montauban, for which our men were now fighting.

Other shells came with an irregular choice of place over the battlefield, and there were moments when those clouds of black shrapnel overhead suggested an immediate dive into the nearest dugout.

I passed across our old line of trenches, from which on Saturday morning our men went out cheering to that great attack which carried them to the farthest point gained that day, in spite of heavy losses. The trenches now were filled with litter collected from the battlefield – stacks of rifles and kit, piles of hand-grenades, no longer needed by those who owned them.

This old system of trenches, in which French troops lived for many months of war before they handed them over to our men, was like a ruined and deserted town left hurriedly because of plague, and in great disorder. Letters were lying about, and bully-beef tins, and cartridge-clips. Our men had gone forward and these old trenches were abandoned.

It is beyond the power of words to give a picture of the German trenches over this battlefield of Montauban, where we now hold the line through the wood beyond. Before Saturday last it was a wide and far-reaching network of trenches, with many communication ways, and strong traverses and redoubts – so that one would shiver at their strength to see them marked on a map. No mass of infantry, however great, would have dared to assault such a position with bombs and rifles.

It was a great underground fortress, which any body of men could have held against any others for all time apart from the destructive power of heavy artillery. But now! . . . Why now it was the most frightful convulsion of earth that the eyes of man could see.

The bombardment by our guns had tossed all these earth works into vast rubbish heaps. We had made this ground one vast series of shell-craters, so deep and so broad that it was like a field of extinct volcanoes.

The ground rose and fell in enormous waves of brown earth, so that standing above one crater I saw before me these solid billows with 30 feet slopes stretching away like a sea frozen after a great storm. We had hurled thousands of shells from our heaviest howitzers and long-range guns into this stretch of field.

I saw here and touched there the awful result of that great gunfire which I had watched from the centre of our batteries on the morning of 1st July. That bombardment had annihilated the German position. Even many of the dug-outs, going thirty feet deep below the earth and strongly timbered and cemented, had been choked with masses of earth so that many dead bodies lay buried there. But some had been left in spite of the upheaval of earth around them, and into some of these I crept down, impelled by the strong grim spell of those little dark rooms below where German soldiers lived only a few days ago.

They seemed haunted by the spirits of the men who had made their homes here and had carried into these holes the pride of their souls, and any poetry they had in their hearts, and their hopes and terrors, and memories of love and life in the good world of peace. I could not resist going down to such places, though to do so gave me goose flesh.

I had to go warily, for on the stairways were unexploded bombs of the 'hair-brush' style. A stumble or a kick might send one to eternity by high-explosive force, and it was difficult not to stumble, for the steps were broken or falling into a landslide.

Down inside the little square rooms were filled with the relics of German officers and men. The deal tables were strewn with papers, on the wooden bedsteads lay blue-grey overcoats. Wine bottles, photograph albums, furry haversacks, boots, belts, kit of every kind had all been tumbled together by British soldiers who had come here after the first rush to the enemy's trenches and searched for men in hiding.

There were men in hiding now, though harmless. In one of the dug-outs where I groped my way down it was pitch dark. I stumbled against something, and fumbled for my matches. When I struck a light I saw in a corner of the room a German.

He lay curled up, with his head on his arm, as though asleep. I did not stay to look at his face, but went up quickly. And yet I went down others and lingered in one where no corpse lay because of the tragic spirit that dwelt there and put its spell on me. I picked up some letters. They were all written to 'dear brother Wilhelm' from sisters and brothers, sending him their loving greetings, praying that his health was good, promising to send him gifts of food, and yearning for his home-coming. 'Since your last letter and card,' said one of them, "we have heard nothing more from you . . . every time the postman comes we hope for a little note from you . . . Dear Wilhelm, in order to be patient with your fate you must thank God because you have found fortune in misfortune.'

Poor, pitiful letters! I was ashamed to read them because it seemed like prying into another man's secrets, though he was dead.

There was a little book I picked up. It is a book of soldiers' songs, full of old German sentiment, about 'The Little Mother' and the old house at home and the pretty girl who kissed her soldier boy before he went off to the war. And here is the sad old ''Morgenlied,' which has been sung along many roads of France:

Red morning sun! Red morning sun! Do you light me to an early death? Soon will the trumpets sound, and I must leave this life and many a comrade with me.

I scarcely thought my joy would end like this. Yesterday I rode a proud steed; today I am shot through the chest; tomorrow I shall be in a cold grave, O red morning sun!

On the frontpage of this book, which I found today at Montauban, there is an Army Order from Prince von Rupprecht of Bavaria to the soldiers of the Sixth Army.

'We have the misfortune' it says, 'to have the English on our front, the troops of those people whose envy for years has made them work to surround us with a ring of enemies in order to crush us. It is to them that we owe this bloody and most horrible war . . . Here is the antagonist who stands most in the way of the restoration of peace. Forward!'

It seemed to me that the preface by Prince Rupprecht of Bavaria spoilt the sentiment in the German folk-songs, which were full of love rather than of hate.

I stood again above-ground, in the shell-craters. Other shells were coming over my head with their indescribable whooping, and the black shrapnel was still bursting about the fields, and the Germans were dropping five-point-nines along a line a hundred yards away.

'Be careful about those dug-outs' said an officer, 'Some of them have charged mines inside, and there may be Germans still hiding in them.'

Two Germans were found hiding there today. Some of our men found themselves being sniped, and after a search found that the shots were coming from a certain section of trench in which there were communicating dug-outs.

After cunning trappers' work they isolated one dug-out in which the snipers were concealed. 'Come out of that,' shouted our men, 'Surrender like good boys.'

But the only answer they had was a shot.

The dug-out was bombed, but the men went though an underground passage into another one. Then a charge of ammonal was put down and the dug-out blown to bits.

This afternoon, while I was still on the battlefield of Montauban, a great thunderstorm broke. It was sudden and violent, and rain fell in sheets. The sky became black with a greenish streak in it when the lightning forked over the high wooded ridges towards La Boisselle and above Fricourt Wood.

'Heaven's artillery!' said an officer, and his words were not flippant. There was something awe-inspiring in the darkness that closed in upon these battlefields and the great rolls of thunder that mingled with the noise of the guns. Artillery observation was impossible, but the guns still fired,

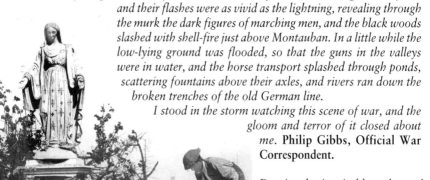

and their flashes were as vivid as the lightning, revealing through the murk the dark figures of marching men, and the black woods slashed with shell-fire just above Montauban. In a little while the low-lying ground was flooded, so that the guns in the valleys were in water, and the horse transport splashed through ponds, scattering fountains above their axles, and rivers ran down the broken trenches of the old German line.

I stood in the storm watching this scene of war, and the gloom and terror of it closed about me. **Philip Gibbs, Official War Correspondent.**

Despite the inevitable sadness that accompanies any battle, there was no doubt that the Pals had won a great

British soldiers excavate an unexploded British howitzer shell at the base of a statue of the Virgin Mary in Montauban village.

victory, and even though the victory was tempered by the news of great losses further north, nevertheless deserved congratulations were not slow in coming to the four City Battalions.

The following is just a selection of those received over the next week or so:

> *The Commander-in-Chief wishes the following wire from His Majesty the King circulated to all ranks. "Please convey to the Army under your command my sincere congratulations on the results achieved in the recent fighting. I am proud of my troops; none could have fought more bravely.* **His Majesty King George V, through GHQ British Expeditionary Force.**

The interior of Montauban Church during the German occupation, showing Allied shell damage.

> *Convey to 30th Division my best congratulations on their splendid work. Lancashire will indeed be proud of them.* **The Right Honourable Earl of Derby.**

> *Please convey to all ranks my intense appreciation of their splendid fighting which has attained all asked of them and resulted in heavy losses to the enemy, nearly 1,000 prisoners having already passed through the cage.* **General W N Congreve VC, CB, MVO, Commanding XIII Corps.**

> *I was deeply honoured by the praise you lavished on my Brigade; it is greatly appreciated by my troops and by me; but I must tell you that we equally admired the superb attitude under fire, the courage and the coolness of our neighbours, the officers and men of the 17th King's Liverpools. In particular, the very cordial relations we enjoyed with their commanding officer, Colonel Fairfax brought to our attention his fine military qualities, and we much regret the absence from the line of this brilliant leader and his magnificent Battalion. I intend to draw the attention of the French military authorities to this superior officer who cooperated so well with the left flank of my Brigade.* **Le Colonel de Coutard, Commandant de la 77e Brigade.**

> *Please convey to all ranks 30th Division my congratulations on their capture and defence of Montauban. They have done excellent work and will be attacking again before long.* **General Sir H S Rawlinson, Bart, KCB, CVO, Commanding Fourth Army.**

This last message from Rawlinson, originally dated 2 July, illustrates quite clearly, the curious way in which a victorious division was rewarded on the Somme in 1916! The 30th Division would not have to wait long before it would attack again.

The death of Trotter

The 18th Battalion, no longer an effective fighting force, was taken out of the new front line, on the evening of 2 July, and moved back to its old assembly trenches on the eastern side of Talus Boise. On the following day, a head count showed an effective strength of just six officers and two hundred and eighty eight men! On 4 July, the 21st Brigade was replaced in the line by The South African Brigade of the 9th Division, and the 18th Battalion moved gratefully into huts and tents at the Bois desTailles. On 5 July the 9th Division also took over the trenches held by the 89th Brigade, and the other three Pals Battalions joined the 18th Battalion at the Bois des Tailles.

On 7 July, however, the 30th Division was ordered to move back into the line to prepare for the next phase of the Somme battle, by securing a defensive line from the Maltz Horn Ridge, through the southern portion of Trones Wood, which was known to be heavily fortified by the enemy. The Pals were not involved in the early fighting to secure this portion of the wood, but on 8 July, what was left of the 18th Battalion was detailed as a carrier battalion for the rest of the 21st Brigade, who had gone into action the previous day. On the afternoon of 8 July Lieutenant Colonel Trotter received orders to move the Battalion forward up Train Alley, and deciding to supervise the move himself, he arrived in Train Alley Trench in advance of the rest of the Battalion.

By about 5.30pm the Battalion had also arrived, and was about to move forward. Trotter was on the point of entering the HQ of the 21st Brigade when the Germans, who dominated the high ground which ran from Trones Wood and Maltz Horn Ridge back to the village of Guillemont, obviously noticed the movement of troops and began to shell their former positions. One shell landed right in the dugout entrance, and killed the Colonel, Second-Lieutenant N A S Barnard, and two other ranks. It also fatally wounded Lieutenant Colonel W A Smith, the Commanding Officer of the 18th Battalion The Manchester Regiment. Smith was himself a former second in command of the 20th City Battalion, and was one of the original Pals officers who had crossed to France in November 1915.

> I was about thirty feet away when it happened. The Colonel was occupying the old German head-quarters. As you came over a hill, there was the headquarters, and there was the German front line over there. The artillery was firing this way, and he[3] fired three shells. One let[4] over there, one let over there, and one let right in the dugout, he must have measured it before he went back, he must have had sights on it. He killed the Colonel, a young lieutenant and two men in that dugout.
>
> I was about thirty feet away when it went up, and you could tell he'd gone. I could see them dragging him out right away. I was with the Colonel's servant at the time, he was a crack shot in the Grenadier Guards, that's how he got the job – he had this little rifle on his arm, for being a crack shot,[5] and he cried like anything, and he got his rifle and he flung it far away, crying like hell!
>
> It hit the Battalion like a brick. It went through the Battalion like that – they all knew in a few seconds – its amazing how the news got round that the Colonel had been killed. **16400 Private S R Steele, 18th Battalion KLR.**

Many obituaries to Great War officers state how much they were loved by their men, but there is no doubt that in Trotter's case, this was entirely true. Since the first day he was posted to the 18th Battalion, he had striven hard to make it the best of the City Battalions, and devoted himself to the training and the welfare of his men. Having seen active service himself, he knew how important it would be for the 'citizen soldiers' under his command to be very fit, and well trained, when the time for battle arrived, and his daily cross-country runs, whilst the Battalion was in England, had earned the 18th the nickname 'Trotter's Greyhounds'. Even some members of the other three battalions accepted, albeit grudgingly, that the 18th was the premier battalion. We have already seen that it was for this reason that Stanley detached the 18th from the 89th Brigade, to swap it for the 2nd Bedfords, in December 1915, and there is little doubt that its professionalism, esprit de corps and physical fitness contributed a great deal to the Battalion's victory at Montauban. There is equally little doubt that any success was directly attributable to the concern, leadership, and fine example shown by Lieutenant-Colonel Trotter.

To him, the welfare of his men was paramount, and two examples of his attitude, both before and after the Battalion left for France, best serve the point.

> The Colonel had an idea that when you got these army boots – they were very crudely made – that you had to have bare feet, put your boots on, and walk quickly for two miles across Salisbury Plain, so your feet got hot, and gradually moulded your feet to the boot. And when you came back to camp, you stood in water, with your boots still on, for half an hour or so, and by that time, he said, your feet were more or less fitted like a glove, into the boots. I don't know where he got the idea from, but it made them very pliable after it, you see. I don't know whether he got it from the South African War or not, but that was all Trotter's idea. **16224 Private T H Brown, 18th Battalion KLR.**
>
> We as a rule, do not very often have too much praise for officers; but if any battalion is blessed with a finer colonel than ours, to put it politely, it must only be in heaven. I could give many accounts of what he has done for the men. We are always his first consideration; but one example, I think, is sufficient. Two of our men went up a wrong communication trench. It was not in use, and had not been for some time, on account of its bad state. To cut the story short, they went into a hole and came out mud and water up to the chest. They managed to return to the battalion

headquarters, and when the CO heard of it, he gave them all dry warm things possible.

But it is necessary to sleep in clothes when without blankets, so he gave up his bed to them for the night, and himself laid on the floor of his dug-out. He is always doing actions of this sort, and I verily believe if he ordered the men into practically certain death I do not think one man would hesitate to go, for they all know he would be there to share it. **16319 Private J F Mills, 18th Battalion KLR.**

The genuine grief at the news of Trotter's death was felt by all ranks, many of whom wrote home with the sad news.

It is with great sorrow I have to tell you of the death of Colonel Trotter, which happened two days ago. I can scarcely realise that we have lost him. It is the blow that has struck the hardest. I am afraid I cannot put to paper our opinion of him. I can truthfully say that you will not find a man on the whole British line like him. No man ever had the welfare of his men at heart as he had – considerate, kind and human: a man who took a personal interest in your private as well as your military life. We would have followed him anywhere, knowing we would be safe in his hands. If ever a monument should be raised in memory to any soldier, it should be to Colonel E H Trotter. **17059 Private G E Jones, 18th Battalion KLR.**

Perhaps the best tribute paid to him was by one who knew him best!

He left a blank that could never be filled, and somehow one would never like it to be filled. His was a unique personality. His discipline was one which sprang from affection which all his men had for him. They all, from their love of him, simply could not do anything but try to please him. His ideals were of the highest. He expected the best out of everyone and by his setting the example himself, he always got it. His love for his Battalion was heartwhole; it always had been so throughout his soldiering – and he made a point of not only knowing all his men, but of knowing their lives and their belongings. Whenever he went on leave, he spent it in the interests of his men, going to their homes, relating to their friends and relations stories of what they were doing, and a hundred other kind actions.

He was more like a father to each individual in his Battalion than anything else, and his love for them was unbounded.

And so died one of the noblest soldiers that ever stepped. His name will never be forgotten in Liverpool and his memory will always be held in deepest respect and affection. **Brigadier-General F C Stanley, CMG, DSO.**

Trotter's death, coming so soon after the losses of 1 July, was a real blow to the survivors of the 18th Battalion, and indeed, many considered that 8 July 1916 was the last day of the old-style 2nd City Battalion. Never again after this date, was there the same spirit of comradeship, togetherness and sense of duty. Something more than their commanding officer had died on that July afternoon.

Lieutenant-Colonel Trotter's body was taken from the battlefield to Maricourt, and eventually buried in what became Peronne Road Cemetery, where he lies today, amongst other comrades, who died on 1 July, and the later battles on the Somme.

Soon after Trotter's death, Major R K Cornish Bowden was ordered to take over command of the Battalion, while it was still in the the line, and then temporary command passed to Major A P White, of the East Surrey Regiment. Eventually, on 23 July, Lieutenant-Colonel W R Pinwill, of the 1st Battalion The King's (Liverpool Regiment) was appointed Battalion Commander.

Trones Wood

After the attack on 1 July had failed to break out into open ground, it was inevitable that the offensive would degenerate into a bitter contest around the natural strongholds that the Germans had so carefully fortified, often before the 'Big Push' began. Thus, the history of the Battle of the Somme after 1st July is often dominated by bitter fighting to control the major wooded areas, which were so easy to defend and

A wartime artist's version of the fighting for Trones Wood.

so difficult to conquer! It is no coincidence, therefore, that the names Mametz Wood, High Wood and Delville Wood are well known to all students of the Great War, for the ferocity of the fighting there, and the unbelievable feats of heroism and endurance which were carried out by soldiers on both sides. The actions in Trones Wood[6] are not so well known, however, but not for any lack of ferocity of fighting. Although the Pals were only engaged in the Wood over a four day period, their casualties were testimony enough to the scale of the struggle.

By the end of the first week of the offensive, Haig was desperate to push forward and outwards along his front, to capture the German second main defensive line. Congreve's XIIIth Corps was given the task of taking this line in an advance which was to run west from Longueval, and this advance would include the Bazentin Woods, Mametz Wood and Trones Wood. It was essential that the latter wood was in Allied hands to prevent the right of the British line, where it met the French, from being left exposed to a possible German counter-attack, which could split the armies in two, and open up a dangerous gap. As a result, it was agreed that when the wood was attacked, the French would also mount an attack on the village of Hardecourt aux Bois.

What was not realised at the time, however, was that Trones Wood was a virtual death trap! Shaped like a pear, it was about 1,400 yards long from top to bottom, and was about four hundred yards wide at its lowest point. Inside, it was choked with dense thicket, which had not been cut down since the war began, and the Germans had fortified it with a veritable warren of trenches, protected by concealed positions, barbed wire and well-sited machine guns. Added to this was the fact that the British approaches to the wood were totally exposed and in the open, apart from one sunken lane, called Trones Alley. Thus, every move could be spotted by the Germans who held the Maltz Horn plateau and the high ground on the Bazentin Ridge. They also had a very strong defensive position in the village of Guillemont, from which they could feed troops into the wood when and where they were needed, virtually undetected by the British or the French. Also, as they had only recently vacated the area themselves, they knew the ranges for directing their artillery fire almost to the inch. Thus the task would not just be to take the enemy positions, but to hold them in the face of the inevitable counter-artillery fire, which would follow any British success.

The original attack was to have been made by men of the 21st Brigade, on 7 July, but because of French pressure, this was eventually postponed until 7.15 am on the 8th. The 90th Brigade, under Steavenson was to be in support, with the 89th Brigade in reserve. We have already seen that the 18th Liverpools, of the 21st Brigade was to be used as a carrier unit, because of the battering it had received on 1 July. The first attack by the 21st Brigade was driven back almost immediately, but another in the afternoon, succeeded in occupying the lower portion of the wood, with the exception of a small, heavily fortified position on the south-east flank. The French attack went according to plan and Hardecourt fell to them. However, once the British attempted to dig in, they came under a furious German bombardment directed from the top of the Bazentin Ridge.

The Germans did not have the monopoly of artillery fire, however, and themselves suffered in the artillery duel which lasted throughout the battle.

On the morning of Saturday, the 8th, at 7.30 am, we began a long bout of firing which went on till 3.30 pm on Sunday, with for objective, the capture of Trones Wood. The whole of the day passed in a series of terrific bombardments directed at that accursed place. At first things did not go well, but by 1.15 pm we had captured the southern half of it, and by 6 pm the whole wood appeared to be in our hands, which brought our men within five hundred yards of the German second line.

But all was not over then. At 11 pm the Hun made a strong counter-attack preceded by an hour of intense bombardment. At the same time he deluged Maricourt and its neighbourhood with tear gas, and after severe fighting, succeeded in retaking a great part of the wood.

Confused fighting went on all night, culminating in another bombardment between 4 and 5am, when we regained some of the lost ground. Even the French words – "Une lutte acharne"[7] – fail to describe what a hell on earth woodfighting in a pitchy black night can be . . .

I forget how many times the centre of the wood changed hands, but we never quite lost our hold of the southern end. The wood used to show up as a dense compact mass on the skyline, but after the appalling shelling it has got it is now only a tortured collection of stumps. The hostile gun-fire has been gaining in severity each day; every night we have tear gas and incendiary shells all round the village, and by day heavy shells of every calibre, even up to three 17-inch. Lieutenant-Colonel N Fraser-Tytler, DSO, TD, CLIst (County Palatine) Brigade, Royal Field Artillery.

An original shell hole burial site at a cemetery known as Squeek Forward Position, which was later totally destroyed by shellfire. The first four names on the left side of the wooden cross, and three others on the right, are of men from the 18th Battalion. See Appendix II to find out where they are commemorated today.

By the early evening, No.2 and No.3 Companies of the 18th Battalion, by this time without their Colonel, had both been sent up to the front line with supplies and trench stores. By about 8.00pm, men of No.3 Company, led by Captain G Ravenscroft, had actually entered the wood to deliver water to the 2nd Battalion The Wiltshire Regiment. Whilst they were in the act of handing it over, the Germans launched a counter-attack, and they had to stay and help to repel it, only returning to their original positions on the following morning. No.1 and No.4 Companies, meanwhile, had also taken bombs, sandbags and ammunition up to the front line. All the companies were eventually relieved during the morning of the 9th, by the 90th Brigade, who were themselves sent forward. In all, in less than twenty-four hours, the 18th Battalion had lost twenty-eight of all ranks killed, wounded or missing.

Wood fighting 10 July to 12 July 1916

In the early hours of the morning of Monday 10 July, following a British bombardment, men of the 90th Brigade had succeeded in taking the German positions at the tip of the wood, but by 10.00 a.m. they had been forced out again. They had sustained no fewer than 800 casualties, most of them from German shellfire. As the 89th Brigade was in reserve, 'A' Company of the 17th Battalion, led by Major G F Higgins, was lent to the 90th Brigade, to mount an attack on the German positions, to relieve the pressure on its troops, many of whom were still pinned down. This attack was made at 2.30 pm on a German position, on the south-west corner of the wood, but the Germans were able to fire on the Liverpool men from their concealed positions, and the Pals were forced to retreat. In the evening another attack was made by 'A' Company and this too, was repulsed. Major Higgins, however, was killed at the head of his men. After this, a second company of the 17th Battalion, also lent to the 90th Brigade, which was about to follow 'A' Company, was   d sent back to the Briqueterie.

It is impossible to establish how many men from 'A' Company died with their commander during this attack, because Company records were confused at the time, and all losses during the fighting for Trones Wood are listed as 'Killed in Action 10–12 July 1916'. There is in existence, however, a hand-drawn map which shows the attack made by Higgins and his men, on 10 July, with an 'X' marking the place where Higgins supposedly fell. It is not known when this map was drawn, or by whom, or even how accurate it is, but it also marks with a '+' the spot where Higgins is supposed to have been buried. This spot is in the ruins of the village of Guillemont, which was in German hands until 3 September 1916. Unless Higgins' remains were found and buried there after this date, then the map can not be regarded as accurate. Certainly his body was not found and identified after the war by the Imperial War Graves Commission, as his name appears today on the Memorial to the Missing of the Somme, at Thiepval.

At 12.30 pm on 10 July 1916, Brigadier-General Stanley received orders from General Shea to the effect that the 89th Brigade was to relieve the 90th Brigade and capture the wood at dawn on the morning of 11 July. As a consequence, all four battalions of the 89th Brigade were moved forward to the immediate south of the wood to prepare for the forthcoming attack. The 20th Battalion began to move into Maltz Horn Trench, at about 10.00 pm on the evening of the 10th, relieving the 2nd Battalion The Royal Scots Fusiliers, of the 90th Brigade. The 17th Battalion was ordered into positions in Trones Alley and Dublin and Casement Trenches, and the 19th Battalion was positioned in Bedford Street Trench and in the old German front line. The 2nd Battalion The Bedfordshire Regiment, the fourth battalion of the 89th Brigade, was positioned in Trones Alley, to the left of the 17th Battalion, with its left flank in the Briqueterie. Stanley had timed the attack to begin just before dawn, at 3.27 am. He had chosen this precise time because 327 was his old school number, and he regarded it as a lucky number! The attack was preceded by a heavy bombardment, and just as it ended, the Bedfords began to move forward along their stretch of Trones Alley. As it became light, they were spotted by the Germans, who held the other end of Trones Alley, where it met the wood, and soon a fusillade of machine-gun and rifle fire began to hit them. Eventually, they overcame this, and forced their way into the wood, which they found to be extremely thick, and virtually impenetrable, visibility being limited to two or three yards in any direction. This not only impeded their progress, but it meant that they could not keep direction, or even stay in touch with each other. The Germans, on the other hand, were able to use this to their advantage, dug in as they were, and with some knowledge at least of the geography of the wood. Having gained a toehold on the edge of the wood, the Bedfords then tried to consolidate their position by digging in.

At about the same time that the Bedfords moved forward to the west of the wood, the 20th Battalion moved along Maltz Horn Trench, north of the ruins of the farm itself and began to bomb the enemy strongholds on the German-held side of the trench. One party, under Second-Lieutenant H A Small, MC, found a fork in the trench which led to the Trones Wood – Guillemont road. He decided to reconnoitre this fork, and there encountered the enemy, whom his party engaged in a fierce exchange of fire. Although the Germans suffered losses, they were backed up by fire from a fortified strong point to the east of the wood, which the Pals

German prisoners from the Somme fighting. Many more were expected.

party were unable to locate. Smith unfortunately, was killed as he tried further to explore the road, and the rest of the Battalion began to fortify their position by digging in, and consolidating what they held of the Maltz Horn Trench.

Another soldier who was killed in the same action, was 22051 Private J Davies, and an account of his death, written home to his parents in Birkenhead, by a comrade, is so typical of many such accounts written home throughout the war, which often only told part of the truth, in an attempt to spare relatives and loved ones the anguish of the whole story.

> *'I was speaking to him a few minutes before his death, and at the time he was directing the throwing of the bombers. He was one of a party of six men in a bombing raid who were taking a redoubt and he had done some glorious work. They had taken the trench and were moving the last few Bosches when he was sniped through the head. He did not suffer at all, dying instantly. He had done his duty nobly, and I trust this thought will help you in your loss.'* **22189 Lance Corporal A S Newell, 20th Battalion KLR.**

Lance-Corporal Newell was himself killed in action just nineteen days later, during the Battle of Guillemont. It would be merciful to hope that he too was sniped through the head, did not suffer at all, and died instantly!

By mid-afternoon the 20th Battalion had still not managed to locate the German strongpoint on the east of the wood, and Stanley decided to send No.3 Company of the 19th Battalion, under Captain H A Smith, up the line and into the south-west corner of the wood, to try and link up with the Bedfords, and locate the strong point. The men immediately came under sniper fire, and although they located the strong-point, from the wood side, they were unable to capture it, and had to fall back to the edge of the wood. Smith himself, and three other officers were badly wounded, and at least sixty other ranks became casualties during the engagement. The survivors, like those of the Bedfords and the 20th Battalion, began to dig themselves in, against the inevitable barrage that they knew would hit them, once the Germans in the wood called up artillery support.

This came at about 6.00pm when the Germans mounted a counter-attack, which although expected, was still virtually unstoppable.

> *'Night was now coming on, and we had been warned that a counter-attack was imminent. Sure enough it came and, alas, they succeeded in getting a bit more of the wood back. A prisoner had been taken that day with the orders on him, saying that the attack was to be delivered when a green light went up, and when the objective was reached a white would be sent up. You can imagine my horror when, after seeing the green light go up, I soon after saw the white one. It was awful!'* **Brigadier General F C Stanley, CMG, DSO.**

At first the 20th Battalion was able to hold out against this assault, which came from the direction of Waterlot Farm to the north-east, and was made by Saxons of 16.Infanterie-Regiment Nr.182, who came from Freiberg and Koenigsbruck. The Pals killed many of these Saxons with especially accurate machine gun fire, but eventually the weight of the German attack forced them to give ground and retire. The Bedfords, who had been in almost continuous action for virtually fifteen hours, were not able to hold out either, and two of their companies were driven right out of the wood. The other two, however, although much depleted in numbers, managed to maintain their positions along the wood's south-eastern edge until help arrived.

This help was in the form of two companies of the 17th Battalion, who were ordered into the fight. 'A' Company, led by Captain H N Brinson, and 'B' Company, led by Lieutenant B S Thompson, entered the wood, established a footing, and managed to beat the Germans back into the position of their original strong point.[8] They were then able to dig in, and establish contact with what was left of the Bedfords. On the following day, Wednesday 12 July, the Commanding Officer of the 17th Battalion, Lieutenant-Colonel B C Fairfax took command of the fighting in the wood, and Fairfax's men and the Bedfords joined up their respective trenches and wired the old gap against further German attack. This proved a very wise move, as the enemy did attack again during the following night, and were completely taken by surprise

PRAISE OF THE PALS

BY BOTH BRITISH AND FRENCH COMMANDERS.

GENERAL STANLEY'S MESSAGE.

EARLIER PERFORMANCE "TOTALLY ECLIPSED."

A magnificent tribute is paid to the City Battalions of the King's (Liverpool Regiment) (the Pals) in a letter received to-day by the Liverpool Seed, Oil, and Cake Trade Association from Brigadier-General F. C. Stanley.

The brigadier states that since the 10th July "the splendid men have done work which has totally eclipsed their performance before that date." Moreover, "they have now not only received the praise of our high commanders, but also that of our Allies."

Mr. Cecil Calthrop, president of the Seed, Oil, and Cake Trade Association, C20, Exchange Buildings, Liverpool, wrote to Brigadier-General Stanley under date 10th July as follows:—

The Committee of the above Association desire me to convey to you their sense of pride and pleasure in the accounts of the excellent work done by the "Pals" Battalions of the King's (Liverpool Regiment), which are so largely composed of employés of the various firms comprising this Association.

Their bravery and tenacity in capturing and holding positions in the late advance is much admired.

I trust that you can make this known to your men.

Brigadier-General Stanley's reply, copies of which have been handed to the Lord Mayor and posted in the Exchange News Room, is as follows:—

Headquarters, 89th Infantry Brigade, 16th July, 1916.

DEAR SIR,—

Your letter was received by me this day, and it will most assuredly be communicated to the City Battalions of the King's (Liverpool Regiment), three of which are in the 89th Brigade and one in the 21st Brigade.

It gives us the greatest satisfaction to know that what we have done has earned the praise of your Association.

I see that your letter was dated 10th July, and I am pleased to say that since that day the splendid men have done work which has totally eclipsed their performance before that date.

No words of mine can express to you my pride in these magnificent battalions which from the date of their formation in September, 1914, have striven to excel at all times, and now have not only received the praise of our high commanders, but also that of our Allies, and your appreciation as well.

Liverpool may well be proud of those whom she sent out so ungrudgingly, and when the war is brought to a successful termination the City Battalions of the K.L.R. will have made for themselves a name which Liverpool will honour for all time.

(Signed) F. C. STANLEY, Brigadier-General, 89th Infantry Brigade.

The President, The Seed, Oil and Cake Trade Association, C 20, Exchange Buildings, Liverpool.

LY 17. 1916.——3

COLONEL TROTTER.

A COMMANDER OF COMRADES FALLS.

"THE FATHER OF HIS MEN."

The Comrades have lost one of their battalion commanders. Colonel E. H. Trotter, D.S.O., has fallen in the advance. We not himself a Liverpool man, the city has reason to mourn one to whom was entrusted the leadership of its soldier sons, and who trained with them and went with them into the great ordeal. He was in many respects a remarkable personality. Lord Derby showed his familiar instinct for the right man when he invited him to command a battalion. From all points it was an admirable choice. Colonel Trotter was a fine soldier; but he was also one of those men whom to know is to honour, such is their uprightness of character and their force of example. Coming from their office desks to the strangeness of the parade ground, the Comrades had before them a pattern of the chivalrous and tactful British officer who did credit to the profession of arms, and who set them from that time onwards the ideal of all that is best in the Army.

Bonds of Mutual Affection.

Brigadier-General Stanley it was who said that "they will worship him before they have done." Never was a truer phrase spoken. Colonel Trotter was the brother of his subordinate officers and the father of his men. Between him and them were the bonds of mutual affection. Gifted with a remarkable memory for names, he knew their antecedents, and was always ready to help them, and he talked with them in camp on terms of intimate equality. Under him the battalion became inspired with a wonderful esprit de corps. Up at Knowsley, indeed, in those happy days when the raw recruits were learning the rudiments of the warrior's trade, there was a genial disputation as to which was the best unit in the brigade. Each of the battalions claimed the title for itself; but there was one which disdained to recognise competitions, and that was Colonel Trotter's. Every effort to shake their proud obsession was doomed to failure, and ultimately the men of the other three units decided, mainly in the interests of peace, to admit that Colonel Trotter's did deserve the palm. Upon that delicate controversy the cautious layman would avoid adjudication; but certain it is that Colonel Trotter's passion for efficiency did leave its impress on his command, and in shooting it stood at the head of Kitchener's Army.

Fine Type of Soldier.

Colonel Trotter was a soldier of the type of John Nicholson and Henry Havelock. He was a man of high and chivalrous feeling, with a devout religious spirit, the soul of genial good nature, and devoted to duty. Ever ready to encourage, he was never harsh when he was compelled to chide, and those who knew him declare they never heard a bad word fall from his lips. He never asked his men to do what he would not do himself. Often on route marches, and in spite of a weakness in the knee that had been caused by a hunting accident, he would march on foot, and leave his horse to be ridden at the rear by a groom. And he was " in the thick of it " on the 1st of July. He did splendidly on the opening day of the advance, and characteristically set the example. Carrying only a walking-stick, and with no more than a dozen men in his immediate entourage, he captured a strong redoubt from the enemy and held it for four hours until relieved. Then, on the 8th of July, just a week after the initial attack, he met his death from an explosive shell. The Liverpool Comrades Brigade has won an enduring name during this battle, but no tribute of esteem could be too high for that Christian gentleman who

when they became entangled in the newly laid wire. The ferocity of wood fighting is well illustrated by the fact that those Germans who were caught on the wire, and attempted to surrender, were simply shot!

Twice during the night of the 12th, the Germans made half-hearted attacks on our line Q A C and southwards and appeared surprised to find wire and organised opposition. On one occasion a party got hung up in the wire, and were disposed of by a Lewis Gun. Lieutenant-Colonel B C Fairfax, Officer Commanding, 17th Battalion KLR.

These attacks were the last ones made on the Pals in Trones Wood, because in the early hours of the morning of Thursday 13 July, the remaining soldiers of the 89th Brigade were relieved by the 7th Battalion, The Royal West Kent Regiment, of the 55th Brigade, 18th (Eastern) Division, and were able to make their weary way back to the Bois des Tailles and rest. The 19th and 20th Battalions had been relieved the previous day. The Pals had not captured the whole of the wood, but they had made sure that the parts they had captured were securely held, which allowed the wood to be cleared by men of the 54th and 55th Brigades of the 18th (Eastern) Division on the morning of 14 July.

An evaluation

Although the part played by the Pals in the capture of Trones Wood had been a vital one, there is no doubt that the whole exercise had been a complete shambles. The battalions involved had never been trained in close-quarter wood fighting, and it is surprising that their losses were not heavier. Brigadier-General Stanley put the 89th Brigade's casualties at five hundred and fifteen of all ranks,[9] to which must be added the twenty-eight which the

18th Battalion lost on 8 July. Post-war figures show that from the four Pals Battalions, a total of eighty nine of all ranks, lost their lives from enemy action, in the period from 8 to 12 July 1916, and this figure does not include those who died later from their wounds.

The ever-professional Colonel Fairfax outlined his views on the fighting, and reached the following conclusions:

CONCLUSION

1. Leadership is the difficulty and control.

2. I am strongly in favour of movement by sections in file. By night, more particularly, there appears to me but slight hope of direction being maintained and objectives reached by troops moving in any other formation than file or single file.

3. I attribute the successful entry of my two companies on the night of 11th July to the adoption of this method of moving, and to the skill of the individual leaders.

4. Men must be lightly equipped to fight, particularly in wood fighting.

5. It is no use soldiers wandering through woods and saying they have taken them – it only leads to annihilation. A firm footing – ie, a trench six foot deep and wire in front must be made and put up first on some definite alignment. From this trench further progress can be made by specially chosen parties. An advance in a wood must be deliberate and cunning. Heroic action probably ends in disaster.

6. A wood, such as TRONES WOOD, should be treated with respect, and slow progress for its occupation expected only.

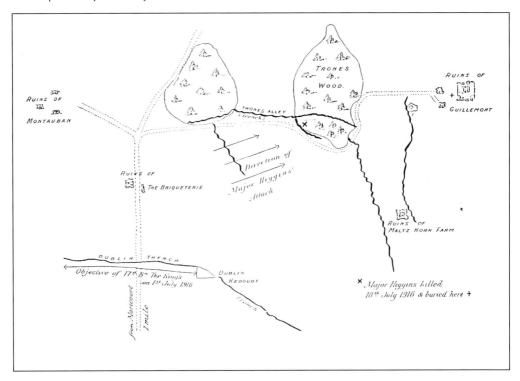

This contemporary map shows the 17th Battalion attack on Trones Wood, on 10 July 1916. Major G F Higgins, who led the assault, was killed, and according to the map was buried near the ruins of Guillemont village. Unless the burial took place after the village fell in September 1916, this information must be viewed with suspicion. Higgins' body was never found and identified after the war and he is commemorated on the Memorial to the Missing of the Somme at Thiepval.

7. Proximity to German infantry in a wood should be a satisfactory position for British Infantry to be in, as they are unlikely to be shelled by German guns.
29-7-16 Lieutenant-Colonel B C Fairfax, Officer Commanding, 17th Battalion KLR.

Despite the knowledge that they had done a good job, the feeling persisted long after the war, amongst some veterans, that the whole episode had been a total debacle. It was even rumoured amongst the survivors of the 19th Battalion, that more than one company of their Battalion had been ordered into the wood by Stanley, and that the Battalion Commander, Lieutenant Colonel L S Denham, realising the impossible conditions that existed inside the wood, had refused to give the order. The veracity of this rumour is compounded by the fact that Denham was replaced as Battalion Commander by Major G Rollo, of the 17th Battalion, on 20 July, and went home to England. No explanation of this sudden change of command was given in either the 19th Battalion War Diary, or in Stanley's post war book The History of the 89th Brigade, 1914–1918'.

Despite any misgivings that might have been felt by the rank and file, however, there is no doubt that the spirit of the Pals was still very high, even under the most terrible of circumstances.

It is said that in the fighting in Trones Wood, east of Maricourt, one of number 4 platoon, 'A' Company of the 19th King's Liverpools, one of the original volunteers and a hard case sailor and soi-disant deserter from the US peacetime navy, had an eye shot out by a sniper's bullet. With the eye salvaged, and in hand, he passed down the communication trench, remarking to the gog-eyed soldiery, 'Havn't you ever seen f.....g Lord Nelson? Stand to attention! **17158 Corporal E G Williams, 19th Battalion KLR.**

NOTES
1. Although Balleine makes the point that Private Redhead was buried in a coffin, this might well have been just for the sake of his mother's feelings. Burials at large military base areas might have been made in coffins, but certainly the standard method of burial in the field and at front line medical stations was in just an Army blanket. Bodies that were subsequently disinterred were often wrapped in chicken wire as well, to keep them as intact as possible for reburial.
2. To keep pace with the growing need to label wooden crosses accurately, special labelling machines were imported from all over Britain, where they normally rested in railway stations and amusement arcades. These cast iron monsters, could be made, for one penny, to punch out in raised letters, any name - or message, on a strip of aluminium, and as such, were ideal for marking grave crosses. Machines of the same type could still be seen in Britain well into the 1960s!
3. 'He', of course refers to the enemy in general, and not a specific person. Great War veterans often refer to the Germans in the third person singular, even if they are referring to more than one soldier. Other similar uses of the singular, were 'Jerry' or 'Fritz', as in 'Jerry attacked', or 'Fritz had occupied the trench'. Officers also sometimes used the derogatory French term for a German, such as in 'Kill the Boche'.
4. The use of the word 'let' in this case means to explode, and is maybe a corruption of 'to let off', as with a firework. This use of the word seems to have been fairly common in the Great War.
5. Marksmen in the British Army were entitled to wear crossed rifles on the left lower arm of the service tunic. Many who were entitled to this distinction, however, often removed the badge in the front line, as, if they were captured by the Germans, they were sometimes shot, on the assumption that they were snipers!
6. The name 'Trones' Wood appears to be a British version of the name by which it is known locally, 'Bois des Troncs', which would translate as '(Tree) Trunks Wood'. It was not uncommon for the British to unwittingly alter foreign place names to what they thought they ought to be! The Official History also calls it 'Trones Wood', which would translate as 'Thrones Wood'.
7. Meaning 'a desperate fight'.
8. Captain Brinson was subsequently awarded the Distinguished Service Order for his conduct during this action.
9. This would also include casualties sustained by the 2nd Battalion, The Bedfordshire Regiment.

Chapter Seven

'Moving off for line tonight – over the top in the morning'

On 14 July the 89th Brigade moved from the Bois des Tallies to Corbie, and from there to its old camp at Vaux sur Somme. On the morning of 15 July 1916, it was addressed by the 30th Divisional Commander, Major-General Shea, who congratulated the men on their performance in Trones Wood, and passed on to them other messages of congratulation from General Haig, General Congreve and Lord Derby. Ominously, he also promised the Brigade that it would shortly attack again! The War Diary of the 19th Battalion records this part of his speech in a little more detail than the others. Its entry for 15 July reads:

> *Major-Genera; JSM Shea, CB, DSO, GOC 30th Division congratulated the Brigade on its excellent performance. The Division had done so well that they could not be spared for more than seven days and their rest would therefore be curtailed.* **Lieutenant and Adjutant W Ashcrott, 19th Battalion KLR.**

Although the tone of the entry might seem a touch sarcastic to a modern reader, it is almost certainly not meant that way by Lieutenant William Ashcroft. The curtailed rest was in preparation for the Battle of Guillemont, which would involve all the 89th Brigade and inflict on them the most grievous casualties of the war so far. The 18th Battalion, also at Vaux, heard Shea make a similar speech to the 21st Brigade. The 30th Division then moved to bivouacs at Happy Valley to prepare for the attack.

Guillemont was a heavily fortified village to the east of Trones Wood, south-east of Longueval and Delville Wood. Waterlot Farm was on the road that connected it with Longueval, and this, too, was heavily fortified. Just to the south of Guillemont, was a high plateau, where there were further fortifications in the ruins of what had been Maltz Horn Farm. From these, the German defences continued to the south-east through the village of Maurepas.

Guillemont was typical of all the German fortified villages on the Somme. They had been chosen carefully way back in 1914, with all-round defence in mind, and the Germans had had virtually two years to fortify and wire them and to site machine-gun, rifle and bombing posts to their best possible advantage. The Allies could not even outflank, isolate or bypass them, as this would have left a determined enemy in their rear! As a consequence, the villages had to be captured, even if this meant head-on frontal assaults.

Thus, it was first planned that two brigades of the 30th Division should take Guillemont on 23 July in a concerted attack with the French. However, the French were not ready to take part at that time, and the plans had to be modified to allow just one British brigade to make the assault. The brigade chosen was the 21st. As the 18th Battalion was still 'licking its wounds' and assimilating reinforcements, it was designated as carrying battalion for the attack, and did not go into action. The other three 21st Brigade battalions did, however, but their attack was a failure. Although the attacking battalions sustained fairly heavy casualties, the 18th King's, waiting in the

The church at Guillemont before the main attack.

The railway station at Combles on 25 June 1916, showing the wreckage of a German ammunition train which received a direct hit from the Allied bombardment.

positions of the old German Glatz Redoubt, was very lucky, losing only one soldier, 32076 Private J Durcan killed, and two officers and 16 men wounded, through shellfire.

One of the reasons that the attack failed was the inadequacy of the tactics that were employed. The three battalions of the 21st Brigade, with insufficient artillery cover, had been expected to cross nearly a mile of rolling countryside, and defeat a well prepared enemy, well dug in, on high ground. Despite the obvious reasons for the failure of the attack, no attempt was made to alter or modify the basic plans for the next attempt to take the village. This was originally planned for the following day, but largely because the French were still not ready, it was postponed until the early morning of 30 July. This would become a date long remembered in the grieving homes of Merseyside.

The objectives for this day were fairly simple. There would be a joint Anglo-French attack. The 2nd Division would attack to the north of the village, as far as the ruins of the railway station, the 30th Division would attack in the centre, to take the village itself, and the French would attack to the south, to take the village of Maurepas. The obvious flaw in the plan was that all three divisions had to succeed at more or less the same time, fighting over different terrain against different fortifications. If just one of them was not successful, then the other two would have at least one exposed flank. In the case of the 30th Division, it was not inconceivable that both flanks might become exposed, with the possible consequence of total annihilation!

From the 30th Division, the 90th Brigade was given the task of taking the village itself, whilst the 89th Brigade's objectives were the German second-line trenches which ran from an orchard, on the south-east of the village, as far as the north corner of Wedge Wood. To get this far, it would be necessary to take Maltz Horn Farm first. The 21st Brigade was to be in reserve, after its drubbing on 23 July.

Ten German Infantry battalions opposed the eight battalions of the 30th Division along its attacking front. They were mainly Saxons, from 5. Koeniglich Saohsisches Reserve Infanterie Regiment Nr.104, 8. Koeniglich Sachsisches Reserve Infanterie Regiment Nr.107 and 9. Koeniglich Sachsisches Reserve Infanterie Regiment Nr. 133. These all belonged to the 24th Reserve Division. In addition there was one battalion of 22 Bayerisches Reserve Infantrie Regiment attached from the 8th Bavarian Reserve Division, who held the southern end of the Pals' attack line, and the line beyond Maurepas.

German troops (wearing pickelhauben), attending a church parade in the village of Combles during the Somme offensive.

Preparation

During the day of 29 July the officers and men of the attacking battalions made their preparations for the following morning. For the 17th and 20th Battalions it was not a new experience any more, having been 'blooded' on 1 July, and again in the fighting for Trones Wood. Most of the men of the 19th Battalion, however, had not yet seen any real action, as they were brigade carriers on 1 July, and only No.3 Company had been deployed in Trones Wood on 13 July. None of them had actually 'gone over the top', in the strict sense, and they were very keen to get into action, and prove themselves to be as worthy as the other three City Battalions!

An example of how the Germans fortified villages on the Somme as strongpoints in their line of defensive trenches.

Before the Brigade left camp, kit that was not needed was stowed away, or left with friends who were not going up the line, goodbyes were said, and last letters home were written. Most of these would be left with a friend, or a trusted person in the Orderly Room, and only posted home in the event of the writer's death. Sadly, there would be only too many of these letters posted home over the next few days. Unhappily, one of the letter writers was Private Ward Evans, of the 19th Battalion, whom it may be recalled, had handed a letter to a railway porter to post home for him, as his Battalion was entrained for Folkestone and France. He had missed the attack on 1 July, because he had been slightly wounded by a shell burst in the front line on 27 June and evacuated to the base hospital at Rouen. He did not rejoin his Battalion until 20 July. On 29 July, he wrote two last letters home to his family in Liverpool, one to his mother and one to his sister, on the only scraps of paper available. These treasured letters have survived the passage of time and are reproduced here. The letter to his mother is all the more fascinating because it was censored at the time, because it contained information concerning the movements of the Battalion, which might have been useful to the Germans had it fallen into their hands. Over seventy years later it has been possible to decipher enough of Owens' original words to understand what he intended to say. The lines which were originally censored are shown in bold print.[1]

France Saturday Evening 29.7.16
  My dear Moth.
   We shall he moving off in about half an hour from this place where we have been bivouacking for the last week, up to the line, and early tomorrow morning I shall be in the first wave when the good old 19th goes over the top to take three lines of trenches from the enemy.
   This will be the first experience of hand to hand fighting for us, as in the past in an attack, the 19th were carriers to the Brigade, so this time the boys are out to fight first in the field – at last.
   Well of course, the future is veiled indeed at a moment like this, but by God's help I hope to fight a good fight and be brought safely thro'.
   It will be idle for me to tell you not to grieve if I should go under – all I would ask you to remember is that the parting is but for a time we shall be together again before long.
   So whatever the days to come

Another German fortified village strongpoint.

Private W B Owens, 19th Battalion

may hold for us Moth we have the assurance that nothing can separate us from the love of Christ that is my sheet anchor for the morrow.

> *God bless you Mother for all you have been + meant to me*
> *Your loving son*
> *Ward*

France Saturday 29.7/16
My dear Gwyn,

I feel I must leave a line for you in case anything happens tomorrow when the fun begins. I should feel much worse about things if I thought Mother would be left alone anytime, and it is a real comfort to me to know that she will always have you to cheer + strengthen her.

Oh Gwyn never give her a moment's anxiety that you can help. You + I will never know all she has suffered this last 18 months a better dearer Mother no son or daughter ever had - what her letters have meant to me out here, I could never put down on paper. May you never have to realise the strength of her love by absence from it.

Now girlie persevere with your school work + music – never mind the failures, the striving is the thing.

And now God bless you + keep you ever in his care.
Your loving brother,
Ward

> **25576 Private W B Owens, 19th Battalion KLR,**

Private Owens also kept a pocket diary, although this was strictly forbidden in the trenches. It was was later recovered from his body, and returned to his family. Its last entry, for Saturday 29 July 1916, is simple and to the point: 'Moving off for line tonight - over the top in the morning'. The rest of the diary is blank.

30 July 1916

The four attacking battalions had left positions near Fricourt on the late evening of 29 July, to arrive in their assembly positions in the British held end of the Maltz Horn Trench in time for zero hour, which was set for 4.45 am. The 19th Battalion was to attack in a slightly north-easterly direction along a line to the south of Arrowhead Copse, whilst the 20th Battalion would advance along a line which was slightly to the south-east, just north of the Maltz Horn Farm. The 17th Battalion was to support these two, 'A' and 'B' Company behind the 19th, and 'C' and 'D' Companies behind the 20th.

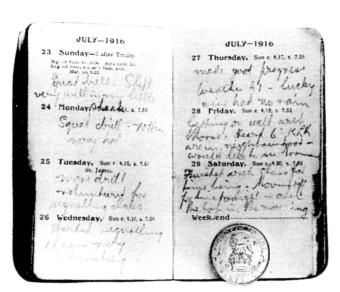

The last diary entry of 25576 Pte W E Owens of the 19th Battalion, who died of wounds received during the Battle of Guillemont. Diaries were strictly forbidden in the trenches, and the size of this one is shown by the sixpence photographed with it.

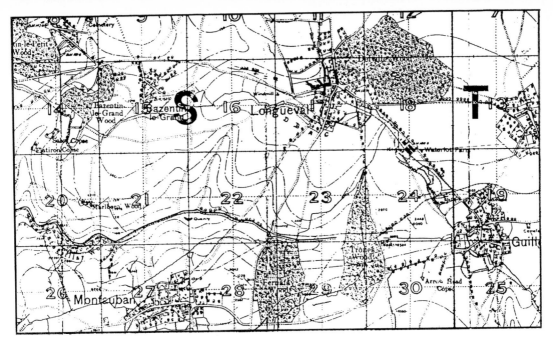

A detail from the 2 June 1916 edition 1:20,000 scale British Army trench map of Montauban, showing the area fought over by the Pals Battalions during July 1916.

The night was very dark and very foggy, but nevertheless the Germans seemed to know that an attack was imminent for they shelled all the approaches to Maltz Horn Trench with gas and high explosive shells, as the units passed south-east of the Briqueterie. The 19th Battalion suffered most from gas, although some of the 20th and the 17th Battalions, including Lieutenant-Colonel Fairfax, the commander of the 17th, also received harmful doses.

Guillemont was a dreadful do, even before we got to the line. We had an officer called Boundy, he was a Chilean, I believe, anyway, while he was talking to us, the Germans were sending some shells over very close, – gas shells, and some of us got a nasty whiff of the stuff. It made you very sick, and I was told your throat closed up, if you got a lot of it. And when we came to go, I fell down about three times, and some others did the same thing, and when we came to take a position in the line, they sent myself and three or four others to a dugout at the back, told us we weren't going over the top, and to stop there. **25561 Private H Redhead, 20th Battalion KLR.**

We moved up on the night of 29th July, and under cover of darkness, went through our old front line to the sunken road which led to the Brick Works on which we had gazed for many months. When we reached the dead horse, (killed on the first day), we left the road to take up our jumping off position in the old German trench of which there was just a little left. Just as we were about to do so, I ran into my brother Harvey[2] who was a captain in our 19th Battalion, who were to be on our left. We only had time for a word, but it was a happy coincidence.

We reached the trench, my platoon in single file behind me, and I was just about to tell the first man to get in, when there was a blinding flash and I found myself on the ground. My sergeant was with me and I suppose I told him to carry on, but I remember little of the next few hours. My batman, Dave Anderson,[1] with help from others got me into the trench, and there I lay. I was hit in a number of places and he did his best to bind the worst wounds with field dressings. There is little pain when one is first hit – that comes as the senses begin to return. Anderson stayed with me – I am sure that if he had left me, I would never have survived. **Lieutenant E W Willmer, 17th Battalion KLR.**

By 2.45 am all four battalions of the Brigade were ready to go, but a dense fog still covered the battlefield, and no one could see more than a few yards in any direction. Fog can often aid an attack, as it shields the attacker, but over such a distance, with the open ground first dipping downwards and then upwards towards the German positions and with the enemy firing on previously calibrated fixed positions, the prospects for success were not good.

It was next to impossible to delay the attack – it was much too big an operation – so forward they had to go. It will give some idea when I say that on one flank we had to go 1,750 yards over rolling country. Everyone knows what it is like to cross enclosed country which you know really well in a fog and how easy it is to lose your way. Therefore imagine these rolling hills, with no landmarks and absolutely unknown to anyone. Is it surprising that people lost their way and lost touch with those next to them? As a matter of fact, it was wonderful the way in which many men found their way right on to the place we wanted to get to. But as a connected attack – it was absolutely impossible. **Brigadier-General F C Stanley, CMG, DSO.**

At 4.45 am, the whistles blew in White Horse Trench, the forming up point for the 19th and 20th Battalions, and the Pals climbed out onto the parapet and began to move off into the fog – in the general direction of Guillemont. At this moment the fog gave help to neither attacker nor defender. The British artillery, firing blindly into the fog, had no way of knowing how effective its shelling was being, as the fall of the shells could not be plotted with any accuracy. The Germans knew just how effective it was, however, and under the cover of the fog many left their frontline trenches and got out into the open of No Man's Land, where they lay concealed in shell holes and depressions, where only a chance shot would hit them.

The first success of the day, however, went to the Allies. In a joint move, the 2nd Bedfords advanced to the east and the French 153me R.I. advanced to the north, and cut off the enemy garrison in the remains of Maltz Horn Farm. The Germans were largely taken by surprise, and many were down in the bottom of the trench, presumably sheltering from the British bombardment. Over sixty of them were killed, and despite some attempts by them to surrender, no quarter was given by the Allied soldiers and only one prisoner was taken – presumably for intelligence purposes

There was a trench just in front of us, with the remains of a farm there, and we had a little plan by which the French made a dash from the south and we from the west, and nipped it off, allowing the line to go forward.

German photograph taken in the summer of 1916 showing the village of Combles, just a few miles from Guillemont, as yet undamaged by the war.

It was most successful and we scuppered all the Boches there, and there were a lot. It took them completely by surprise and they were all lying down in the bottom of the trench. Of course there was a lot of 'Kamerad'[4] business, but that is not thought much of now, and one only was taken prisoner. **Brigadier General F C Stanley, CMG, DSO.**

The early capture of Maltz Horn Farm was essential to the rest of the attack, as it was on high ground and overlooked the surrounding area. Once it had fallen into Allied hands, it was possible for the rest of the 89th Brigade attack to move forward.

Whilst this was happening, the 19th Battalion, followed up by 'A' and 'B' Companies of the 17th Battalion, was stumbling forward into the early morning fog. This fog began to clear in pockets, but visibility was still very poor and good communications impossible. The advancing Companies became separated, and as the fog began to clear became perfect targets for those Germans who lay concealed out in front of their lines. Many of the Pals were caught out in the open with no cover at all, and were quite simply mown down by rifle and machine-gun fire.

No.3 and No.4 Companies, intermingled because of the fog, soon came across a small German strongpoint on their line of advance, just to the south of Arrowhead Copse, and after a fierce hand-to-hand encounter, they captured it. Two soldiers friends from Chester wrote home separately to describe their experiences.

Major (later Lieutenant-Colonel) G Rollo, Second-in-Command of the 17th Battalion until July 191G, and then appointed Commander of the 19th Battalion until the end of the war.

We were in reserve for a week just behind the line waiting for our third attack. On the night of 29th July we moved up to our position just ahead and on the right of Trones Wood. Here we took up our position in shell holes just behind the 19th, and dug ourselves in for safety, awaiting early morning when the advance was due to take place at 4.45. We were in our stations, myself being with Ossy Eyes.[5] Frank[5] and his men were quite near, also Sam's[5] gun team.[6] We were under constant fire, but not heavy, being mostly gas shells. It would be towards one or two o'clock when poor old Sam met his fate. Our sergeant had just given us our rum ration and gone to the shell hole where the gun team were, and here, unfortunately, one gas shell found its mark, landing in the centre of the gunners. Poor lads, it wiped the whole of them out.

It was a bad start for us, but at 4.45 the boys were up – into the mist they went, headed by our section commanders. We ploughed along taking shelter here and there, for they poured one continual rain of lead at us. We were suffering terrible losses but the boys kept on. When we first started the attack, I saw Frank leading his section on. He was on our right, but he disappeared into the mist. We kept pushing forward and were then held up by a German advance trench (a strong point). Here we fought for three-quarters of an hour, when the enemy saw their chance was hopeless. They downed arms, hands up, and cried like children for mercy.

We took up our position in what was once the German trench, – only three of us out of our section, our NCO, Ossy, and myself. Getting lost, we attached ourselves to the 19th. Here we met another of our Pals who had also got lost. He was one of Frank's section. Then he told us the terrible news. Frank was leading his section in the charge, and unfortunately was shot through the heart. The sights were bad enough, but the shock of losing Frank and Sam as well! I can't describe my feelings – its heart-breaking. They were two fine fellows, so very popular in the company, and not only were they excellent soldiers but thorough gentlemen too! **15971 Lance Corporal H Foster, 17th Battalion KLK.**

Ossy Eyes and I who managed to come out of this without a scratch are badly cut up over this terrible affair, and I know you, and all those who know those dear comrades of ours will be badly

The human debris of war on the Somme.

upset about it. Poor old Charlie[5] met his death while performing a very brave act. Our battalion were in the first wave that went over, and when we advanced so far, we got orders to get down and Charlie, who noticed a German machine-gun a few yards ahead of him, charged with one or two others to try and capture it, but was fatally hit by a bomb. Frank Pierce was sniped through the head and Sam Thomas was knocked out by a gas shell while going up to the trenches the night previous to the attack.[7] Gordon Pinches[5] was killed with a bullet, but where it caught him exactly, I could not say. **21646 Private C J Wright, 17th and later 18th Battalion KLR.[5]**

For his part in the same action, another Chester man, 21495 Lance Corporal A J Edwards was awarded a field promotion to Sergeant, and the Distinguished Conduct Medal. However, not long after this, as the fog swirled around them, the so-far victorious men of the 17th and 19th Battalions also came under fire which pinned them down and they were unable to move in any direction.

By this time the 19th Battalion Commander, Lieutenant-Colonel G Rollo, in Battalion Headquarters in Maltz Horn Trench, was wondering why no news of the attack had come back to him, and determined to find out for himself how things were going. Ordering his Adjutant, Captain W Fraser, and Lieutenants Lloyd and Lewis to accompany him, he left his dugout and led his party over the parapet. Almost immediately he was hit and fell to the ground. Fraser was hit soon afterwards, followed some forty minutes later by Lloyd. Stretcher-bearers were called and the Commanding Officer and Lieutenant Lloyd were evacuated to the rear, but Captain Fraser died, unfortunately, before help arrived. This left Lieutenant Lewis temporarily in command of the Battalion, without knowing anything at all of the situation up ahead.

However, No.1 and No.2 Companies, led by Captain A L Dodd and Captain W Nickson, on the left of the 89th Brigade attack, managed to move forward through the enemy fire and arrive at the main objective at the edge of the orchard south-east of Guillemont village. There they began to dig in, to try to consolidate their position. However, any hope of their holding the captured position depended totally on the success of the troops on their left flank. These were men of the 2nd Battalion of The Royal Scots Fusiliers, of the 90th Brigade, whose task was to capture and hold the eastern edge of the village. Although the Scots managed to fight their way in, their numbers were so depleted that they were simply too few of them to hold their gains and they were eventually overwhelmed by the German defenders and all but wiped out. The only survivors from this Battalion were the wounded, who were later recovered from shell holes.

The fate of the Scots Fusiliers had catastrophic results on the Pals, as it then meant that their own position was totally untenable, enfiladed, as they were, with the Germans able to pour murderous fire onto them from the safety of the village fortifications. As these men from the 19th Battalion had received no word from the rear or from either flank, they believed themselves to be totally isolated thus, in the early afternoon, they decided to withdraw and try to consolidate something of their advance, by falling back and digging in. An estimation of the effective fighting strength of the 19th Battalion alone, at midday, was put at seven officers and 43 other ranks.

The 20th Battalion, supported by 'C' and 'D' Companies of the 17th Battalion, had also gone into the attack at 4.45, their objectives being the German trenches from the north corner of Wedge Wood, up to the orchard area to be taken by the 19th Battalion. They did not, however, get that far. Like the other attackers, all six companies became intermingled in the fog and uncertain terrain, and were similarly caught out in the open when the fog began to clear.

We have already read reports of the 20th Battalion's part in the victory of 1st July from Lance-Corporal J Quinn. Quinn himself was to be a casualty of 30 July, and this report of his death, sent home by a comrade some weeks after the battle, outlines the early stages of the 20th Battalion's struggle.

Joe was in command of the Section adjoining mine, and we were together the whole of the previous night, 'going over the top' together in the morning. There was a very heavy mist, in which it was practically impossible to distinguish individuals at a distance, and I, busy keeping the men extended, was unable to keep Joe in sight. About halfway across No Man's Land, whilst waiting in a shell-hole for one of our own barrages to lift, I became aware of the fact that Joe was in the next hole to mine, and we smiled encouragement to each other. Enemy machine guns were sweeping the whole place with exposive bullets, and there was a fearful noise, and speech was impossible. The streams of death whistled over our shell-holes, coming from the left flank, and Joe's hole being to the left of mine, he received the bullet in his side. He slipped away quietly – just a yearning glance, a feeble clutching at space, and then a gentle sinking into oblivion, with his head on his arm.
21956 Corporal G E Hemingway 20th Battalion KLR.

Oxton Cricket Club.

OXTON,
BIRKENHEAD.

We have learned with profound regret that ATHERTON was killed in action on July 30th.

He has been in the employ of the Cricket Club for fifteen years, and earned the esteem of all – as Cricketer, Groundsman, and Professional.

He was one of the very first to join the colours on the outbreak of War, although a married man; and he leaves a wife and four little girls, the eldest aged 7 and the youngest 2½ years.

The Committee have decided to appeal for a Testimonial for the benefit of the Widow and Children, and it is hoped that the necessary sum will accrue to enable them to carry out a proposal to hand over monthly, a sufficient amount to tide the Widow over the next six years, or until the children are in a position to assist the home by their own earnings.

We realise that there are many calls at the present time, but feel that this one will appeal very directly to all members and friends of the Oxton Cricket Club.

Donations should be sent to, and will be gratefully acknowledged by :—

G. E. GODWIN, *President.*
48, SHREWSBURY ROAD, OXTON.

D. N. HEBBLETHWAITE, *Hon. Treas.,*
32, BIRCH ROAD, OXTON.

B. K. STRATTON, *Acting Hon. Sec.,*
MOUNT PLEASANT, OXTON.

The first report of the Battalion's progress received at 20th Battalion Headquarters, also in Maltz Horn Trench, by its commander, Lieutenant-Colonel H W Cobham was timed at just after 6.00 am. It was from Lieutenant R E Melly, who commanded No.1 Company, and stated that he had captured the German-held end of Maltz Horn Trench, killing many Germans. About half an hour later another message was received from Second-Lieutenant C P

A notice published at Oxton Cricket Club, Birkenhead to raise financial assistance for the widow of 22311 L/Cpl S Atherton of the 20th Battalion who was killed in action during the Battle of Guillemont. Atherton was the groundsman and professional for the cricket club for fifteen years before the war.

Moore, who was at a position about two hundred yards east of a sunken road which led to the outskirts of the village. He was the only officer left, in command of a party of about one hundred and fifty men, and he was cut off from anyone else and under fire. A second message from this officer about three hours later revealed that he had already lost half his strength, and although he had sent two patrols forward, neither had returned!

Further forward from Melly's position, a 'mixed bag' of about two hundred and forty men from the 17th, 19th and 20th Battalions had managed to cross the bullet-swept downward slope, and despite losing some two hundred of their number, had reached a second sunken road, which connected Guillemont with Hardecourt, and which allowed them some shelter, although they were still under enfilading fire from Guillemont. They were very gallantly commanded by a young Second-Lieutenant, J W Musker, who also managed to get messages back to Battalion Headquarters which told of his plight. All attempts to reach this party with fresh orders failed, as the runners were cut down, and eventually, with his command dwindling in number, Musker himself was killed by a sniper, and command passed to another subaltern, Second-Lieutenant J H Worsley.

At 11.00 am men from the Headquarters Company, with two companies of the 2nd Bedfords occupied and consolidated the newly captured German trench which ran through Maltz Horn Farm. In the afternoon, a further trench was then dug to link up with Arrowhead Copse, and Second-Lieutenant Moore and his party, still holding out further forward, were withdrawn to this new position. The new trench was then extended by the Bedfords to link up with the French on the right, and this then became the new Allied front line. At 9.30 pm Second-Lieutenant Worsley arrived back at Battalion Headquarters with the remnants of his morning command, about two hundred men, in all. This was the last organised party to return to the safety of the British lines! For their gallantry in action on 30 July 1916, both Worsley and Moore were later decorated with the Military Cross. Melly and Musker got nothing, probably as it was not usual, at this stage of the war, to recommend posthumous awards, other than that of the Victoria Cross.

An evaluation

By nightfall on 30 July the Allied line had been pushed forward about three hundred yards, on a line roughly from the eastern edge of Trones Wood to the eastern side of Maltz Horn Farm. In fact, the 89th Brigade now held the Maltz Horn Ridge, which would allow the Allied artillery a slight advantage in the continuing battles to capture Guillemont. However, there was no escaping the fact that the Brigade had not taken and held any of its objectives, and had sustained unbelievably high casualties.

The total killed, wounded, missing or prisoner, from the three Pals Battalions alone was estimated at the time, to be forty-two officers, and 1,073 other ranks. The breakdown of these figures shows fifteen officers and two hundred and eighty-one other ranks from the 17th Battalion, eleven officers and four hundred and thirty-five other ranks from the 19th Battalion and sixteen officers and three hundred and fifty seven other ranks from the 20th Battalion. Post war published figures indicate that of these, five officers and one hundred and eight other ranks from the 17th Battalion, nine officers and one hundred and eighty-four other ranks from the 19th Battalion and nine officers and one hundred and thirty-seven other ranks from the 20th Battalion, were killed in action or died of wounds actually on 30th July. Although these are not so reliable as the cemetery returns of the Imperial War Graves Commission, they always err on the light side, so the actual dead from the three Pals Battalions on 30 July 1916, would actually number well over four hundred and sixty! The losses of the 2nd Bedfords were fewer, as they did not have to endure the punishing time out in the open that was suffered by the Pals.

Furthermore, unlike the victory of 1 July, when all the objectives were captured, after 30 July, those wounded who lay in No Man's Land between Guillemont and the new British front line had little hope of being picked up, and many must have died simply because they could not get any form of medical help. Only those who were fortunate enough to be wounded early in the assault stood any chance of survival. One of these was Lieutenant E W Willmer, whom we have already seen, had been wounded by shellfire whilst waiting to go over the top.

Much of the time after I was hit, I must have been unconscious, but I have a vague recollection of the platoon going over, and that we were subjected to some unpleasant shelling, when several men were killed. I was told afterwards that it was a misty morning and that the French did not think an attack could succeed in such weather. We suffered heavy losses, the 17th coming out with about one hundred and twenty men, and the 19th with about eighty. My cousin was killed that day with the 19th,[8] and my brother Harvey was wounded, but carried on and as the senior officer surviving, brought his battalion out, next day. He was awarded the MC.

Just before the ambulance reached Corbie, I asked where we were, and was told that we were just passing through Heilly-sur-Somme. I thought to myself that at any rate we were out of range now!

I was not kept in Corbie, which was just a Clearing Station, and as casualties were so heavy at the time, everyone was cleaned up and sent on as soon as possible. I was whisked away to No.8 General Hospital in Rouen, where I was operated on for the wound in my leg, which was

The effect of the long hot summer of 1916 on July corpses can be easily appreciated in this photograph.

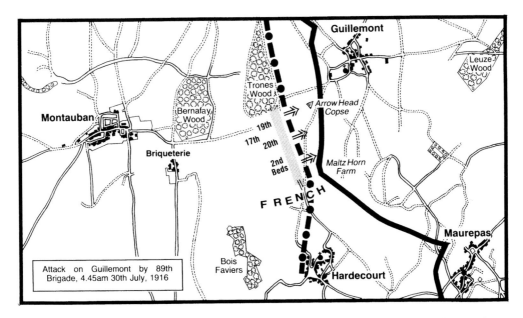

Attack on Guillemont by 89th Brigade, 4.45am 30th July, 1916

a hole through the muscular part of the thigh. Fortunately that splinter did no vital damage, but the medical people were very concerned about the wound in my shoulder, though at that time I did not know why.

There was no delay in Rouen either. The demand for beds was so great that men had to be moved on to England as soon as they were fit to travel. But as it was apparently important that I should be kept as still as possible, I was sent by barge down the river to Le Havre, where I was loaded onto a hospital ship which in private life, had been a ferry between the two islands in New Zealand. We were told that the ship would cross by night as there were U Boats in the Channel and I suppose I slept. When I woke, it was early morning, and I asked one of the orderlies if we were in Southampton, but he told me that we had not been able to start as there were submarines outside the harbour and that we would have to wait for nightfall to make the dash across.

Hospital ships were not immune from the torpedo, and they had been forced to sail without lights for some time, just like a combatant ship. It seemed a very long day, but all was well and we made the trip safely and were disembarked at Southampton in the early morning. The stretchers were loaded onto a hospital train, part of which was to go to Manchester, where I was to be at a school building in Whitworth Street which had been converted into a military hospital, mostly for surgical cases. Before the train started, two pretty girls came down the corridors between the stretchers and gave anyone who wanted to smoke, packets of cigarettes of any kind they fancied. I was told that they had met every hospital train since the war began. It was a wonderful thing to do. **Lieutenant E W Willmer, 17th Battalion KLR.**

Lieutenant Willmer eventually recovered from his wounds, but was never fit enough to serve on the Western Front again.

For well over five hundred men whose bodies lay on the slopes before Guillemont, or who would die of their wounds over the next week or so, there would be no return to England. The bodies which lay in No Man's Land decayed in the intensity of the August sun, or were blown apart by the bombardments which criss-crossed the area until the village finally fell to the British on 3 September. After this, the locations of any bodies that were found were marked, or the remains themselves were retrieved from the battlefield, for later interment in the nearby burial ground, to be named Guillemont Road Cemetery. For most of the dead, however, there would be no burial, identified or not, and today their names are inscribed on the Memorial to the Missing of the Somme at Thiepval. Many years later, even after Guillemont Road Cemetery was closed for burials, the bodies of dead soldiers were still being discovered

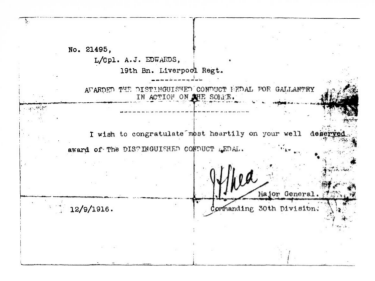

No. 21495,
L/Cpl. A.J. EDWARDS,
19th Bn. Liverpool Regt.

AWARDED THE DISTINGUISHED CONDUCT MEDAL FOR GALLANTRY
IN ACTION ON THE SOMME.

I wish to congratulate most heartily on your well deserved
award of The DISTINGUISHED CONDUCT MEDAL.

Major General.
12/9/1916.   Commanding 30th Division.

Above and below: The original citation for the award of the DCM to L/Cpl A J Edwards of the 19th Battalion for distinguished conduct during the battle of Guillemont. He was also rewarded with a field promotion to sergeant.

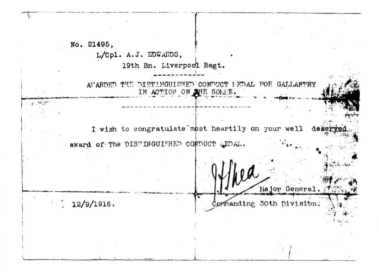

No. 21495,
L/Cpl. A.J. EDWARDS,
19th Bn. Liverpool Regt.

AWARDED THE DISTINGUISHED CONDUCT MEDAL FOR GALLANTRY
IN ACTION ON THE SOMME.

I wish to congratulate most heartily on your well deserved
award of The DISTINGUISHED CONDUCT MEDAL.

Major General.
12/9/1916.   Commanding 30th Division.

on the nearby battlefield. Captain William Nickson, of the 19th Battalion, who was killed on 30 July, is commemorated on the Thiepval Memorial and also has an actual grave, ten miles away in Serre Road Cemetery No.2. As the Thiepval Memorial was unveiled in 1932, his remains must have been discovered after this date and reburied in one of the few Imperial War Graves Commission cemeteries still admitting burials at that time.

What went wrong?

So what did go wrong to make 30 July 1916 the blackest day in the short history of the Liverpool Pals and plunge all of Merseyside into mourning? Certainly there was no lack of zeal and courage on the part of the attackers, motivated, as they were, by a total belief in their cause and confidence in their leadership. After 1 July they were also confident in their fighting skills, and not even the debacle in Trones Wood had dented this. It would be tempting to blame the failure of the battle on those perennial of Great War scapegoats – the Generals – but if they were responsible, which ones, and at what level?

After the abortive attack of 23 June, there had certainly been no attempt to change the plan of attack, even though it had shown itself to be totally inadequate. On that occasion, one complaint had been the lack of covering artillery fire, but any extra fire that might have been used on the early morning of 30 July was largely negated by the weather conditions. There is no doubt, as we have already seen, that the whole attack was at best too ambitious, depending, as it did on an assault across a mile of open undulating country, and relying for its success on further successes on each flank. The Official History hints at where the blame ought to lie.

It is little matter for surprise that the attack made on 30th July should have taken almost the exact course of action of the 23rd: the conditions under which it was delivered were practically the same. After the first experience it seemed to the local commanders that an assault against the Guillemont positions from the west – up the exposed shallow trough which marked the termination of Caterpillar Valley – and from the south-west, over a crest and down a slope, both devoid of cover, had little chance of success. This was particularly the case if the right were not supported by an attack in force across the heads of the Maurepas ravine. **Captain W Miles, late Durham Light Infantry.**

Thus, it would appear that the 'local commanders', presumably those who commanded the infantry brigades, were not happy about the planned attack. We have already seen that Stanley was far from happy with the situation on the morning of 30 July, but was powerless to do anything about it. It is therefore inconceivable that he would not have expressed his disquiet to Shea, who should have been able to have the attack plan modified, but to what? It is also difficult to see how a modified plan would have made any difference to the local conditions encountered that foggy morning.

Despite the gas and high explosive shelling which disrupted the troops on their way to the forming-up point on the night of the 29th, there is no doubt that the thick fog which covered the battlefield, and did not clear fully until some four hours after the battle had started, was the real barrier to success. Even though Stanley realised that the fog might mean disaster, he was powerless to halt, or postpone the attack, as it was on such a large scale. Although some of the Pals undoubtedly survived the assault because they could not be seen in the fog, the fact remains that for the most part, the Germans did not need to see them. They knew that the Allies would have to attack across open ground to reach their strongpoints, and they knew the exact direction from where they would be coming. Machine-gun and artillery fire on these axes were bound to hit their targets! Furthermore, the Allied artillery did not have this advantage, because although they knew where the German trenches were and were able to range on them, they could not see the effects of their shelling, nor could they see that many Germans had left their front line and were sheltering in comparative safety, out in No Mans's Land. It was these troops who would create so much carnage amongst the Pals, once the fog began to lift.

It was also most important for an attack involving three divisions to keep in touch with its flanks at all times, and there was no way that the fog allowed this. Sections and companies became entangled as well as battalions, and no one had any real idea of how any other troops were faring, or how the attack was progressing. This also meant that even when the fog did clear, total confusion prevented any link up of units or joint consolidation of positions. As a result, not only did troops suddenly find themselves at the mercy of the Germans with very little cover, but they were not able to count on vital artillery support, to help them hold what they had won, because the Allied artillery had no idea where its own troops were. This point is proven by the men of the 19th Battalion, who had actually taken their objectives on the edge of the orchard, but eventually withdrew in the early afternoon because they thought they were totally unsupported.

Thus, overall, the fog benefited the Germans more than the Allies. The only areas where this was not so were the few strongpoints, such as Maltz Horn Farm, or the fortified trench traverses that were not far from the Allied start lines. Because they were fairly close to the Allied front lines they were vulnerable to sudden armed assault under the cover of the fog and could be rushed and taken. There was no chance, however, that the defenders on the outskirts of Guillemont could be surprised in the same way.

Therefore, if the fog was the root cause of so many losses, it is interesting to speculate if the attack might have been successful had there been no fog at all. The inescapable answer, however, is that had there not been thick fog, it is entirely possible that the 89th Brigade might have sustained even more losses. The only Allied advantage that might have been gained had there been no fog, would have been that of artillery cover for the attack, but it is doubtful if this could have been very helpful to the men who had to cross a mile of unknown undulating land. Although effective fire from the German front line trenches could have been neutralised by Allied artillery, the Pals would still have had to cross virtually a mile of open, and they would have been easy targets for the German Foot Artillery firing from the

A German sentry on the Somme.

129

This concrete obelisk stands in a field on the road to Hardecourt sur Bois past the crucifix marking the original site of Maltz Horn Farm. The inscription commemorates two French soldiers, Marcel Boucher, and Romeo Lepage of the 153rd Infantry Regiment, who fell on that spot on 28 July 1916. It was, of course, the French 153rd Infantry Regiment who were to the right of the 17th Battalion KLR on 1 July 1916, and who helped the 2nd Bedfords capture Maltz Horn Farm in the early morning of 30 July 1916.

high ground in the rear of the second defensive line.

However, if one accepts that it was essential to the greater outcome of the Somme battle, that Guillemont had to fall, then despite the inevitable losses, there does not seem to have been any other way, at the end of July 1916, to reduce it than by infantry assault. Perhaps if the French had managed to capture Maurepas, then an outflanking attack from there might have isolated Guillemont, but Maurepas did not fall either! The only obvious way that Guillemont could have been captured with few losses to the infantry would have been as part of a tank attack – the country would have been ideal for tank warfare – but tanks were at least six weeks in the future and another year away from real success in open warfare!

The inescapable fact remains, then, that 30 July 1916 was to the Liverpool Pals Battalions of the 89th Brigade what 1 July 1916 was to most of the other northern Pals Battalions and the 18th Battalion of the King's Regiment. For the most part the carnage on each of these dates wiped out the concept of the original Pals Battalions, and their fervour and patriotism was diluted to such an extent that it was difficult to resurrect. Now, also, the 19th Kings, 'the good old 19th,' had certainly been 'blooded', and they too, like their comrades, understood the full realisation of the horrors and losses of battle.

Few returned from the 2,000 yards approach in the early morning mists across the flat open country to the enemy trenches, strongly held by fresh troops just deployed for counter-attack. Few even reached them. Later on, the next morning, under a blazing sun and blue skies, we marched away after a pitiful roll call, leaving behind at our Happy Valley bivouacs four large mounds of loaves of bread and other rations, – rations intended for four companies, and for which there were no partakers. The old comrades had diminished in numbers to one quarter of their strength in January that year. Here if ever, was a moment of truth! Gone was the excitement, the hope of breakthrough, the early savage satisfaction of getting our own back on Jerry with heavy barrages and bayonet, after being for so long at the receiving end without means of reply. Only remained the clear realisation of the stark picture of the war. The dreams carried over from civilian life (how many eons ago that seemed) were discarded with the footballs and drums and blithe spirits which the first crossing of No Mans Land would take. Only a grimmer spirit to face the future remained. **17158 Corporal E G Williams, 19th Battalion KLR.**

When the grim realisation of the scale of the losses began to dawn on those still left in the business offices and financial institutions of Liverpool, a hollow numbness descended over all. Such was the closeness of the community that when Mr R G Williams, father of Corporal E G Williams and Private T S Williams, realised that both his sons had survived, when the sons of so many of his friends and associates had perished, he felt almost ashamed that he was not able to share in their grief.

NOTES

1. It might interest readers to know how the censored portion of the letter was revealed. A rectangular section was cut from a piece of card to provide a 'window'. The censored part of the letter was then laid over the 'window', and illuminated from beneath by the light of an overhead projector. An ultra-violet light was then shone onto it from above, and the varying amount of light thus transmitted and reflected allowed the major portion of the words in the censored section to be read. This method works particularly well with the old indelible pencil used at the time, although it will

An original entrance to a German dugout near the old orchard on the outskirts of Guillemont.

also work with modern graphite pencils The author is indebted to Miss G Owens, sister of Private W Owens, for permission to reproduce these letters.

2. This was, Captain Harvey Thew Willmer, 19th Battalion KLR, who was wounded at Guillemont, and subsequently awarded the Military Cross for his actions during the battle. He survived the war.
3. 15804 Private R D Anderson, 17th Battalion KLR, survived the war.
4. This German word literally means 'comrade*, but was used widely by German troops in the act of surrendering, presumably to signify that they accepted their sudden change of status. Many British soldiers, however, thought that the word actually meant 'I surrender'.
5. 'Ossy Eyes' was 15985 Lance Corporal I O Eyes, 17th Battalion, himself killed in action at Flers, on 12 October 1916. 'Sam' was 15938 Private S H Thomas, 17th Battalion, 'Frank' was 16016 Lance-Corporal F A Pierce, 17th Battalion, 'Charlie' was 21518 Private C. Heath, 19th Battalion and Gordon Pinches was 21589 Lance Corporal N G Pinches, 19th Battalion. All were childhood friends from the Chester area, and played football for Sealand United Football Team pre-war. Lance-Corporal Foster was wounded in the fighting around Flers in October 1916, recovered, and was eventually captured on 29 April 1918, during the German March offensive. He spent the rest of the war in a German Prisoner of War Camp. Private Wright was not so fortunate. Having won the Military Medal, and been transferred to the 18th Battalion, he was hit and killed by an aerial bomb on 8 October 1918, just a month before the Armistice.
6. This 'gun' would have been a Stokes Mortar, often referred to at the time, as a Stokes Gun.
7. The accounts of the action written by Lance-Corporal Foster and Private Wright differ in one or two respects, which, apart from anything else, illustrates the inevitable confusion which was ever-present during a battle.
8. This was Captain Walter WiUmer, of No.4 Company, 19th Battalion KLR, who was hit in the head by a machine gun bullet. He, like his cousin, came from Birkenhead, Cheshire.

Chapter Eight

The attack was a shambles – many men killed and taken prisoner'

Reaction at home

By August 1916, people in Great Britain were beginning to feel uneasy about the euphoric way in which the press was still reporting the war, and beginning to realise that the Somme was not the great breakthrough that had been promised. Despite the reports of great gains made and great deeds of valour performed, the lists of dead and wounded that appeared in both local and national newspapers after 1 July began to tell their own story. With so many of their men under arms, Merseysiders had been used to casualty lists since 1914 but after the losses of the 18th Battalion on 1 July the news of the losses of the other three Pals Battalions at Guillemont just over a month later finally brought home the harsh reality of war. The same kind of spirit that had bound the Pals and their families together in joy and duty in 1914 was now to bind them together in sorrow and mourning – just two years later. Grief is always a very personal emotion, but many families, even those who had been bereaved were at least proud that their menfolk had 'done their bit', and could no longer be described as 'Derby's Lap Dogs'.

In January 1916, the following poem had appeared in the Birkenhead News, a local newspaper owned by the Willmer family, whom we have already seen, had provided three recruits for the Pals, all officers.

Lord Derby's Pals Brigade

'D'ye call these 'ere chaps soldiers? They're only "knuts", no more,
They'll never do no fightin' and they'll never see no war.
They call them "Derby's Pets", wot thinks they'll take the gals by storm.
That's why they're dressed in puttees and a khaki uniform.'

Such words as these were good enough – or any slanderous thing,
When Derby's lads chucked up their jobs for Country and for King.
And this is true – "twixt me and you – they hadn't to be pressed,
The thought of each on 'listing was – "I mean to do my best."

For many weary months they drilled and toiled and trenched and marched.
At Prescot, Knowsley, Grantham, oft hungry wet and parched.
No "stackers" these, they did their work, and so they would again;
No finer men had Kitchener when they marched from Salisbury Plain.

And now they're out in Flanders, and giving of their best,
Doing their bit as Britishers – not "standbacks" like the rest.
Of those who still remain at home, afraid to do their share,
Their duty is to help us, but they feel "much safer" here.

I'd like to meet the scoffers now, I'd rather like to know.
Do they say the same of "Derby's Pets" as they did a year ago?
Good luck to each and all of them – the boys are not afraid.
And when they return Set's do what we can for Lord Derby's Pals Brigade.'

A F Cotton Birkenhead, 19 January 1916

A F Cotton was better known as Arthur Cotton, Solicitor, and Deputy Coroner for the Borough of Birkenhead, and the tone of his poem already suggests disillusionment with those men who chose to stay at home rather than join the colours. Unfortunately, the Cotton family was to pay a high price for its devotion to patriotic duty. Arthur and Maud, his wife, had three sons. Their eldest son, Lieutenant A E Cotton, of the 13th Battalion The Cheshire Regiment, was killed in action on the Somme on 7 July 1916. The 13th Cheshires had been raised in a similar way to the Pals in 1914, but by Lord Leverhulme, in Port Sunlight village, site of Lever Brothers soap works. They were known locally as "The Wirral Pals". Their second son, 15209 Private E B Cotton had joined the 17th Battalion KLR in

British prisoners taken on the Somme, marching into captivity.

October 1914, and was reported missing, presumed killed after Guillemont. Eventually, at the beginning of August 1916, any hopes that he might have survived the battle were dashed, with a letter from his Company Quartermaster-Sergeant.

It is now quite definite that Ernest has been killed. It happened on the 30th July. He is known to have been buried, and the place will be notified to you by the War Office. I do not know where exactly, nor could I tell you were I in possession of the information. There is no one who can give you any more details than those contained in this letter, because nearly all the Company is gone.

I was with him a good deal before the 30th, and shook hands with him before moving off during the evening before that day. We two or three times discussed the letters he received from home, regarding your other son.[1]

As to who saw him last, it is impossible to say. The move took place in a heavy mist which hung on for some hours after the attack. I can only say that the last duty he performed was to endeavour to obtain stretcher-bearers for a wounded man. It is with sincere regret that I have written the foregoing. I ask you again to accept my sincerest sympathy. **15910 Company Quartermaster-Sergeant E R W Potter, 17th Battalion KLR.**

Private Cotton's body was never found and he is commemorated on the Memorial to the Missing of the Somme at Thiepval. To complete the family's agony, their third and only surviving son, Second-Lieutenant/Flight Cadet T E Cotton, of the Royal Air Force, also died tragically after the war, on 25 May 1920, of injuries sustained in a flying accident at Shoeburyness.

Although Arthur Cotton's poem represents the feeling in Merseyside of January 1916, maybe his own feelings altered somewhat after the loss of two sons on the Somme. However, curiously enough, an altered version of his poem, this time called 'After Two Years', and attributed to a 'WAH', appeared in the October 1916 edition of 'Progress', the house magazine of the previously mentioned Lever Brothers. It seems to be an update of Mr Cotton's earlier work, and reproduces, almost exactly, portions of some of Cotton's verses, yet adds more patriotic reactions to the part played by the Pals in the Somme battle, – albeit with little regard for historical accuracy.

After Two Years

D'ye call these 'ere chaps soldiers? They're only "knuts", no more,
They'll never do no fighting and they'll never see no war;
They call them "Derby's Pets", wot thinks they'll take the gals by storm,
That's why they're dressed in puttees and a khaki uniform.

Such words as these were good enough – or any slanderous thing —
When Derby's lads chucked up their jobs for Country and for King.
And this is true, 'twixt me and you, they hadn't to be pressed,
The thought of each on 'listing was – 'I mean to do my best.'

The time for Britain's push came on, to clear the Huns from France;
You'll think it hardly possible, at least a slight romance,
When told that out of all the troops at General Haig's command,
To drive the foe from Montauban, he chose Lord Derby's band.

And when their object they'd achieved, like warriors of old,
They showed their British grit and cried, "What we have got we'll hold."
They smashed the German counterblows, and fought the Prussian Guard,
Resolved that every man would die before he'd yield a yard.

They've made their name those "Swanky Pals", but the toll they had to pay,
Too many of their bravest chums now lie beneath the clay.
I'd like to meet the scoffers now, I'd rather like to know, Do they think the same of
"Derby's Pets" they did a year ago.'   WAH.

It is interesting to speculate whether or not Arthur Cotton ever came across WAH's version of his poem, and what his reactions were if he did. Probably, by October 1916, his sentiments would have been better represented by those expressed in a poem written by Alice, the sister of 26540 Lance Corporal Joseph Quinn, of the 20th Battalion, also killed at Guillemont. She wrote these words, very much in the style of the times, exactly one month after her brother's death in action.

In Memoriam
My Darling Brother, Lance-Corporal Joseph Quinn (Killed in Action, France 30 July, 1916) RIP.

Dear Soul, dost thou live in Eternal Light,
While thy corse lies cold 'neath the sod?
Brave Spirit, passed from the mire-stained trench,
To the Great White Throne of God!

Great Heart, so spent in the stubborn fight!
Hast thou ceased to beat for aye?
O tender eyes with the old lovelight!
Thou wilt sometimes glance our way?

Dear one, thou must know of our loneliness,
And pity the tears we shed,
We note no beauty in earth or sky,
Since through whom we loved are dead!

Yet brave, true Heart, thou dost bid us hope,
That, when Life's grim fight is done,
We shall meet at the muster of Christ's array.
When the final Victory's won!' Alice
2 Alexandra Road, Waterloo, near Liverpool. 30 August 1916[2]

There would be over five hundred other families consumed by similar grief that August after the Battle of Guillemont, and many more men from the Pals would tread the slow and painful road to recovery from their wounds.

Givenchy and back to the Somme

Whilst the relatives of fallen Pals were trying to come to terms with the enormity of their losses in Merseyside, out in France, Brigadier-General Stanley was trying equally hard to come to terms with the practical realities of the same losses! He estimated at the time, that the 89th Brigade's casualties from all causes were approximately 1,450.[3] Apart from heavy losses amongst the commissioned ranks, the three City Battalions from the Brigade had also lost very heavily at non-commissioned officer level. This had left the 17th Battalion with only ten sergeants and twenty-one corporals, the 19th Battalion with eight sergeants and seven corporals, and the 20th Battalion with eight sergeants and twelve corporals. An approximate war-time strength of NCOs in an infantry battalion in France in 1916 might be sixty sergeants (excluding warrant officers) and one hundred corporals (including lance-corporals). At this stage of the war, it was far easier to replace commissioned officers than it was to replace experienced NCOs;

On 31 July, the 30th Division was pulled out of the line and moved back to the Citadel Camp, south of Fricourt. On 2 August, the Pals Battalions entrained for the encampments around Abbeville, where they began a period of rest and reorganisation. One officer who did not go with them, however, was the Commanding Officer of the 17th Battalion, Lieutenant-Colonel B C Fairfax. He had not recovered at all from the gassing he had received on the night of 29 July, and despite his protestations, he was ordered to report sick, sent to No.38 Casualty Clearing Station at Heilly, and from there, was evacuated to England on 8 August, on board the hospital ship 'Asturias'. It was a blow both to himself, and to the whole Battalion, as he had been a guiding influence on it and the whole Brigade since 1914! Although he was to survive the war, he never returned to the Pals. His place as Battalion Commander was taken by Major J N Peck, who was promoted Lieutenant Colonel, upon appointment.

Because of the mauling it had received, since 1 July, the 30th Division was ordered to be temporarily attached to the XVII Corps of the First Army, which was stationed further north. As a result, on 3 August 1916, it left Abbeville for the area around Givenchy, where training, receiving of new drafts and re-organisation continued. Presumably, as the 18th Battalion had not been involved in the fighting at Guillemont, it was judged fit for front line service earlier than the others, and entered the line on 11 August. The other three Pals Battalions followed on 26 August. Although the Givenchy sector was not as hectic as the Somme, it was by no means quiet, and because the nature of the warfare there was more static, both sides had engaged in mining and counter-mining since 1915. It was a new experience for the Pals, and one which they did not find too comfortable. The trenches there were also quite close together in places, so both sides were very active with artillery, trench mortars and rifle grenades. Night patrols and trench raids were made by both sides to gain information about enemy movements, and to keep up the 'offensive spirit'. A typical patrol was undertaken by men of No 4. Company of the 20th Battalion on the night of 31 August/1 September.

It was there that I went on a night patrol, but luckily it was only to try and observe where the Germans were sending up very lights from, because sometimes you had orders to go into his line and try and take a couple of prisoners and then get back with them if you were lucky enough to

escape. Then sometimes if you took more than two – if you took four and they only wanted two – I believe you'd have to take a couple and shoot them to get rid of them, and bring two back only!

It was a very bad bit of the

A German trench in front of Guillemont showing the effects of the British bombardment on the village, before its final capture.

line; it had been that way for twelve months and when we went over, it was wet and slimy, and we had to go over on our stomachs. We couldn't stand up, we had to crawl through our own wire and then make for this point in No Man's Land. There were half a dozen of us, and in the end, the corporal decided he could give a good inclination as to where they were coming from, and he gave us orders to go back, and we turned round and went back on our stomachs. There were a lot of sticks – I think they must have been from rifle grenades – as you were crawling along you had these iron sticks sticking up out of the soil, all over the place, – I think they were from grenades.

We were trying to get to a point which was about half way across No Man's Land, and the lights kept going up, and every time one went up near to us, we had to keep 'doggo', and keep our faces down so that they wouldn't show white. When we reached this point, we stopped there for a few minutes while the corporal was deciding what to do, and then he told us to go back. It was quite dark there apart from when the lights were going up, and there were lots of holes and there seemed to be lots of sticks and lots of rubbish in No Man's Land, although I didn't see any bodies. We had to crawl back to this particular point where we went through the wire to get back into our own trench – the password was 'whisky', – I remember that. **25561 Private H Redhead, 20th Battalion KLR.**

On 4 September some men from the 19th Battalion made a trench raid to try and identify the German unit opposite them, but as they reached the enemy wire they were discovered and fired upon, one man being hit. His cries then attracted more fire, and the party was forced to retire in confusion. On 13 September a similar raid was attempted, with similar unfortunate results, during which Lieutenant R G Lloyd MC, 17800 Company Sergeant Major E Concannon, 52076 Private T Jordan and 5440 Private A R Worsley were killed, and Second-Lieutenant N L Taylor and eighteen other ranks were wounded. Taylor subsequently died of his wounds on 18 September 1916. On that same day, the 30th Division was ordered to rejoin the Fourth Army. It was originally intended that the Division would serve as part of the XIVth Corps, but this was then changed, and it was ordered to XVth Corps instead. As a consequence, on 18 September the 18th Battalion entrained for Doullens and then took up billets at Naours, and the other three Pals Battalions left for Gezaincourt, which was about three miles from Doullens, and then took up billets at Vignacourt. The Division then carried on further training whilst in a reserve role. In all, over the thirty-nine day period in which they served under fire in

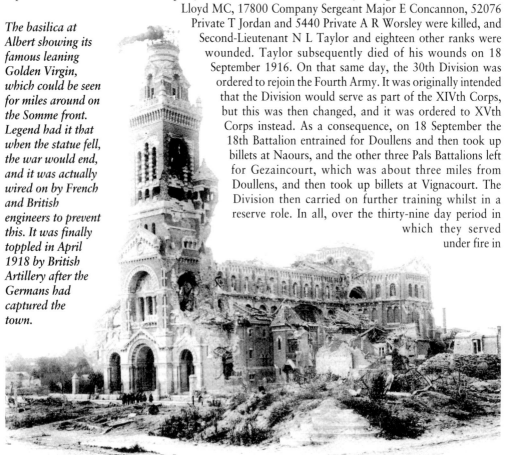

The basilica at Albert showing its famous leaning Golden Virgin, which could be seen for miles around on the Somme front. Legend had it that when the statue fell, the war would end, and it was actually wired on by French and British engineers to prevent this. It was finally toppled in April 1918 by British Artillery after the Germans had captured the town.

The devastation on the Somme.

the Givenchy area, the four Pals Battalions lost a total of three officers and fifty-four men killed, or died of wounds.

It was whilst the 19th Battalion was stationed at Vignacourt, on 29th September 1916, that its temporary Commanding Officer, Major Robert Kneale Morrison was accused of drunkenness and tried by General Court Martial. He was convicted of the charge, and the sentence of the court was that he should be dismissed the Service, and returned to England![4] He actually left the Battalion for home on 16 October, one day after the official date of his dismissal. Morrison, who had been with the Battalion since the beginning, had taken over temporary command after Lieutenant-Colonel Rollo had been wounded at Guillemont, and his fall from grace, under such circumstances, perhaps illustrates the pressures of front-line command. Although it was a great disgrace for an officer to be dismissed the service – under any circumstances – in principle at least, Morrison's sentence was a lot more lenient than could have been expected by an other rank convicted of a similar offence. A soldier found drunk on active service could expect imprisonment and/or a fine, although at the front this was usually commuted to Field Punishment Number One, which we have already seen could often take a most distressing form! Although it is likely that a cashierd officer would be shunned by all his circle of contacts at home, and might well have to start a new life elsewhere, at least it could be said that he was alive and out of the firing line — a situation that many at the front would have envied. It is interesting to speculate whether or not Morrison was within the age group for conscription when he was returned to England, and whether or not he was subsequently forced to endure the indignity of further military service in the ranks.

In early October the comfortable billets were left behind and the inevitable terrors of the Somme battlefront beckoned again.

> On the 4th of October, a party of commanding officers and myself went off to the line where we were to go in. We passed through many places that one knew so well by name, and some of them by sight – Montauban, Longueval, Bernafay Wood and High Wood, and Flers. It was impossible to describe the absolute state of desolation of all the country. As for the ground, it was so pitted with shell holes that you could not walk a yard without being able to touch one with your stick, and there were some terribly gruesome sights to be seen. **Brigadier-General F C Stanley, CMG, DSO.**

> The thinned and shattered ranks had been filled with new men who had been given experience in the Givenchy trenches, and then we moved south. Once more we reached our landmarks of December 1915 when in the trenches for the first time at Fonquevillers. Eastwards from the comparative quiet of our own billets, the Somme battlefield still thundered and prayed on the troops fed into it, in its last weeks of active offensive. Further south, the leaning virgin of Albert still held her son in outstretched arms, and the Somme river still flowed by poplars and osier beds beside Bray sur Somme, now in temporary peace. Once more we marched along the Albert to Peronne highway with its double-rank trees between Mametz and Carnoy, and once more we entered the ruined village of Maricourt and turned north towards Longueval, Flers and Bapaume, still in enemy hands.

Was this the countryside we knew only a few months or ages ago, – was it indeed the cushy[5] bit of line where we had fished in the millraces of Bray sur Somme and in the lily pad studded etangs[6] of the river marshes, and where we had travelled at night when taking up the rations to the battalions in the trenches, with only the occasional shell salvoes, and the spent bullets singing their night songs?

Now in October 1916, it had become a muddy rutted way, and a grey and dull landscape, the passageway of many who had gone in to the terrible arena of the battlefield. As we marched along that road in a long column of route, we came across an encampment, a bivouac of brown tanned canvas tents set out in orderly rows on the brown trodden earth on a level of ground below us. Browner than the brown earth, the little tents stood empty with their side flaps open for airing, and showing neatly set out kits and bedding.

The men were away, but there remained one lone tall Scottish piper, adorned with green and black tartan kilt, and plaid and bonnet. He played on the pipes as he paced rhythmically round. His stance and the vibrant music fell strangely and excitingly on the ears of the Sassenachs passing by. We heard it a far as we approached, and it faded into the background mutter of the war, as we marched on.

We passed through Maricourt's red brick ruins and out into the dun terrain of the old French lines battlefield of 1st July, and rested in our ranks on the south sloping stretch of ground from whence we looked back over the churned ochre mud of the sunken road, to the shredded trunks of what was once Trones Wood, standing pale and stark in the half sunshine of an October afternoon.

Strangely, against this background, appeared a group of marvellously uniformed staff officers — khaki and gold braid, and red in tabs and visage. They stood around one slight khaki clad figure with a boy's face gazing sombrely at us. His chest bore an array of rainbow ribbons, – Edward, Prince of Wales, and his entourage. They looked at us and departed, – and so did we, but up to the battlefield of Flers, a few kilometres away to the north. **17518 Corporal E G Williams, 19th Battalion KLR.**

Flers – 12 October 1916

Although veterans of the Pals always refer to the Somme fighting in October 1916, as 'The Battle of Flers', a more appropriate title for it would be the Battle of Ligny-Thilloy, as Flers had already fallen on 15 September, during the first tank attack in history. In point of fact, the October fighting around Flers was part of the Battle of the Transloy Ridges, and directly involved the 89th Brigade on 12 October, and the 21st Brigade, on 18 October.

By the end of September 1916, nearly all the high ground on the Somme battlefront was in Allied hands. Only on the ridges above Thiepval, Le Sars and Le Transloy were the Germans still in superior and commanding positions. However, the fine weather of July and August was beginning to break, and it became vital to force the ridge above Le Transloy as soon as possible, and from there to take Bapaume, which had originally been one of the objectives of 1 July! However, despite fine weather on 1 October, after the attack commenced it rained incessantly for the next six days, which severely hampered progress and meant a temporary postponement of the next phase of the attack until 7 October, when the weather cleared a little. Although some progress was made then, total

British troops of the Royal Warwickshire Regiment in reserve on the Somme.

After a very hot summer, the weather eventually cracked and broke on the Somme.

success was still hampered by the foul weather.

Not only did the constant rain make sudden infantry attacks an impossibility, it also rendered the barrage and counter-battery work of the British artillery all but useless. This was for two separate reasons; firstly, the 'eyes' of the artillery, the spotter planes of the Royal Flying Corps, were rarely able to fly, in what was virtually total cloud cover, and secondly, because of the state of the churned-up mud, the time taken to replenish the exhausted stocks of artillery ammunition meant that some batteries were not able to lay down effective covering fire during an infantry assault. Only local artillery batteries, with line of sight and secure supply routes, were able to support any attacks, and they could not hope to put down
• enough fire to ensure complete success.

As a result, despite fierce fighting, little progress was made, and the assaults faltered as the impetus was lost. Haig then ordered that the main attack should continue on 12 October, which would involve the whole of the XVth Corps in the area south of Ligny-Thilloy. This included the 30th Division, who would attack with the 12th Division on its right, and the 9th Division (from the 3rd Corps) on its left. As a result, Shea's Division moved up to relieve the 41st Division in the line north-west of Flers, on the night of 10/11 October. Its orders were to continue the attack on the afternoon of 12 October, by capturing and consolidating the German trenches on its immediate front. Zero hour was set for 2.05 pm, with the 90th Brigade on the left and the 89th Brigade on the right. The 21st Brigade was to be in reserve trenches near the ruins of Flers. Within the 89th Brigade the 17th Battalion on the left, and the 2nd Bedfords on the right were to lead the attack, with the 20th Battalion in support and the 19th Battalion in reserve.

During the night of 10/11 October, the 17th Battalion moved into assembly trenches which ran north of Flers, and north-east of the ruins of the village of Eaucourt l'Abbaye. At mid-day on 11 October, the British artillery began its preliminary bombardment, and the Germans immediately replied, killing Second-Lieutenant N Smith and seven other ranks, and wounding Second-Lieutenant J H Fearon and twenty other ranks. It was already obvious that the Germans knew that an attack was coming, and from which direction it would be mounted. On the evening of the 11 the 20th Battalion moved up the line, and dug two deep assembly trenches behind the 17th Battalion's positions for the attack the following day. The 19th Battalion also moved into its reserve positions known as Flers Trench. Although the rain had stopped, the ground was like a morass, and with all the natural vegetation destroyed, it was difficult to tell exactly where the objectives lay.

On the afternoon of the 12th, at exactly 2.05 pm, the attack began along the whole Corps line, covered by the local batteries of the Royal Field Artillery which still had line of sight.

The bombardment continued all the night of the 11th and the morning of the 12th, the attack being timed to start at 2.5 pm. On such days I always observed from the ridge behind the guns, where all the other batteries have their OPs', from where one gets a general if rather distant view; it would be useless to attempt front-line observation, as it would be impossible to keep a line going, owing to the Hun counter-barrages.

Just before zero hour, everybody comes up to Switch Trench [8]. It makes a splendid grand-stand,

The ruined village of Flers, on the Somme, captured by the New Zealanders in 1916 with the aid of tanks, which were being used for the first time. The area was lost when the Germans launched their March offensive in 1918 and subsequently regained at the end of August of that year. The ruined building is all that is left of the village church.

and as the batteries have already all the endless lifts and alterations in range, we at the OP are simply spectators. I have seen many of these zero hours, and they get more stupendous each time. Often there is a lull during the last five minutes; then at the appointed second the whole world seems to explode. It is impossible to exaggerate what Hunland looks like on these occasions, erupting as it were, in one vast volcano! Then the endless Hun SOS rockets ascend and down comes a Hun counter-barrage, followed by a period in which there is nothing to be seen except whirls of flame-stabbed smoke, incendiary shells bursting, and more rockets. **Lieutenant Colonel N Fraser-Tytler, DSO, TD, CLIst (County Palatine) Brigade, Royal Field Artillery.**

As the whistles blew, the 17th Battalion left its trenches to move forward. At the same time, No.1 and No.2 Companies of the 20th Battalion moved forward and occupied those trenches vacated by the 17th Battalion. As they too went over the top, No.3 and No.4 Companies took their place and waited their turn to follow. No.2 and No.3 Companies of the 19th Battalion, meanwhile, moved up to occupy the assembly trenches dug the previous night by the 20th Battalion.

As soon as the attacking waves left their trenches the enemy artillery began to register on them, and at the same time, the defending infantry commenced a murderous rain of fire. These men were Brandenburgers of 4, Brandenburgisches Infanterie-Regiment Nr.24, from Nueruppin, and Bavarians of 16, Bayerisches Reserve Infantrie Regiment from Passau, and 21, Bayerisches Reserve Infanterie-Regiment from Furth. Although their numbers had been depleted to some extent by the British bombardment, they were trained and experienced soldiers, well dug in on high ground, and for the most part, looking out on uncut wire. As such, it was virtu-

Aerial view of the German Switch Line Trenches near Ligny-Thilloy, October 1916.

ally impossible for them to miss the City Battalion men struggling to advance in the mud towards them.

The 17th Battalion, on the left, was particularly badly hit, as its portion of No Man's Land contained a slight rise in the ground, and as the troops emerged onto it they were silhouetted against the sky and became easy targets. Those on the left of the attack, who managed to avoid the hail of bullets and make it to the German wire, then found that it was totally uncut, and thus trapped, they too became easy targets, to be picked off almost at the enemy's will. It was hardly surprising that, seeing the first waves being wiped out, some of the following waves turned back and made for their start lines. These lines were now packed with other waves of troops, however, and the fleeing men added to the congestion already there, and became easy prey for the German gunners. There is some evidence also, to suggest that at this stage, the British trenches were also being hit by their own heavy artillery shells which were falling short.

On the 12th, we had more difficulties about the exact location of our line. It seemed very hard indeed to actually fix it on the map. The result was that we suffered very considerably from the effects of our own heavy artillery fire. We were warned when we came in first of all that there was a great deal of short shooting, but we certainly never expected anything like what we got at that time. It seemed absolutely useless to complain and to ask the heavy gunners to increase their range. The only answer that one always got was that they were not shooting at all. I know from practical experience that they were our own guns which were shooting, and which were causing us quite a considerable number of casualties. The fault lay at that time from the fact that the heavy gunners would not send their FOO's[9] far enough forward, but were content to observe from right back.' **Brigadier-General F C Stanley, CMG, DSO.**

25561 Pte H Redhead, of the 20th Battalion.

The view that British shells were falling short was also shared by many who were underneath the barrage. One private in the 20th Battalion, waiting his turn to go over the top, in the assembly trenches dug the previous night, was similarly certain. He, however, considered that the weather was to blame.

What happened was that we'd had such a lot of rain and the ground was so bad that they couldn't get the guns as near as they should have done, and they were firing short and our people were getting very badly hit. I heard that the Colonels kept sending word back to notify the batteries to stop firing or alter their range or something, because the Germans weren't being touched badly by our shells at all! **25561 Private H Redhead, 20th Battalion KLR.**

It is more than possible that the fault for the guns firing short did lie with with the weather, but not necessarily for the reasons that Private Redhead had considered. It was not always easy to lay, and keep intact, a land line all the way from a heavy battery to the front line, and as a result, the Royal Flying Corps often spotted for heavy batteries who were firing from the rear. We have already seen that the foul weather made flying virtually impossible and exceedingly dangerous, as the cloud cover was so low, and this was demonstrated in no uncertain terms by the shooting down of a spotter 'plane not long after the attack began. The War Diary of the 20th Battalion reports this tragic event with typical clinical thoroughness: 'At 2.40 pm one of our contact aeroplanes fell on our lines having been hit – direct hit by hostile shell fire.'

Private Redhead also witnessed the event, however, and was a little more graphic in its description.

I was in the third wave, and we were to go later, to bomb dugouts, take prisoners and that sort of thing, and while we were waiting to go over, there was one of our 'planes, a monoplane, going backwards and forwards. As I watched him, I saw what looked just like a newspaper floating out, as part of his wing went west, and then it went back, and then the other went west, and he fell like a stone in No Man's Land, only a couple of hundred yards away from where we were.' **25561 Private H Redhead, 20th Battalion KLR.**

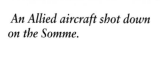
An Allied aircraft shot down on the Somme.

The aircraft was a Morane-Saulnier Type P two seater of No.3 Squadron, Royal Flying Corps based at the airfield at Lahoussoye, which was located between Albert and Querrieu. Its crew was Second-Lieutenant L C Kidd, MC, from Bromsgrove, and Second-Lieutenant F E S Phillips, of the 3rd Battalion The Devonshire Regiment, attached to No.3 Squadron, Royal Flying Corps. Both men were killed and, as their bodies were never recovered, they are commemorated on the Royal Flying Corps Section of the Memorial to the Missing at Arras. Phillips' date of death, however, is shown as 13 October 1916, presumably in error.

Some ground, however, about one hundred and fifty yards in all, had been gained by the 17th Battalion and the 2nd Bedfords, on the right of the attack line, and although there was still confusion on the left, two companies of the 20th Battalion were sent forward to help the 17th and to try to consolidate the new position. No.1 Company led by Captain H Beckett reached the new front line with few casualties, as did No.2 Company, but its commander, Lieutenant R D Paterson, was killed on the way. Once there, the two companies managed to reach the beleaguered men of the 17th Battalion, and despite the ferocity of the small arms fire, they were able to reinforce their position. As darkness began to fall No.3 Company, led by Captain G S Button, and No.4 Company led by Captain G A Brighouse, also moved up the line, and Brighouse's men were detailed to dig a strongpoint in front of the old front line trench. At the same time, contact was established with the rest of the survivors of the 17th Battalion and the 2nd Bedfords. Private Redhead, of No.4 Company, although knocked down by a shell, was able to take part in the consolidation.

Then we got the word to get ready to go over, and we filed down to take our position, and the next thing I knew, I was on my face, blown down by a shell, with lots of black smoke, – I couldn't see for a moment or two, because of the black smoke of that shell. I got up and looked around and half a dozen men were sprawled out behind me. The chap just behind me said 'I can't go on, I've been hit in the shoulder.' I thought 'Well, I'll have to go on on my own', and so I turned round and went round the traverse, and there were a couple of bodies, one on top of the other, so I had to clamber over them. Beyond them there was a length of trench and there wasn't a soul in it, so I ran along this trench and around the corner, and found masses of our troops all congregated together in two trenches that met.

Nobody knew what to do, or what was going to happen, and apparently the first two waves were more or less wiped out with machine-gun fire. The German wire wasn't touched, it wasn't cut as it should have been, and when the waves got up, they were simply mown down, and got into shell holes. It was evidently decided it was no good our going to have the same fate, and so we stopped there until after 4 o'clock. It was October, and it got dark soon after 4 o'clock, and we waited until after it got dark, and we got word to go over our front line, and dig another trench in front of our front line trench.

So we went over, – I forget how we got the shovels, I think they brought those up, – and some were so tired that they couldn't dig deep enough, but anyhow, some only got down a couple of feet, and when morning broke, they had to keep down low, because it was daylight. **25561 Private H Redhead. 20th Battalion KLR.**

During that night of 12/13 October, the 17th Battalion, who had taken the brunt of the fighting and of the casualties, was relieved by men from the 19th and 20th Battalions. As this meant that all four companies from the 20th Battalion were now forward of the original British front line, the Battalion Commander, Lieutenant-Colonel H W Cobham, was ordered to take command of the Brigade front. The 20th itself was not relieved until nearly twenty-four hours later, by the 18th Battalion from the 21st Brigade, and consequently its men had to endure another day and night in the newly dug front line!

> *Our officer came along at dawn, and gave us our rum ration. He'd given me mine, and a minute later they sent word to say 'Grennan's got it through the head', – he'd been shot through the head and was killed![10] All that day we waited. We didn't know if we would have to attack or retreat, or what, and there wasn't much shelling at all, although the Germans must have seen the freshly dug soil that was thrown up. I don't know what happened, and he didn't do anything, and it was pretty quiet. Then night broke, and in the middle of the night the bombardment started again. The sergeant roared 'Get ready, he's coming over, get ready!', and we all had our rifles ready to meet Jerry, but he never came. I think it must have been a false alarm, started by the bombardment. In the morning, we were relieved.* **25561 Private H Redhead, 20th Battalion KLR.**

The 19th Battalion, which had played a reserve role and had thus not sustained the same heavy level of casualties, stayed in the front line until the next attack.

Flers 18 October 1916

It is probable that the attack of 12 October was made in daylight to enable the attacking infantry battalions to see their objectives, and the FOOs of the artillery to see the fall of their shells. As it had turned out, the weather and the faceless nature of the landscape had almost certainly taken away any possible advantage that these might have given and swung the balance of the attack in favour of the Germans – with grim consequences for the British. As a result, Rawlinson decided that the next attack on the Le Transloy ridges, to be made on 18 October, would be made at night.

This time, in the XV Corps sector, the 89th Brigade was to be in support, but would take no active part in the fighting, whilst the 21st Brigade would attack with the object of seizing and consolidating the trenches in what remained of the German-held Gird Lines, south-east of the ruins of Eaucourt L'Abbaye. The 18th Battalion was to be in the centre of the attack, with the 2nd Battalion, The Yorkshire Regiment (Green Howards) on its left, and the 2nd Battalion, The Wiltshire Regiment on its right. The 19th Battalion, The Manchester Regiment, the fourth Battalion of the 21st Brigade, was to be in reserve. Zero hour was to be set at 3.40 am, about two hours before daylight, and two tanks were allocated to the Brigade should they be needed. As the attacking troops would not be able to see their objectives, it was arranged that they would all be in position in prepared assembly trenches on the evening before the attack, and their attack lines would be clearly marked out for them in advance by white tapes. The bombardment to cover the attack actually began on 13 October.

Captain RP Heywood, MC and Two Bars.

The 18th Battalion had been in support or reserve trenches since 13 October and on the night of 16/17 October, was moved from its reserve lines to the assembly trenches, carrying its own trench stores and equipment for the forthcoming attack. In the quagmire conditions, this move was exhausting and took a long time to execute. The assembly trenches were not deep, and gave no protection from shelling and little from the heavy rain which began to pour down from the early evening of the 17th. Men already wet, and worn out from carrying duties and lack of sleep were then subjected to enemy shelling, which, if not totally effective, was at least most demoralising.

The Military Cross and two Bars awarded to Captain R P Heywood of the 18th and later 13th Battalion KLR. Military Crosses were never officially named, and Heywood's cross was privately inscribed, but the engraver has erroneously given the date of the action near Flers as 19 October. Heywood's first Bar was won for his gallant conduct at Passendale (see page 172) and his second Bar was won on 31 August 1918. whilst he was serving with the 13th Battalion. There were only 169 awards of the Military Cross and two Bars during the Great War.

In the late afternoon of 17 October orders were received from Brigade which altered the objectives to be attacked by the Battalion, which, at that late stage, gave no time for any further reconnaissance of the area. Because of the narrowness of the trenches and their general congestion, it was not easy for Company Commanders to acquaint their men with these changes in plan. It was not a good omen.

At 6.30 pm the commander of No.3 Company, Captain G Ravenscroft, and 16984 Sergeant J S Orret went out into No Man's Land to tape the start line for the attack. At about 7.30 pm the Germans, realising that an attack was imminent, must have become nervous, as an SOS rocket was sent up, calling for artillery fire support. The subsequent bombardment came down on the British fire and communication trenches, causing inevitable casualties there amongst the packed bodies of men, and also killing Ravenscroft and Orret,[11] who were probably on their way back to their own lines at the time. It also destroyed most of the tapes they had laid, and Lieutenant R P Heywood was detailed to try to repair them. He managed to re-lay what remained of them, and for this, and his subsequent conduct during the attack, was awarded the first of three Military Crosses he was to win whilst serving with the King's Regiment.[12] His own diary entry for this action is very modest and matter of fact: 'Over the top on the 18th to attack GIRD TRENCH at 3 am (won MC)!' The citation for his award, published in the Supplement to the London Gazette of 10 January 1917, states: 'For conspicuous gallantry in action. He assumed command of and led his company forward on three separate occasions. On the same occasion he marked out under fire the assembly position for his battalion.' Presumably, the recommendation for the award did not differentiate between the laying of tapes on the evening of the 17th and the subsequent action on the 18th.

No.3 Company was so badly shaken by the German shelling, and the loss of Guy Ravenscroft, its much respected Commander, that the Battalion Commander, Lieutenant-Colonel W R Pinwill, pulled it out of the front line and replaced it with No.2 Company, which was originally intended to be the reserve Company. This changeover inevitably took time and was only just accomplished before zero hour. For many of the Battalion it meant they would have to go over the top and face the enemy's fire, cold, wet, dispirited and exhausted, having had no sleep at all for the previous two nights.

At exactly 3.40 am the whistles blew, and the Battalion left its assembly trenches, in three waves, approximately fifty yards apart, and began to cross No Man's Land. Almost immediately, the German barrage fell on the first wave and halted its advance, so that the second wave soon caught up with it. This was not a great problem at first, and the two combined waves were able to advance together for about 300 yards, whereupon they encountered the German Grid Trench system. On the right of the advance, it was found that the wire was largely intact, apart from a few gaps, and the Germans bombed and machine gunned these gaps, which prevented any further progress. Elsewhere along the trench, however, the wire was cut and there did not seem to be any serious opposition. Nevertheless, the men hesitated to jump down into the German trenches, and instead, began to filter back across to the safety of their own lines.

By this time the third wave had caught up, as had a fourth wave, which had been detailed to mop up

Reinforcements moving up towards Flers, crossing the front line German trench which had been captured on 15 September 1916.

any opposition once the trenches had fallen, and all four waves became intermingled, which added to the confusion. No less than three attempts were made to try to get the men to go forward again, but each attempt became markedly less successful than its predecessor, and eventually the attack came to a stand-still. Although the British assembly trenches had received the attention of the German guns, the attackers in No Man's Land had not come under any great intensity of fire up until this point. However, once it became obvious to the Germans that the attack was disorganised and faltering, they began to fire into the massed men from the flanks. It was probably this that finally settled the issue and convinced the Pals that they could no longer gain the enemy trenches, and all four waves, now merged into one, began to retreat to their own lines. The whole attack had been an abysmal failure, and no ground had been gained at all.

It was a debacle on the night of the 17th/18th at Flers, where the tanks' first attack had done little. There were many of them lying around blasted by shells. They had come there just a week or two before. The trenches were little more than shell holes linked up. We didn't know where the line, if any, went. We didn't even reach the German trenches. We didn't even know where they were, and I don't think our officers did. The attack was a shambles, many men killed and taken prisoner. Our Captain was taken prisoner, with all the Company's money – our wages – having wandered into their lines![13]

We were hiding in a so-called trench just behind an old tank. I was wounded in the left arm, this time by a bullet, and was glad to be out of the carnage. I had to walk to the railhead – I can't think of its name, but it must have been five miles away. There were no stretcher bearers – there was no proper line there at all! Eventually I was sent to Rouen, and then home. Altogether, I spent fourteen months in hospital. **17122 Private W Gregory, 18th Battalion KLR.**

What went wrong?

The simple cause of the failure of both of the attacks at Ligny-Thilloy was the weather. The popular image of the battle of the Somme amongst the public is one of overburdened soldiers valiantly struggling to achieve their objectives through glutinous mud, but as far as the Pals Battalions are concerned, the attacks on 12 October and 18 October were the only ones attempted in anything approaching these conditions. The two attacks were exactly the same in plan, the only significant difference being that the latter one was made at night, but we have already seen that the weather was the real arbiter of any success on both occasions. On 12 October, it took away any advantage that daylight might have given the attackers, and on 18 October it lowered the morale of the troops to such an extent that the cover of darkness actually hindered the progress of the attack!

Low morale was not a factor which had affected the Pals before October 1916, in fact the reverse had

17122 Pte W Gregory of the 18th Battalion, photographed in 'hospital blues' at Staveleigh Hospital, Blackburn, in December 1916, whilst recovering from wounds received at Flora on 18 October 1916.

22031 Acting Corporal J D Buxton. of the 20th Battalion.

more often been the case, but in both attacks, when the impetus of the assault was lost, even though victory was still at least possible, the men faltered, and then turned to run back to their own lines – even though some ground was captured and held on 12 October. It is hardly surprising, however, that morale was low on both occasions. Firstly, all four Pals Battalions had been in virtually continuous action since 1 July – the penalty for their success on that date – and had fought some desperate actions since. Nearly all the original officers and senior NCOs had been lost to the Battalions through death or wounds, and the few surviving originals amongst the men who had crossed the Channel in November 1915, were certainly suffering from what would be described in later wars as 'battle fatigue'. Although a significant number of the 'originals' would eventually return to their own battalions after recovering from their July wounds, many of the men who fought the October battles were their replacements, hurriedly trained and without the dedication and esprit de corps of the silver badge bearers.

When enemy forces are evenly matched, it is usually the side with the better morale that wins the day, and this fact alone gave the Germans the edge. Despite the batterings they had received since 1 July, they were still dug in, in deep, fairly safe positions, and as they occupied the Le Transloy Ridges, the worst of the atrocious weather flowed away from them down onto the British lines. Even though they had to suffer the terror of bombardment and the nervous exhaustion of the inevitability of impending attack, they were not subjected to the rigours of exhausting carrying parties, the possibility of being shelled in shallow ill prepared trenches and the quagmire moonscape of No Man's Land. We have also seen that the worst of the weather also spared them from the worst of the British bombardment. This was especially so during the fighting on 18 October, when, according to the British Official History[14], even one of the defending Saxon infantry battalions, the 15. Infanterie- Regiment Nr. 181, attributed their success to the fact that 'the mud was an enemy of the British, for it prevented a rapid surprise advance, after rendering the preliminary bombardment comparatively ineffective.'

After the battle was over, The Commanding Officer of the 18th Battalion summed up the reasons for his Battalion's failure on the 18 October, and many of his conclusions would have been equally valid for the failure of the earlier attack.

Reasons for Non-Success

It is considered that the non-success of the operation was due to loss of morale brought about by the following causes:

(a) The Battalion had been subjected to persistent shelling ever since it arrived in the forward area, i.e. in FLERS TRENCH as well as in the front line. It was also subjected to an attack by gas shells during the period.

On the night of 16/17th October, i.e. the night before the attack, the Battalion was hurriedly moved from FLERS TRENCH to fresh quarters in GOOSE ALLEY: this move was tiring and took a long time.

Also, carrying parties were employed up to 2 am 17th October. This meant that a great number of the men got no sleep either on the night 16/17th or 17/18th October.

(b) The various objectives and duties required of the different lines could not be adequately explained, owing to the congested state of the trenches and the short time available.

(c) The German barrages put up about 7.30 pm on the 17th shook many of the men: there were several casualties and many men were buried.

(d) The state of the weather was very adverse. The men were soaked

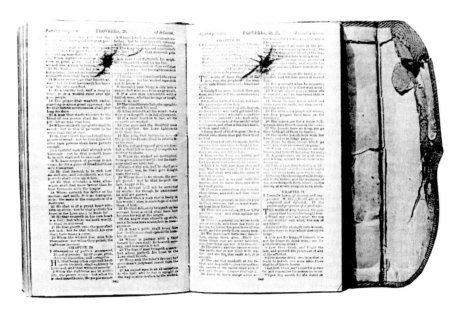

Corporal Buxton's field service Bible pierced by a German bullet which wounded him.

through and chilled the whole night. Rifles and Lewis Guns were clogged with mud and this undoubtedly unnerved many of the men.

Suggestions

(a) Units about to attack to be given adequate rest before-hand: no long and arduous working or carrying parties being demanded.

(b) More time for reconnaissance, issue of orders and for explaining objectives etc., to all ranks.

(c) Night operations on the scale attempted require very special preparation if direction is not to be lost, particularly in a country like that encountered, where land marks are few and inconspicuous. No opportunity for practising similar movements has been available for a long time.

Lieutenant-Colonel W R Pinwill Commanding 18th Battalion KLR.

During the attack on 12 October the 17th Battalion's casualties amounted to five officers and eighty-nine other ranks killed or died of wounds, and five officers and about two hundred and thirty other ranks wounded and missing. The 19th Battalion, which was in support, lost seven other ranks killed, and two officers and twelve other ranks wounded or missing. The 20th Battalion lost one officer and twenty-nine other ranks killed, and five officers and ninety-three other ranks wounded or missing. The 18th Battalion's attack on 18 October cost it four officers and twenty-seven other ranks killed or died of wounds. Taken as a whole, from the start of the Transloy Ridges operation on 11 October to the time when the last Pals Battalion was taken out of the line on 22 October, Pals losses amounted to thirteen officers and two hundred and thirteen other ranks killed, and twenty-one officers and five hundred and forty-six other ranks wounded or missing. Some of those recorded as missing from all four Battalions would rejoin their Battalions, or later prove to have been taken prisoner, but others had been killed, and later information would include them in the total tally of dead.

Soon after the withdrawal of the Pals from the line, the Battle of the Somme ground to its inevitable conclusion, as the weather became the final victor, and Haig ended the offensive. For the men of the City Battalions it would mean the end of full-scale fighting in 1916, but not the last some of them would see of the Somme. The Somme had been their introduction to active service, their baptism of fire and the graveyard of their best!

NOTES

1. Presumably CQMS Potter is referring to the death in action of Lieutenant A E Cotton.
2. All the references to Lance-Corporal Quinn used in the text, are taken from a booklet privately published by his family, in 1918, to commemorate his life and death. The author is indebted to Mrs Eileen Quinn of Liverpool for permission to quote from this booklet.
3. These numbers would naturally include casualties from the 2nd Bedfords.
4. The author is indebted to Miss J P Adams and the staff of the Lord Chancellor's Department, for undertaking a time-consuming and difficult search to ascertain the nature of Major Morrison's offence. Apart from 'special cases', all Great War Courts-Martial files dealing with officers have been destroyed 'in accordance with the relevant schedules', and as a consequence without the help of this Department, it would have been impossible to complete this point of historical detail!
5. 'Cushy' was army slang picked up in India from the hindi word 'khush', meaning pleasant. In the Great War, it meant anything that was easy or non dangerous.
6. 'Etang' is French for pond or pool.
7. OP is the usual artillery abbreviation for observation post.
8. The Switch Line system was south of the ruins of Flers.
9. FOOs were Forward Observation Officers, artillery officers whose job it was to come forward to the front lines and observe the fall of shells, and then correct the ranging of their own gun batteries, usually by field telephone, until their shells were falling on target.
10. Lieutenant G L Grennan is listed as being killed in action on 12 October 1916, but Private Redhead's account would indicate that he was killed in the early hours of the following morning.
11. Although both Ravenscroft and Orret are officially listed as being killed in action on 18 October, there is no doubt that they met their deaths on the evening of the 17th. This fact is confirmed by the Battalion War Diary, and a letter written to the author in 1974 by former 17122 Private W Gregory, of the 18th Battalion.
12. He was awarded a Bar to his Military Cross for his gallant conduct during the first day of the Third Battle of Ypres, Belgium, on 31 July 1917, and a second Bar, whilst serving with the 13th Battalion of the Regiment, on 31 August 1918, near Ecoust, in France.
13. No.4 Company's Commander, Captain E B Beazley. was captured that night, but it is not known whether or not he was carrying the Company's money! He was finally repatriated to England on 14 June 1918, via Switzerland.
14. See History of the Great War, Military Operations France and Belgium 1916, Volume II, page 446.

Chapter Nine

'The Hun asked for war, and now at last he is getting it in good measure, heaped up and overflowing.'

A lull in the fighting

At the end of October 1916, the battle-scarred and battle-weary 30th Division moved north of the Somme sector and took over a portion of the line opposite Bienviflers and Berles. Although this meant moving back into the front line again so soon after Ligny-Thilloy, it was still very quiet compared with the Somme, and very 'cushy' in comparison to what had been expected and dreaded, – a move further north still to the Ypres Salient!

Despite the inevitable casualties always sustained during front-line duty, the area provided a much-needed opportunity for rest and recovery of lost strength, and apart from trench raids made by the 17th and 20th Battalions in December, the time spent in the area was fairly uneventful. The period of time spent there also gave opportunities to rebuild shattered morale when the troops came out of the line. A great part in this rebuilding process was played by the City Battalion Comforts Fund and by Battalion and Brigade entertainment troupes.

The Comforts Fund had been established by Lady Derby and her sister-in-law, The Honourable Mrs Ferdinand Stanley, in the first week of November 1915, prior to the departure of the 89th Brigade for France. They had organised a committee to seek donations, money and 'comforts and luxuries' from local firms and businesses to be distributed amongst all four City Battalions on active service. So successful were they at fund raising that the Pals rarely went short of these items in the front line for the rest of the war — even in the dark days of March and April 1918. To give some idea of the value of their work, which was purely voluntary and totally unaided by the state, by March 1918 the following sets of items, selected at random, from a much larger list, had been distributed to the troops: 29,121 pairs of socks, 4,320 towels, 6,444 pairs of mittens, 3,330 tins of sweets, 5,220 tins of vaseline, 29,616 lbs of candles, 2,933 tooth-brushes, 49,950 packets of cigarettes, 52,044 newspapers and magazines, 1,860 mouth organs and whistles, 4,874 pairs of boot laces, 1,852 shaving brushes, 2,181 shirts and 160 periscopes. It must be remembered, in relation to these distributed amounts, that the total combined numbers of all four Pals Battalions probably never exceeded 4,500 men throughout the war.

The entertainment troupes, or concert parties, were of particularly high quality. Initially, in England, the 17th Battalion had raised a troupe of pierrots called The Optimists', and later in France, the 19th Battalion had raised a similar band called The Duds', and the 20th Battalion had raised The Very Lights'. Some of them had been professional entertainers in civilian life, but most were total amateurs, merely extending talents discovered through the Victorian tradition of parlour entertainment. Although The Duds' and The Very Lights' remained within

'The Duds' concert party of the 19th Battalion. By the end of the war, Cornish. Harrison and Webster had been killed.

the control of their battalions, The Optimists' were judged to be so good that in the summer of 1916, they were ordered by Brigadier-General Stanley, to become Brigade Entertainers, and thereafter they were established at Brigade Headquarters, which took sole control of their 'bookings' and performances. At the same time the performers still carried out their military duties and made it a point of honour that they did as much work as everyone else; All of them, in fact, lost a high percentage of their members during the course of the war through enemy action.

The excellence of the entertainment 'The Optimists' provided, in pre 'ENSA'[1] days, was proved by the number of requests that the 89th Brigade received from other brigades who wanted to hire their services. This came in very useful when Stanley needed to augment the Comforts Fund, and was especially beneficial on one occasion, when an old barn burnt down, which had contained all the troupe's kit – including a much prized piano.

It must not be thought that the troops were always in a state of despondency and alarm. Most of them reacted well to the release of line tension when out at rest, and occasionally there were concerts, and sometimes a pay-day. The local estaminets[2] did good trade and many a humble dwelling responded to the request for 'caf', or 'des oeufs, pommes de terre frites'.

A concert in a huge barn by the Divisional Concert Party ended late one cold clearless snow-whitened night in a huge blaze, and all the troops were turned out of bed to help. Even the ancient hand-pumped, hand-drawn fire engine was brought out, but failed to stand up to the strain. 'Buckets, canvas, troops for the use of, conveyed an inadequate supply of water along a long line of men from the village pond over which about four inches of ice had to be broken. **17518 Corporal E G Williams, 19th Battalion KLR.**

To pay for new equipment, however, Stanley was able to offer the services of 'The Optimists' up and down the line, and very soon, they had earned enough money to make up all their losses!

Inevitably, when soldiers came out of the line to rest, they came across ordinary civilians, who were happy to provide food and entertainment – of all varieties, but usually at a hugely inflated price! The experience of British soldiers during the Great War with civilians, and particularly with French civilians, was not always a good one, and after living amongst local villagers for any length of time, many thought that Britain was fighting the wrong nation! This opinion seems to have filtered through every level of soldiery.

It is so extraordinary in France how, in one place, you are welcomed with open arms, and five miles away from it, you are looked upon with suspicion and every obstacle put in the way of friendly relationship. They did not seem to realise that we had come from our country and given up everything to save them. The fact of the matter was that a good many of them had not felt the pinch of war at all, and in some of these districts all the peasants think of is how to get money out of British pockets. **Brigadier-General F C Stanley, CMG, DSO.**

'In one place in particular, the lady there took the handle off the pump. We arrived in this field in pitch black dark, and some of the lads went up with their dixies to get water, only to find there was no handle. One of our fellows said "Hey, there must be a handle to it", so he knocks at the door, and this old lady comes out and raises hell! 'Oh no, no, no, no, anglais soldat no bon!' she says, and shuts the door!

Then one of the lads said 'Hey – w' better have the handle or else!' So the other fellow that was with him says 'I'll go and get it', and he went back and got his gun, and he knocked at the door, and when she answered he said 'Listen missus, see this, you – handle – quick!' but she still wouldn't give it to him, and they sent for the interpreter in the finish. The interpreter just went to her and said something, and out came the handle, but she was determined that she wasn't going to give it to us! I firmly believe that if the interpreter hadn 't got the handle, that lad would have shot her! **24644 Private W Stubbs, 18th Battalion KLR.**

Others, however, reported good friendly relations between soldiers and locals, despite the language barrier!

The men had gone, and the women were doing nearly all the farmwork. I remember one day when the cow had got out in the lane, and this chappie dashed up to Madame, and he couldn't speak French properly. So he dashes up to Madame and he says, "Madame, madame, du lait parti.
    'Pardon Monsieur?'
    'Du lait parti, du lait parti.'
    Then she knew that the milk had gone!

16224 Private T H Brown, 18th Battalion KLR.

As 1916 drew to a close, Christmas gave a good example of the hardened attitude to the enemy that had been brought about by the brutality of the fighting on the Somme. Now, there was no thought of any kind of fraternisation between the British and the Germans, of the type that had been the hallmark of Christmas 1914. It had been replaced with a grim desire on both sides to kill each other as efficiently as possible! In fact, on Christmas Eve, Lieutenant H Foster-Anderson of the 19th Battalion demonstrated his own version of Christmas spirit by leading a ten-man patrol out into No Man's Land, armed with two Lewis Guns. When the patrol was close to the German lines, the men settled themselves in shell holes, and began to sing Christmas carols to the Germans in the hope of luring them out into the open, where they could be shot! Disobligingly, the enemy did not fall for this particular gesture of good will, but on Christmas day a sniper from the 19th was able to shoot a German soldier who was foolish enough to attempt a festive walk along the top of his trench line.

The 19th Battalion came out of the line in time for New Year, when, paradoxically, on New Year's Eve they celebrated Christmas with a full dinner of goose, plum pudding, wine and English beer, and a concert by 'The Duds'. The 17th and 18th Battalions, which were in the front line at the change of year, reported a quiet entry into 1917, and the 20th Battalion, billeted at Berles, reported a similarly quiet time. 'The new year was ushered in with the usual compliments from other units in the Brigade, and the Brigadier, but not from the Boche.'

In all, during 1916, sixty-four officers, and 1,274 other ranks, from the four City Battalions gave their lives for King and Country. This figure, for the first full year of active service, represents nearly fifty per cent of the combined total of dead lost by the Pals throughout the whole of the war!

Agny and the Hindenburg Line

On 7 January 1917, the 30th Division was relieved in the line, and moved back to the area around Halloy for a month's training. This should have been, in effect, a month's rest, and the first real rest that the Division had had since the start of the Somme battle. However, any thought of rest or training was dispelled in the second week of the month, when the weather worsened and snow, intense cold and driving wind, made living conditions extremely difficult, even in billets. It was, in fact, the coldest winter in living memory. At the same time the Division was ordered to find over 1,000 men per day for work parties needed to widen the line between Doullens and Arras. The only consolation lay in the fact that at least the men were out of the front line, and sports activities and concerts helped to make it a fairly pleasant time for most, despite the inevitable exhaustion of the fatigues!

At the end of January, orders were received to move up to the front again, and on 4 February

A German equivalent of a concert party performing at the opening of a cinema for the troops at Longueval 1916.

the 89th Brigade began to move to the trenches around Agny, just south of the ancient town of Arras, birthplace of the French Revolutionary leader, Robespierre, and equally famous for its mediaeval tapestries. The 17th Battalion was actually billeted in the town itself, going from there into the line on 12 February. Its first tour of duty in the new sector was fairly uneventful and followed the normal pattern of trench life. On 22 February, however, the normal tragedy of death in the trenches was made even more appalling by the death of 15568 Private G Smith, of 'C' Company. He was not killed by the enemy, but accidentally shot by of one of his comrades, 49005 Private W Houghton, of the same company. Private George Smith, from Wavertree, Liverpool, was one of the original 1914 Pals and had crossed to France with the 17th Battalion on 7 November 1915, and had continued to serve with it all through the fighting on the Somme, without so much as a scratch;

After the shooting, Private Houghton was put under immediate arrest, and eight days later, on 2 March 1917, was tried by a Field General Court Martial, having been charged, under Section 40 of the Army Act, with 'Neglect to the prejudice of good order and military discipline' in that he negligently discharged his rifle, thereby mortally wounding a comrade. He was found guilty, and sentenced to one hundred and twelve days imprisonment. This sentence was clearly impractical in the front line, and was later commuted to ninety days Field Punishment No.l, by Lieutenant-Colonel H S Poyntz, DSO, of the 2nd Battalion The Bedfordshire Regiment, who was commanding the 89th Brigade during the absence of Brigadier-General Stanley. It is not known if the punishment was ever carried out, although after the death of Private S B Heyes, also of 'C' Company, during Field Punishment No.l a year earlier, and the questions that his death had prompted at the highest level,[3] it would seem unlikely. On 2 August 1917, Private Houghton was himself hit by enemy gunfire in the right ankle and left hand, and a week later was evacuated to England.

Although the incident of Private Smith's death does not rate a mention in the 17th Battalion War Diary, the post-war HMSO publication 'Soldiers Died in the Great War' states that he was killed in action, which is the way it must have been officially represented on the Battalion casualty returns. However, his next of kin, his sister Mrs Mary McCann, was obviously told the truth about his death, because on the family gravestone in Toxteth Park Cemetery, Liverpool, where he is commemorated, the inscription states that he was: 'Accidentally Killed in the Trenches'. Smith was buried in Agny Military Cemetery, where his remains lie today. The only other incident worthy of recording during the Battalion's stay in the trenches occurred the day after Smith's death, when there was an exchange of fire with an enemy patrol, after which two wounded soldiers from the 3. Sachsisches Reserve Infanterie Regiment Nr.102, were captured.

On 4 February 1917, the 18th Battalion had split into two, some men moving to Saulty for training at the 30th Divisional School there, and the rest moving to billets at Beaumetz and eventual trench duty at Agny, by the middle of the month. The 19th Battalion took over the line at Agny from the 9th Battalion The Rifle Brigade, on 6 February and had a similarly uneventful time, until the 18th, when the Germans began to bombard its positions with minenwerfer projectiles.[4] This fire continued intermittently for the next five days, and on the 22nd a chance round hit the roof of one of the company headquarters' dugouts and exploded, collapsing the position, and burying Captain C R Bolton and Lieutenant H Foster-Anderson – the same officer who had been so 'unseasonal' on the previous Christmas Eve. Although rescuers eventu-

ally got them both out from the splintered wreckage, Bolton was found to be dead. He too, is buried in Agny Military Cemetery, not far from Private Smith of the 17th Battalion, killed on the same day. Foster Anderson, although wounded, eventually re-

An aerial view of a section of the Hindenburg Line. The thick belts of wire, (seen here as dark rectangles), were designed to force attackers into 'killing zones'.

Details of the special
equipment used on an 18th
Battalion trench raid – 15
March 1917.
(See Appendix III).

2. DETAILS OF SPECIAL EQUIPMENT AND STORES.

	Officers	Other Ranks	Bombs	Bomb-bags	Knobkerries	Torches	Wirecutters	Ladders	Stokes Bombs	R.G.M. Bombs	"P" Bombs	Bangalore Torpedoes
1st WAVE.	1	1										
Right Bombing Party		6	60	4	1	1	-	-	-	-	-	-
Left "		6	60	4	1	1	-	-	-	-	-	-
Right Blocking Party		5	60	5	1	1	-	-	-	2	6	-
Left "		5	60	5	1	1	-	-	-	2	6	-
Cleaners		16	30	4	8	5	10	6				
T:M. Personnel		4	-	-	-	-	-	16	-	-	-	
Engineers		2	-	-	-	-	-	-	-	-	-	2
2nd WAVE.	2	2										
Right Bombing Party		6	60	4	1	1	2	-	-	-	-	-
Left "		6	60	4	1	1	2	-	-	-	-	-
Right Blocking Party		5	60	5	1	1	-	-	-	2	6	-
Left "		5	60	5	1	1	-	-	-	2	6	-
Right Road Block		4										
Left "		4										
Cleaners		14	30	4	4	5	5					
T.M. Personnel		2	-	-	-	-	-	8				
Headquarters.	2	10										
Artillery Liaison	1	2										
Lewis Gunners		12										
Total	6	117										

covered. The 20th Battalion commenced front line duties near Agny on 7 February, and subdivided, its troops alternated between training at the 30th Divisional School and trench duties, for the rest of the month. By then reports had been received from positions further south of Agny that the Germans were preparing to retire from their trenches, but despite especial vigilance by patrols, no evidence was found of any withdrawal – at that time.

However, the Germans had begun, in mid February, to enact a plan for a withdrawal from the front line, to a set of carefully prepared fortifications, later called the Hindenburg Line by the British, from where, once more, they would be able to command the high ground lost to them on the Somme. To the Allied press their withdrawal would be hailed as a retreat, and thus a considerable victory for the Allied cause, but to the Germans, professional as ever, it was a sensible and necessary military move to better positions. It also allowed them to shorten their line and thus save valuable manpower, which had become an acute problem for them. They had, in fact, been engaged in the constant drain of almost continual warfare since their attack at Verdun over a year previously. Even though a withdrawal would deliver a huge boost to Allied morale, and a corresponding blow to its own, the German Army would still be in occupation of nearly all of Belgium, and most of northern France. Moreover, as it still held the high ground, it also held the key to Allied political and military strategy; for if the Allies were determined to win the war in the field, they would still have to evict the Germans from France and Belgium, and the only positive way to do this was by frontal assault on the German fortified positions.

Thus, from the last week in February, on a line from south of Arras, to east of Soissons, the Germans had began to pull back their troops, sometimes as far as twenty miles, but on average about twelve miles. The Hindenburg Line, known as the 'Siegfried Stellung' to the Germans had been carefully constructed by pioneer/engineers over a period of some five months to make use of natural geographical features and fortified villages, and was a much tougher nut to crack than the old front line on the Somme had been in July 1916. The whole operation was supposed to have been carried out in secret, but the Allies inevitably found out what was happening, but were powerless to do anything about it without mounting a huge offensive. Once the 'Siegfried Stellung' had been completed, the Germans had given themselves five weeks to move back into its carefully prepared fortifications. Although the first front-line trenches were evacuated on 22 February, the main withdrawal was scheduled to begin on 16 March.

The newly prepared defences were solid and seemingly impregnable, protected by yards and yards of wire entanglements in front, and well constructed support, reserve and communication lines in the rear. They were also built on the reverse slopes of hills, which meant that they could not be seen by attacking infantry, who themselves would be vulnerable to artillery fire, whilst they struggled up the forward slopes. The Germans had also devastated the countryside on their line of withdrawal, partially to slow down the Allied follow through, but also to deny its new occupants any benefit at all from the land. Not only were buildings blown up, and booby-trapped, but wells were poisoned and even fruit trees uprooted and burned, in a positive orgy of destruction often quite incidental to any military purpose!

As the withdrawal was carried out in different sectors at different times it was necessary for the British to send out frequent patrols to try and ascertain exactly what was happening. On 15 March five officers and one hundred and five other ranks from the 18th Battalion, led by Captain R W Jones, made a very carefully planned night raid on the enemy trenches opposite Agny. Its purpose was to gain prisoners for

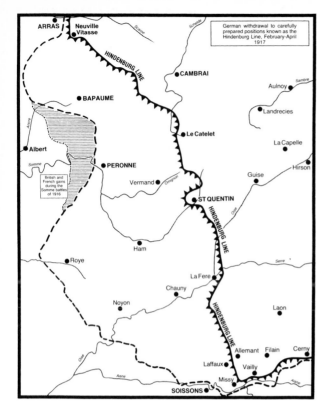

German withdrawal to carefully prepared positions known as the Hindenburg Line, February-April 1917

British and French gains during the Somme battles of 1916

identification purposes, to inflict casualties on the enemy and to destroy his dugouts and emplacements. Despite all the careful planning, however, the raid was not a total success, because the Germans were obviously already reducing their numbers in the front line, and after initially opposing the Pals' assault, they then retreated to their reserve lines, leaving no one behind to be taken prisoner.

Then we went up to the Arras front, and we'd no sooner got there than Jerry started straightening his line, – he went back, so of course we had to follow him; but before we followed him, I remember that our fellows had to do a bombing attack, and they went over with their faces blacked, and two or three of our signallers went up to run the line. I stood by in case the line was broken, and when they got there, they found there was no Jerries – he'd gone! **16400 Private S R Steele, 18th Battalion KLR.**

Nevertheless, during the raid, 16566 Corporal J Eardley from Liverpool was killed and seven other ranks were wounded. It is interesting to note the close attention to detail in the planning of this raid, however, and details of this are reproduced in Appendix III. the 18th Battalion also received a new Commanding Officer just after this raid, when Major W L Campbell of the 2nd Battalion, The Royal Scots Fusiliers took over from Lieutenant-Colonel Pinwall. Lieutenant-Colonel J W H T Douglas of the 2nd Battalion The Bedfordshire Regiment had taken over command of the 20th Battalion, on 6 March.

By 18 March it was obvious at Agny that the Germans had begun to retire, as burning villages and previously manned positions had been seen in the distance, and the enemy artillery was obviously firing from extreme range.[5] Reports had also been received that British troops further down the line had entered unoccupied German front-line trenches. The 17th and 18th Battalions were both out of the line at the time, but the 19th and 20th held trenches opposite Agny, and in the early morning of the 18th both sent out numerous patrols to investigate the enemy positions. Although these discovered that the trenches were still manned at 1.00 am, by 3.00 am it was obvious that the enemy had gone! Both battalions then began to move forward across No Man's Land, and into the first three lines of German trenches. Their advance was cautious at first, especially after one soldier from the 19th Battalion stepped on a trench board, and a carefully placed bomb went off underneath, and wounded him. However, it soon became obvious that the Germans had definitely gone, and throughout the day both battalions moved steadily forward. Eventually, taking into consideration the advance of those troops on either side, the two battalions halted for the night in open country.

The advance then continued over the next three days, when both battalions were pulled out of the advance for rest, but not before the 20th Battalion had taken part in a skirmish near the village of Neuville Vitasse on 21 March. Believing the village to be deserted, a ten man patrol led by Second-Lieutenant R E Green had had the misfortune to encounter a German patrol there, which numbered some forty men, probably from the already mentioned 3. Sachsisches Reserve Infanterie Regiment Nr. 102. Being so outnumbered, the British stood no chance and Green and Corporal R McArdle were killed, and five other ranks were captured.[6] By the time that the Pals had been taken out of the advance, the Germans had reached their appointed positions in the Hindenburg Line, and were firmly entrenched in Neuville Vitasse, and on the ridge in front of the valley of the River Cojeul.

The move forward across open country to catch up with the the Germans was a novel one to soldiers used to static trench warfare, and at least one part of the Signals Section of the 18th Battalion became lost in the open country, looking for their new signals station to which they had been ordered to lay a land line. One former signaller recalled a poem written by two members of the section about an over zealous signals sergeant called Sam, and a frustrating hunt for a signals station in unfamiliar territory.

There were these two men, one was Grey, H Grey, and the other was Max, – Pemberton, I think. They were both linesmen, and they got together and wrote this little poem.

Near The Hindenburg Line
The scene – the front line near Neuville Vitasse,
The hero a linesman considered first class,
The victims three men and an NCO.
They tramped over and over, the language did flow.
The faithful drum reeled steadily out,
The victims paused, they were in doubt.
Who knows where is this blinking station?
'My Colonel wants communication.'
When someone comes this tale to tell,
Old Sam on the right of Mercatel.

You often read such things in fiction,
But never was heard such malediction,
Upon the head of one so 'Great',
Rained thick and fast the Hymns of Hate.
'Communications at all cost,'
Sam murmered, he could not be lost.
Then after further communication,
The victims turned in desperation,
Consigned the 'Hero' down to hell,
And turned their backs on Mercatel.

The grave of 15568 Pte G Smith of the 17th Battalion at Agny Military Cemetery, near Arras. Smith was accidently shot and killed by a comrade, 49005 Pte W Houghton. on 22 February 1917. The headstone erroneously displays the White Horse of Hanover.

I don't know what made them write it, but those two were always writing poetry, they used to write in the front line, even when the shelling was on. We were moving so fast that they hadn't time to get some of the telephone stations out there. He was moving back so quickly, and that's why Old Sam was supposed to know where the station was, but we couldn't find it, and that's why they got so fed up. It was a form of open warfare, you see, because he was moving back all the time, but he knew where to stop! **16224 Private T H Brown, 18th Battalion KLR.**

It was only when the Pals took stock of the land that they had just crossed, however, that they realised the utter devastation that the Germans had left behind them.

It is impossible to convey to anyone who has not seen it the state to which the enemy had reduced the whole of the country. As regards his trenches, he had blown up every single dug-out, smashed in all his machine-gun emplacements, OP's, etc. He had demolished everything except his wire

General Robert Nivelle.

entanglements, and of these there were belts without number of extraordinary width and thickness. These were intact, and I can only thank Heaven that we did not have to attack through this wire. It would have taken us months of wire-cutting.

Apart from his trenches, he had set himself to destroy everything. What remained of the villages had been blown up; there was not a single house standing higher than about three feet. Every tree had been cut down; he had not even spared the gooseberry and currant bushes. The roads and bridges were all blown up. In fact, we have always given him the utmost credit for his thoroughness and this wholesale destruction was one of his masterpieces. He even went so far as to arrange all kinds of booby traps with bombs and explosives of various kinds. There was no souvenir hunting on this occasion – it was far too dangerous an amusement. **Brigadier-General F C Stanley, CMG, DSO.**

The Arras offensive

In November 1916, the seeds of the Arras offensive were sown like the seeds of the Battle of the Somme had been nearly a year earlier, at the French General Headquarters at Chantilly. In discussing concerted war plans for 1917, the Allies agreed to mount a series of almost simultaneous offensives on different fronts, which would keep the Germans tied down and unable to fortify one front for fear of weakening another. By adopting these tactics, it was hoped that the Germans would be ground down to inevitable defeat. After the reasonably successful outcome, for the Allies, of the Somme battle, the German Army had been forced into a salient formed by the River Scarpe and the River Ancre, and Haig's first priority for 1917 was to attack both shoulders of this salient and cut it off, with resultant losses to the enemy. Once this had been accomplished, he hoped to turn his attention to the great salient in the north, the Ypres Salient, where he perceived that the war could actually be won.

However, events in Europe overtook these plans, and by the end of March 1917, not only had the Germans retreated to the Hindenburg Line, but the first Russian Revolution had taken place and there had been dramatic changes in the command structure and policy-making of the Allied armies. The British had a new Prime Minister, David Lloyd George, who had taken up office on 6 December 1916. He had never believed that the war could be won on the Western Front, and had already begun to interfere in military policy making in a way which would ultimately starve France and Flanders of much-needed British manpower. The French military leader, Joffre, had been replaced by one of the heroes of Verdun, General Robert Nivelle, who promised that he could deliver the military 'holy grail' sought by all governments – a rapid victory with few casualties. Such was his eloquence and charm that politicians in France and Britain, justifiably appalled by the carnage of 1916, were taken in by his specious but rash promises. As a result, Lloyd George engaged himself in shameful duplicity by trying to trick Haig into accepting Nivelle as Commander-in-Chief of all the Allied Armies on the Western Front, and when this failed, Haig was forced to accept the compromise of a subordinate role to that of Nivelle in the forthcoming offensive.

Field-Marshal Haig, Prime Minister Lloyd George and General Joffre.

The new plans were for a large-scale offensive to be made by the French (which suited Lloyd George) in the Champagne/Aisne area of France and for the British to open a feint to this offensive in April 1917 with the attack already envisaged by Haig. It would be an attack made along a front of about fifteen miles, from Vimy Ridge in the north to the village of Croiselles, south-east of Arras, in the south. The German withdrawals of February and March did not alter the planning for this significantly, except that it meant that the southernmost five miles of the British assault would be made against the northernmost end of the newly constructed fortifications of the Siegfried Stellung.

The area all around Arras just below the subsoil is composed of almost pure chalk, and centuries of previous persecuted or oppressed inhabitants of the region had taken advantage of this to dig lots of tunnels and hideaways. Some of these still existed in relatively good repair in 1917, and when the Army discovered this, they decided to adopt the same idea and excavate the chalk to provide miles and miles of tunnels, passages and dugouts. These then gave shelter in comparative safety to the tens of thousands of men preparing to attack the German lines, who were able to move around underground, their intentions invisible to the enemy on the surface.

Arras of course, was a good place to stay because we had lovely places underground there. There was chalk subsoil there. There was a little town outside called Agny, and we were there all in chalk. I know it was chalk because when we came up in the morning, everyone was white. Their uniforms had gone white in the night with the chalk. **16224 Private T H Brown, 18th Battalion KLR.**

Scenes of destruction in Arras 1917.

The attack was scheduled to begin on 9 April which was Easter Monday in 1917, and it was preceded by a three-week bombardment which was designed to cut the enemy's wire and wreck his strongpoints and lines of communication. Artillery lessons had been learned from the Somme debacle, and although the wire remained uncut in places where it was laid in very dense entanglements, overall the barrage was a success which would contribute greatly to the infantry successes of the first day.

From the 31st March till the 8th April we were desperately busy every night getting up ammunition. It meant that practically every horse and man had to be out each night working over very congested and bad roads, and fighting through continual blizzards of snow and hail.

We have often shifted 2,000 rounds in a night, so what with shooting all day, and ammunition fatigues at night, all ranks are pretty dead beat. Beside that, for our own sake we had to do repair work such as filling up craters in roads etc., as if we waited for the RE,[7] things would never be done in time. We built quite a substantial bridge for our waggons across the Cojeul River, and we had a pretty hot time of it, too, while doing it, as the Hun now continually shells all important points like bridges and roads.

In order not to disclose our positions, no gun-pits had been dug until the guns got in, but by 5 am (on 4th April), we were more or less ready and started the bombardment of the Hindenburg Line, which had been hitherto out of range. The bombardment had continued right up to today (8th April) the eve of the battle. It is a very unsatisfactory job trying to cut wire with HE[8] at this long range, in fact an impossible one for our 18-pounders, and only the heavy howitzers are doing the job successfully.

On the afternoon of the 8th, which for a change was very fine, orders came in for the great attack on the 9th. This meant a very complicated barrage, needing sheets of different angles for each gun. The reason was that the division to the north of us were to open the attack, which would then gradually extend south, eventually reaching our position on the extreme right, and consequently our set task would begin at zero – plus six hours nine minutes, lasting to plus twelve hours and forty nine minutes. **Lieutenant-Colonel N Fruscr-Tytler, DSO, TD, CLIst (County Palatine) Brigade, Royul Field Artillery.**

The division to the north of Fraser-Tytler was the 56th (London) Division, and although the 30th was not 'on the extreme right' of the attack (the 21st Division was actually beyond it and further south) the position of Shea's Division was certainly precarious. Not only did the Division face a seemingly impregnable part of the Hindenburg Line west of the villages of Neuville Vitasse and St Martin sur Cojeul, but its frontage constituted a salient, and it was thus vital that the attacking divisions either side were successful before it too moved forward. As a result, although the main offensive further north was timed to begin at 5.30 am, the 30th and 21st Divisions were not scheduled to attack until the afternoon, or at least until Neuville Vitasse had been captured by the 56th Division.

The 18th Battalion was to be on the left of the 21st Brigade's assault, whilst on the 89th Brigade's front the 20th Battalion was to be on the left of the attack, the 19th Battalion on the right, and the 17th Battalion in Brigade Reserve. Despite the careful preparations, however, Brigadier-General Stanley was not totally happy with the situation.

I was very uneasy about the wire being cut and so was Shea. In fact, we had a long talk about it and eventually, on the night of the 8th, I sent in a telegram stating we were not satisfied it had been cut properly, and asked for all the 'heavies' possible to be turned on to it. This was done, and we had more favourable reports before we started to attack. **Brigadier-General F C Stanley, CMG, DSO.**

First Battle of the Scarpe

At exactly 11.38 am the 18th Battalion left its assembly trenches south of the village of Neuville Vitasse, and began to move forward, on the left flank of the brigade. Its objectives were the German front-line trenches known to the British as Lion Lane and Panther Lane. These trenches included a strongly fortified position known as 'The Egg'. No Man's Land at this point was at least 2,000 yards wide and was totally open, and covered by German machine guns sited in sunken roads. As the battle was over six hours old by this time, any element of surprise had been lost, and the advancing Pals came under fairly heavy artillery and machine gun fire. Those who managed to get as far as the German front line then found that the wire was virtually intact. It was a familiar story. Although Second-Lieutenant H F Merry then led a bombing party forward to try and exploit two gaps in the wire, all but three of his men were killed in the attempt, and no breakthrough was possible. The remnants then began to dig in, to try to consolidate their position, and despite constant enemy fire of all calibres, they managed to hold on until 3.00 am the following morning, when they were relieved by the 19th Battalion The Manchester Regiment, who had been in support. The Battalion took no further part in that first phase of the battle of Arras, later officially termed The First Battle of the Scarpe. During the fifteen and a half hours they were in action, the 18th Battalion lost two officers and fifty-nine other ranks killed, and eight officers and an unspecified number of men wounded. The two dead officers were Second-Lieutenant H G Ewing, and Second-Lieutenant F Ashcroft. Frederic Ashcroft was the first of three officer brothers serving in different Pals Battalions to be killed in the war.[9] Second-Lieutenant Merry was later awarded the Military Cross for his gallant conduct in front of the German wire.

With the advance of the 21st Brigade held up on the German wire, the chances of success for the 89th were very slim. Lieutenant-Colonel Fraser-Tytler saw the 21st Brigade assault, and then watched the 89th Brigade, as its Battalions formed up, and eventually moved forward to attack.

About noon the 56th Division streamed across the ridge on our left, the left half of the attack getting through the wire, but the right eventually had to come back and line a sunken road some four hundred yards from the Hindenburg wire. Our counter-battery work was most vigorous, and the enemy only put up a slight barrage, although our infantry must have presented a perfect target during their long advance across the open. About 2pm the Brigades of the 30th Division commenced their attack, advancing down the slope and passing right over the holes in which we were standing. They were naturally surprised to see us there, and much chaff was in the air about our having secured front seats for the show. They had a long advance in full view the whole time, and did not launch the actual attack on the Hindenburg Line till 3.30 pm. Previous to their final attack they had had very few casualties, which shows how non-plussed the Hun was for the moment.

While the 30th Division were advancing along the bottom of the valley the 21st Division attacked all along the high ridge of Henin Hill on our right flank. During the last few days we had all reported time after time that the wire across the bottom of the valley was quite untouched, and sure enough the infantry found this was the case and had to dig themselves in along a stretch of dead ground about four hundred yards from the wire. However, in spite of this attack failing, our losses were not heavy, and I really think that the wire was so dense and strong that the Hun could neither see nor shoot through it from his front line.

The 21st Division did better and got through by two gaps on the extreme summit of the hill, and disappeared from view. But while we, the right-hand pivot of the attack, were being held up to some extent, the Divisions which had attacked earlier in the morning to the north of us began to press down almost in the rear of the enemy on our immediate front; from our perfect OP, we could see great confusion on the roads behind Heninel, and our set programme being over at 5 pm, we were able to engage some glorious targets – roads congested with gun limbers, retreating Huns and vehicles of every sort, all went into the hash together. The Hun asked for war, and now at last he is getting it in good measure, heaped up and overflowing. **Lieutenant Colonel N Fraser-Tytler, DSO, TD CLIst (County Palatine) Brigade, Royal Field Artillery.**

The fate of the 89th Brigade was very similar to that of the 21st. Both of the attacking battalions, the 19th and the 20th moved forward just after 3.00 pm, up the gradual slope that was No Man's Land and onto the crest of a small hill from which they disappeared from sight. To the anxious observers, it seemed that they must have captured the enemy front line. However, they had not even reached the German wire, and were in fact caught in the open, in a narrow dip in the ground. The German front line was about another hundred yards away, on the reverse slope of another small rise in the ground. What the British could see from their own front line, and had imagined to be the German front line, was in fact the support line, which was further back still. Moreover, as with the 18th Battalion attack, despite Stanley's extra bombardment, the wire was virtually uncut, probably because it had not been seen by the heavy artillery FOOs.

I found out afterwards that on most of our front the wire was so well sited that, although we thought we could see the front line, it was the 'support' line we were looking at, and the front line was just behind the crest of a little rise in the ground. **Brigadier-General F C Stanley, CMG, DSO.**

A British tank practising against barbed wire entanglements of the type which fronted the Hindenburg Line.

Casualties from both battalions began to mount, and both began to seek whatever cover was available from shell holes and natural folds in the ground, to wait for nightfall. It was obvious that no progress could be made whilst the wire remained intact, and just before midnight both battalions were withdrawn to the safety of sunken roads behind the forward positions, so that the heavy artillery could attempt once more to flatten a path through the wire.

During the day of 10 April false reports were received that first Monchy le Preux, and then

Heninel, both villages behind the German lines, had fallen to the British, and patrols from the 19th and then the 20th Battalions were sent forward to consolidate what was hoped to be abandoned German trenches. In both cases, however, these reports were untrue, and unneccessary casualties were caused to both Battalions when German machine guns opened up on them. However, following the promised bombardment of the German wire by the heavy artillery, it was reported that two lanes either side of the River Cojeul had been opened up, and that the way to Heninel was clear. As a result, two companies of the 2nd Battalion The Bedfordshire Regiment were ordered to proceed through these lanes at 6.00 am on the morning of 11 April. Two companies of the 20th Battalion were also ordered to follow them forward through the left hand lane. At the allotted time, as the British bombardment lifted from the enemy trenches, the Bedfords advanced through the wire and were immediately hit by machine-gun fire from the German front line. They, and the supporting Pals immediately went to ground and tried to dig in where they were, in what was a most vulnerable position. The situation looked grim, when suddenly, literally out of nowhere, two tanks suddenly appeared over the hill and made straight for the German lines. Brigadier-General Stanley takes up the story.

'Then we heard that the Bosches were running like hares, whereupon I turned all the guns on to where they were, i.e., Heninel. We must have done a lot of harm to them, because they were seen by all the FOO's. After the Tanks had played about a bit, our gallant lads made another effort, but with the same result. As a matter of fact, the 20th Battalion made six efforts, but each time with the same result. What happened was that each time the Tank approached, the machine gunners got down into their dug-outs, and when it had passed they immediately reappeared. The Tanks reported that they had been repeatedly fired at by machine guns but could not locate them. Well, the Tanks walked up and down their lines, but the result was always the same, whenever we tried to get across the wire, up came the machine guns.

Then I sent a message to one of the Tanks, (by this time we had four of them playing about on our front), that he was to walk up and down the wire. This he proceeded to do. Then we heard that the front trench was simply teeming with Bosches, so being much afraid of a counter-attack, I had to turn on all the guns, telling then to be careful of our gallant Tank. The way they went was marvellous. As the Tank came along, each battery in turn lifted and he was never touched. It was a beautiful bit of work. **Brigadier-General F C Stanley, CMG, DSO.**

Although the Bedfords and 20th Battalion King's could not get into the enemy front line, the tank action contributed to the eventual victory. By 1.00 pm the 56th Division on the left had managed to penetrate the Hindenburg Line, and was pouring troops left and right through the gap. At 2.00 pm the 17th Battalion sent a patrol forward into the enemy lines north of St. Martin sur Cojeul, which met up with men from the 56th Division, and by 5.00 pm the whole of the Hindenburg Line as far south as the River Cojeul to a depth of about eight hundred yards was in British hands. The portion south of the river fell to the 21st Division on the following morning, as the German position there was also untenable, once the 30th Division had occupied their portion of the Line.

Only when the Line had fallen did the troops realise how formidable an obstacle it had been. The trenches were found to be extremely deep, with huge dugouts which literally went on for miles. The great bands of wire

German concrete fortifications in the Hindenburg Line.

which fronted the Line were virtually impenetrable, as the Pals had found, and for the first time, concrete emplacements, or 'pill boxes', were encountered, which intelligence officers had interpreted as trench mortar pits when they had seen them on aerial photographs. It is no wonder that the machine-gun fire had been so complete.

> *'After breakfast, my captain and I rode forward to look for new positions, and then went on to inspect the far-famed Hindenburg Line. It is rather disappointing in some ways, being merely a very wide traversed trench, only half-completed and with very few fire steps. The wire in front of it is of course marvellous. The machine-guns are either in between the front and support trenches, or in the support trench, with just a few right forward in front of the wire. Here and there were marvellous concrete machine-gun forts. The dug-outs were all in the support line; in fact the entire support line has a passage below it thirty to forty feet deep at this spot, running from Heninel over the hill to Fontaine, a distance of about 3,500 yards. No wonder the Hun was annoyed at losing all this so quickly.* **Lieutenant-Colonel N Fraser-Tytler, DSO, TD CLIst (County Palatine) Brigade, Royal Field Artillery.**

Although the Pals themselves had not actually stormed the trenches and wrested them from the enemy, their part in the victory should not be diminished. Theirs was by far the hardest part of the line, and it was known in advance that they could not succeed unless those on either side of them were victorious. Their stoicism in attack had allowed the assaults either side of them to continue, and their casualties bore testimony to their commitment.

By the time that the 89th Brigade was relieved on the afternoon of 12 April, Stanley assessed the casualties of the three Pals Battalions at about five hundred. Of these, the 17th Battalion lost two officers, Second-Lieutenant B S Davies and Second-Lieutenant A R Carr, and twenty-two other ranks killed or died of wounds, the 19th Battalion lost Lieutenant G W Mason and thirty-three other ranks killed or died of wounds, and the 20th Battalion lost thirty-one other ranks killed or died of wounds. Of the others who were wounded, the vast proportion were from the 19th and 20th Battalions, who bore the brunt of the fighting. For the gains made, these casualties must be considered quite light in Great War terms.

The Second Battle of the Scarpe

The first phase of the Arras offensive was an overwhelming success for the British Army, although few people today have ever heard of the fighting, apart from the much publicised success of the Canadian troops at Vimy. This is probably because colonial troops often received a much larger portion of media attention in the Great War than their numbers deserved. In many ways this attention was quite justified, because an initially European war was not really their fight, and yet they volunteered to leave their homes and loved ones on behalf of the British Empire. Their flamboyance of style and indifferent attitude to discipline made them natural targets for media attention, also, they always seemed to make a habit of being present when victories were achieved. This naturally tended to take the 'limelight', however, from the steady relentless pressure put on the enemy positions by the British county troops who had often stood firm against all odds until the colonial troops arrived. The second phase of the Arras offensive, later known as The Second Battle of the Scarpe, was not such a success for any British troops, however, and would ultimately make the whole offensive one of the costliest of the war, in terms of the number of days the battle lasted and the casualties sustained.

One of the reasons for the successes of the early phases of the battle was that the Germans were not able to bring up reserves quickly enough to mount the necessary counter-attacks to drive the British back, and Haig would have been more than happy after 12 April to

Victorious Canadian troops on Vimy Ridge in what was the first truly British and Empire success of the war.

Men of the Manchester Regiment sheltering beside a disabled British tank.

consolidate his gains and conserve his troops for the coming battles of the Ypres Salient. However, Nivelle's offensive on the Aisne was postponed by bad weather until 16 April, and as a result Haig was forced to renew his attack near Arras on 14 April, to keep German reserves pinned down. By this time, however, the initiative had been lost, the Germans had replaced many of their casualties and the positions to be attacked were even harder to penetrate than before. The next phase of the attack was planned for 21 April, but this too, had to be postponed until 23 April because of bad weather. As a result, the 30th Division was ordered back into the line on 22 April.

Shea's main assault was to be made by the 90th Brigade, with the 21st Brigade in support, and the 89th Brigade in reserve. The objective was the enemy positions on top of a ridge west of the village of Cherisy. The 90th Brigade assault began at 4.45 am but was a failure because of the ferocity of the German defence and the fact that the Brigade lost direction, and by the late afternoon, the remnants of the attacking troops were back at their own start lines. Thus, the 21st Brigade was ordered to renew the attack at 6.00 pm, the 18th and 19th Battalions of the Manchester Regiment forming the main assault, and the 18th King's in support as 'mopper uppers'.

The Manchesters actually took their objectives, but the part played by the Liverpool Pals was a complete shambles. Although the attack was scheduled to begin at 6.00 pm, orders for it were only received at Battalion Headquarters at 4.30. The troops were moved forward immediately, but there was too little time to prepare for action, and the company commanders only rejoined their companies at 5.45 pm, after Major Campbell, the Battalion Commander, had given his briefing. Guides were not available to show the men the exact locations of the Manchesters, and all the company commanders had to work with was a map reference, in an area where all geographical features looked the same. By this time 6.00 pm had passed and as the Manchesters attack was moving forward, the Germans began to shell the whole of the front area with a retaliatory bombardment. As the Pals moved off to try to catch up with the Manchesters, they inevitably fell under this bombardment and rushed for the cover of shell holes. As the company commanders went forward to reconnoitre, they realised that they would never be able to complete their task, and at nightfall ordered a gradual withdrawal to the old British line. In the early hours of the following morning the Battalion moved forward again, and upon learning that the enemy had vacated his front line trench, took and occupied it at 8.30 am.

For the next few days the men held this position, until relieved on 28 April by the 6th Battalion The Northamptonshire Regiment. During the period it had been in the line, the Battalion had lost Second-Lieutenant C D Calcott and thirty other ranks killed or died of wounds, and three officers and seventy-one other ranks wounded. On 26 April, Major W L Campbell relinquished command of the Battalion to Major M Owen of

A British trench near Arras.

the Rifle Brigade, who was unfortunately wounded as he was making his first visit to the companies of his new command. It would seem fairly reasonable to assume that Major Campbell's demise was connected with the debacle on the 23rd. Major G S Clayton then assumed temporary command until 8 May, when Lieutenant Colonel W R Pinwill took command once more.

Meanwhile, the 89th Brigade, in brigade support had also returned to the front on 23 April, where the 17th Battalion supported the 21st Brigade in the line and the 19th and 20th Battalions were engaged in carrying, working and burial parties, until all three were relieved on 28 April. During this time Second-Lieutenant L Band of the 17th Battalion and a total of twenty-one other ranks from all three Battalions were killed. Early in the morning of 29 April 1917, the Pals Battalions left the Hindenburg Line at Neuville Vitasse, and marched to Arras, where they entrained for Petit Houvin, near Vaulx, and rest. Thus, they were not present when the fighting at Arras ground to a halt on 17 May. Although the battles at Arras had played their part in tying down troops to allow Nivelle's offensive a chance of success, that too was finally called off on 20 May, having achieved few of the General's promises and ultimately caused the French Army to mutiny.

One final 30th Division casualty of the Arras offensive was its popular and professional commander, Major-General J S M Shea. In late April 1917, the GSO(l)[10] of the 30th Division. Lieutenant-Colonel W H F Weber, who had been with the Division since its arrival in France, was dismissed from his post. The exact circumstances surrounding his dismissal are not known, but were sufficient to make General Shea consider that Weber had been very badly treated. He wasted little time in making his feelings about the affair quite plain to his superiors, whereupon he also incurred their wrath, and he too was dismissed from his command on 30 April 1917. This was a double blow to the 30th Division, for Shea was much loved and much respected, and had been with the Division for over a year. More importantly, he had helped the Division grow from a collection of amateur citizen soldiers to the most successful Division in the British Army on 1 July 1916, and had then nursed it through its heavy losses in the later Somme battles. He was replaced by Major-General W de L Williams, who remained in command for the rest of the war. Shea himself was too good a general to remain in disgrace for long, however, and on 6 August 1917 was given command of the 60th 12nd/2nd London) Division, and eventually covered himself with further glory serving under Allenby in Palestine. His loss to the 30th Division was soon diminished, however, when, in mid-May, orders were received to proceed north to the dreaded Ypres Salient.

GSM Creasy and l6113 Colour Sergeant A Randies of the 17th Battalion.

Senior NCOs of the 17th Battalion in L917. They are: 16013 CSM J Shaine, 15408 RSM A Lovelady and 15755 C/Sgt.E E Bryan.

Notes
1. 'ENSA' stood for Entertainments National Service Association, and was the body of men and women raised to provide entertainment for service personnel during the Second World War.
2. An estaminet was a French cafe. which sometimes offered entertainment, where the soldiers could spend their money, usually paid to them in local currency. The most popular items of estaminet fare were a strange concoction of low strength beer, red wine, and food, when available. This latter was most commonly egg and chips, and the french order for this, 'des oeufs, et pomnes de terre frites'.

Senior NCOs of the 17th Battalion in May 1917. They are: 15421 Sgt E A Wray, 15790 CQMSJ H Bracher 15779 C/Sgt J A Jack and 15444 Sgt J Glover. Note the square black 17th Battalion indentification patches worn on their shoulders (see page ??).

was delivered by the British soldiery in many different forms, all of which they were quite sure was perfect French! Common amongst these linguistic gems was 'Deyzuff Bombadier Fritz.'

3. See Chapter 4 for a full account of this incident and its ramifications.

4 A minenwerfer, (literally mine-thrower), was a trench mortar fired by specially trained German troops. It came in several calibres, and was much feared by the British for the devastation it could cause, and particularly for its ability to penetrate deep dugouts. The slow almost lumbering procession of its projectile through the air could often be seen, and could usually be heard, but it was virtually impossible to predict exactly where it might land! The British troops referred to the projectiles as 'Moaning Minnies', or 'Minniewerfers'.

5. Even the least experienced infantryman could tell this from the sound of the passage of the shells through the air, and their correspondingly reduced accuracy. Another crude method often employed by the troops was to read the time graduations on the fuzes of shrapnel shells that had exploded nearby. Each shell was fuzed to burn for a set number of seconds before exploding, depending on the distance between the gun and the target. As a result, the longer the time for which the fuze was set to burn, the further away was the target.

6. The Battalion War Diary states that Green was captured, and presumably this was reported by the four members of the patrol who escaped, and made it back to the British lines. However, nothing was ever heard of him again, and he must either have been killed at the time, or died before he reached a German prison camp.

7. Royal Engineers.

8. 'HE' were high explosive shells, as opposed to sharpnel shells.

9. Lieutenant W Ashcroft of the 19th Battalion was badly wounded on 9 April 1917, and had only just returned to the Battalion, when he was killed in action near Roupy on 22 March 1918. Lieutenant E H Ashcroft of the 17th Battalion died on 12 May 1918, in German captivity, of wounds received two weeks earlier.

10. 'GSO(1)' stands for General Staff Officer (1st Grade) whose job it was in each division to liaise between the division and the General Staff, to make sure that the orders laid down by the staff were interpreted correctly.

Chapter Ten

'The City of Fear'

The Salient and 'The Big Bang'

The ancient Flemish town of Ypres[1] first came to the attention of the British in mediaeval times, when its geographical position made it ideal for the marketing of woollen cloth. British wool at that time was acknowledged to be the best in Europe, and many Englishmen followed the route across Flanders to Ypres with trains of heavily laden pack horses. The town's geographical position on the edge of the Flanders plain, and on the route to the Channel coast, inevitably also made it the natural focus of military attention, and every time that war swept across the so-called 'cockpit of Europe' the town suffered accordingly. The British Army had fought in the area under Marlborough during the War of the Spanish Succession and, a century later, had also been in the area during the Napoleonic Wars.

In 1914, after the battles of Mons and the Marne, when both the German and the Allied armies raced to get to the Channel, the British had stopped the German advance at Gheluvelt, just outside Ypres, and had then fallen back to defend the town itself. This was later called the First Battle of Ypres. In April and May 1915, the Germans had tried once more to wrest the town from Allied hands, and almost succeeded after their devastating first use of poison gas, but once again the defence held firm, although the noose around the town was further tightened. This became known as the Second Battle of Ypres and resulted in the German Army stabilising its positions in a semi-circular bulge around the east of the town. This bulge in the Allied line, known as a salient in military terms, was to become the graveyard of a generation of British soldiers. To have to defend a position in the middle of a salient is bad enough, because the enemy can fire at you on three sides of the bulge, but at Ypres it was worse, because the salient was bound by three ridges of high ground upon which the German Army was firmly established. To the north of the town was Pilkem Ridge, to the east of the town was Passendale Ridge, whilst to the south was Messines Ridge. Thus the Germans were able to see every single British movement in and around the town, and shell it almost at will. Not only that, but virtually the whole of the Flanders Plain upon which Ypres stands, is land reclaimed in modern times a lot of it below sea level. Just north of the Salient at Nieuport, in 1914, a resourceful lock-keeper had opened the lock gates that held back the River Yser and flooded the area, so preventing the German advance. All over the Ypres area, thick clay is just below the surface, and as a result, there is no natural drainage of the plain to take away rain and flood water. Over the centuries the Flemish farmers had contrived artificial drainage methods, but these were soon destroyed by the shelling in 1914 and 1915, which consequently reduced the whole area to little better than a morass. Thus, in some areas, topsoil was just not present, just glutinous clinging clay, and in any case, water was encountered virtually everywhere, in any hole dug deeper than eighteen inches. This, of course, was not so drastic a problem to the Germans, who, as ever, held the high, well drained ground.

The tragedy of Ypres for the British is, of course, that not far away from the town are a series of hills, the Flanders Hills, which were held by the Allies for most of the war. Had the British Army pulled back to these and aban-

The city of Ypres in ruins.

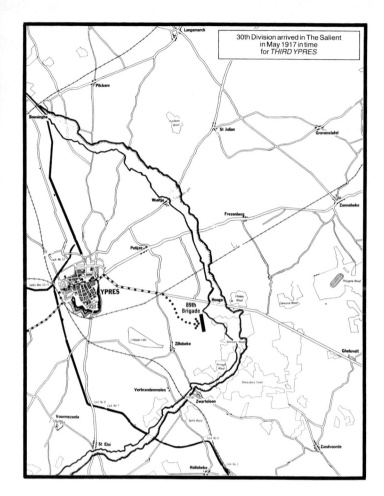

30th Division arrived in The Salient
in May 1917 in time
for *THIRD YPRES*

doned Ypres to its fate, its position would have been virtually impregnable. The Germans would have had to have followed up, and accepted the low ground, where they would have been dominated by the British-held high ground. If they had remained determined to take the Channel ports, they would have had to have dragged the British from off the hills first, because they certainly could not have afforded to have left them in British hands in their rear. So why didn't the British Army pull back to the Flanders Hills? The answer is really a political one. The British High Command favoured a withdrawal, but this would not have been popular with the public at home, who had been bombarded by the emotive propaganda of 'the rape of little Belgium', and would probably not have been at all sympathetic to giving up, especially when British lives had already been lost in the defence of Ypres. This also suited the French, who were quite happy to see the Germans tied down in the exhaustion of continual warfare in the north, whilst they held the line further south and tried to recover from the severe wounds inflicted upon them in 1914. Thus, the decision to defend Ypres was made by politicians, not generals, and their decision was to consign the British Army to nearly four years of unceasing warfare, unspeakable conditions and almost indescribable horror. The Pals first arrived at Brandhoek, in the Salient, on 28 May 1917, and were billeted in different parts of the vast camp there, which all bore Canadian names given to them by their original builders and occupants. They had left Vaulx on the 20th and had marched all the way, at a fairly leisurely pace, in fairly good weather, halting occasionally. Whilst the marching column of the 89th Brigade was passing near St Hilaire, on 22 May, Lord Derby, who, at this time, held the position of Secretary of State for War, drove past them in his motor car and took their salute. On arrival at Brandhoek, 'A' and 'D' Companies of the 19th Battalion were sent into support lines south of Hooge, where men of another Battalion of the King's Regiment, the 10th Battalion (Liverpool Scottish) had distinguished themselves in June 1915. One soldier from the 19th Battalion, clearly remembered the first time he saw the dreaded town of Ypres, although he was wrong about the date. It was, in fact, the end of May 1917, and not the beginning of June.

In June 1917, our division marched through Poperinghe and entered the dreaded Salient for the first time. We were to hold a piece of line south of Hooge. Behind the German trenches way back along the Menin Road was Gheluvelt. It was here in the early days of the war that the enemy's great drive for Calais and the coast was checked. Every inch of the intervening ground had been fought over, and rumour had it that another big Flanders battle was shortly to take place. It was an eerie

This famous photograph shows the stark reality of living and fighting conditions in the Ypres Salient.

experience marching through the deserted streets on our way up the line. The Cloth Hall was flooded in bright moonlight. Black shadows lurked amongst the ruins; our footfalls on cobbleways sounded strangely unreal and it seemed as though they belonged to other feet than our own. Perhaps we imagined a ghostly echo? The steady tramp of thousands who had gone before, never to return. If only we could move along faster – run – anything to get away from that brooding silence amongst the ruins. One felt that any moment the calmness of the beautiful June night would be changed into a raging inferno, yet nothing happened. We passed through the city of fear without a shot being fired. **17519 Private T S Williams,[2] 19th Battalion KLR.**

Although the 19th Battalion initially only held this position for 24 hours, it suffered quite heavy artillery fire, and Second-Lieutenant F.J. Mackie was killed and four other ranks wounded. The Battalion was still in the line on 7 June, when the Messines Ridge was captured by the British in one of the most successful operations of the war.

As a preparatory operation to what later became known as the Third Battle of Ypres, (more popularly known as the Battle of Passendale), Haig decided to try to capture the Messines ridge, which ran south of the Salient. Because of the fairly static nature of the trenches near Ypres, both sides had become accustomed to making war by tunnelling under each other's positions and packing underground galleries full of high explosive, which would then be exploded to blow up large sections of each other's trenches. Taking this type of warfare a step further, the British decided to undermine the whole of the German positions on the Messines Ridge and blow the ridge sky high, to enable them to capture what was left of the enemy lines. From as early as August 1915, engineers began to dig tunnels, to lay twenty-one huge mines under the ridge, from Hill 60 near Ypres itself, to just north-east of Ploegsteert Wood. At exactly 3.10 am on 7 June 1917 these mines were detonated, and the top of the ridge virtually evaporated.[3]

At 3 o'clock this morning we exploded two enormous mines – one at Hill 60 and the other at Messines.[4] I believe they are far bigger than anything that has ever been attempted before, and have taken months and months to prepare. I watched it from a point of vantage, and it was really a most remarkable sight. It was just getting light when suddenly there were two immense sheets of flame – very much like a very red sunset. Then there was a distinct kick to the earth, and two seconds afterwards the whole rocked violently.

I believe that a mass of Boches were blown up, as Hill 60 was packed at the time. This was the signal for the general attack, and in seconds you couldn't imagine what a row there was. Every gun was firing for what it was worth. One couldn't see anything much, because it was just too dark, but the roar was simply deafening and splendid, and somehow gave one a feeling of intense pride, when one thinks how we started from practically nothing, and now have grown to this enormous extent.

After that we soon began to hear that the objectives had been reached all the way down the line. We really were not taking part in the attack at all, but nevertheless we came in for a very hot fight. **Brigadier-General F C Stanley, CMG, DSO.**

A typical photograph showing the Flanders 'moonscape' terrain with German concrete blockhouses on the skyline.

A sniper's bullet hit the wire just in front of our post, and the whine of the ricochet had scarcely died away before another sound was heard – a sound which one would least expect in such a situation. It was the sound of a bird. I could scarcely believe my ears, but there was no mistaking that passionate burst of music, it was the nightingale's inimitable crescendo. One could not have wished to hear a more fitting requiem for those who had made the supreme sacrifice in this gruesome moonlit battlefield. During the nights that followed, I listened intently for the lovely contralto bird voice. Every night the songs sounded sweeter and louder as the rifle and machine-gun fire grew in intensity. The more active the snipers, the more vehement became the notes of the hidden bird.

On the 6th night of June, I heard the nightingale for the last time. Early the following morning, our guns were strangely silent. Everything was calm and peaceful when the first faint light of dawn came. Suddenly, and without any warning, two huge sheets of flame surmounted by columns of red smoke rose from the ridge away on the right. The concussion was terrific. The little advance trench which we were holding at the time started to rock. A deafening roar reached us, and at the same moment, it seemed that every gun on the whole Flanders front had opened up. The German position on Hill 60 had been blown sky high, and the attack on the Messines Ridge had commenced. 17519 **Private T S Williams, 19th Battalion KLR.**

Immediately after the explosions four patrols from the 19th Battalion were sent out to ascertain if the Germans still held the trenches in front of the Battalion's positions. One patrol, led by Second-Lieutenant J Ross, got to within forty yards of a German front-line position known as Jackdaw Trench, and Ross went forward on his own, only to find that the trench was full of Germans, and after firing two rounds at them from his revolver, he hurried back to his men, and brought them back safely to the British lines, with only one man wounded. Second-Lieutenant G W Sharpies led another party into No Man's Land, which suddenly came under rifle and machine gun fire from a group of about two dozen Germans, who had left their trenches and were lying in wait out in the open. Sharpies' patrol scattered and Lewis gun fire from the British trenches dispersed the Germans. However, it was not until the stragglers finally made it back to the British lines, after dark that evening, that they reported that Sharpies and two others had been killed. Another patrol then went out to look for their bodies, which were eventually found and recovered about forty yards from the German wire. The other two patrols sent out that morning met with heavy small-arms fire and were unable to move forward any further.

The enemy opposite the 19th Battalion had not been directly affected by the blowing up of the Messines Ridge and throughout the day of 7 June made several counter-attacks, but few of them got through as far back as the Battalion's positions, and those that did were soon dealt with. British artillery fire finally dispersed the rest. On 10 June the Battalion was relieved in the line by The 2nd Battalion, The

A German MG'08 machine-gun team

Royal Scots Fusiliers, and marched back to St Lawrence Camp at Brandhoek. The 17th Battalion, which had been in support, was also relieved and proceeded to Erie Camp near the same village. The Battle of Messines had been a great victory for the British, which, unfortunately, was not followed up quickly enough, with an attack on the main German positions on the other two ridges, and this was a mistake for which the British would pay dearly in the months to come.

On 16 June Brigadier-General Stanley left for England to begin three months leave, and his place was taken firstly by Brigadier-General W W Norman, and then by Brigadier-General W W Seymour. Such was Stanley's reputation, however, that when he was offered the leave, it was on the strict understanding that he could resume command of the Brigade on his return to the front. This was a most unusual arrangement after such a long leave in England and illustrates the value that his superiors evidently placed upon him.

Captain N G Chavasse. VC and Bar, MC, RAMC.

The build-up to the main attack on the Salient had begun by this time, and all four Pals Battalions alternated between training, trench duties and working parties, with the inevitable casualties that went with front-line service. Particularly costly were reconnaissance patrols, which were a necessary part of the planning for any major attack, and in the course of one such patrol, on 4 July 1917, Lieutenant A Chavasse of the 17th Battalion was fatally wounded and Second-Lieutenant C A Peters was killed trying to rescue him. Aidan Chavasse was the brother of Captain N G Chavasse, VC and Bar, MC, of the Royal Army Medical Corps, who was the Medical Officer of the 1/10th Battalion The King's Liverpool Regiment, (Liverpool Scottish). He had won his first VC at Guillemont in August 1916, and died of wounds received in winning the Bar to the Cross at Wieltje, on the Salient, exactly a month to the day after Aidan's death. The Medical Officer of the 17th Battalion during its service on the Salient, who also searched in vain for Aidan's body, was another Chavasse brother, Captain F B Chavasse, RAMC. On the day after Aidan Chavasse's death, Lieutenant-Colonel T S Rendall assumed command of the 17th Battalion. General Williams, the 30th Divisional Commander, was obviously dissatisfied with its former commander, Lieutenant-Colonel J N Peck, and had him sent home to England. Rendall would share the same fate just over three months later. For all four battalions most of the rest of July was spent out of the line, training for the big attack which was to take place at the end of the month.

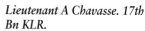

Lieutenant A Chavasse. 17th Bn KLR.

The Third Battle of Ypres

Since the end of 1914, Douglas Haig had been convinced that the British Army had the capability to win the war by beating the German Army in Flanders. German forces occupied some twenty-eight miles of the Belgian coast line, and Haig reasoned that if this area could be taken back from the enemy then the German Navy would be denied access to the sea and would not be able to disrupt British movement in the Channel. Once the coast was captured, it would be possible to make a two-pronged attack from inland, and from the coast, which could crush the German Army between its twin jaws. The basic flaw in this plan was that the British Armed forces did not have the facilities to mount such an amphibious assault, especially as it would mean crossing the land around the River Yser, flooded in 1914. It was possible to drain off this land, but not without giving the Germans the clearest possible indication that an attack in the area was imminent. Haig then reasoned that the next best thing would be an attack outwards from Ypres itself, to capture the ridges which surrounded it. Once accomplished, the Allied forces could turn towards the coast, and if a smaller amphibious landing was then made, it could join up with the advancing army, so cutting the German forces in two. Throughout the Somme campaign, this idea had gestated in Haig's mind, and although Nivelle's plausible promise of sudden and decisive victory for the Allies once again pushed the idea into the background of the planners' strategy, the failure of the Nivelle offensives

The German U-Boat offensive took its toll of Allied shipping in 1917, and also led to the Battle of Passendale.

gave it new credibility. The turn of events in the war also gave it new support.

As an island nation, Britain's survival depended on her ability to import and export goods to and from the Empire, and to protect them on their journies. The existence of the mighty Royal Navy bore testimony to this, and the sudden building and growth of the Imperial German Navy had been one of the challenges to British power which had helped to bring Britain into the war. Ironically, however, it was not the German surface fleet which eventually caused so much trouble in Britain's sea lanes, but the menace of the U-Boat. Before the convoy system became effective, which gave a measure of protection to merchant ships on the high seas, German U-Boats were able to score huge successes against British and Allied shipping, which almost brought victory to the Central Powers. In the first three months of 1917 nearly 650,000 tons of British ships were sunk by U-Boats, which seriously dented the ability of the Mercantile Marine to continue the carriage of vital war goods. In April 1917 alone, a record number of one hundred and fifty-five vessels were lost to U-Boats, with a total tonnage of 516,394. It was thought at the time that many of the U-Boats were operating from the Belgian ports of Ostend and Zeebrugge, and the fear began to grow, especially in naval circles, that the war would be lost unless these two ports, and perhaps the rest of the Belgian coast, were taken from the Germans.

Although by mid 1917 the tide of the U-Boat war was probably turning in favour of the Allies, and few U-Boats actually operated from the Belgian coast anyway, these facts were not known at the time. Thus, the seemingly critical situation at sea suddenly brought new and unexpected allies for Haig's plans. Haig was also able to argue, after the Nivelle offensive, that French morale was so shattered that the French Army would collapse entirely if the Germans were allowed to make a concerted attack in the French sector. An attack on the German Army in Flanders would prevent this possibility. The British General Staff, after a little wavering, was also in favour of an offensive which would take the British Army away from the impossible military position it faced at Ypres, with an average of two hundred and fifty to two hundred and seventy-five casualties per day.[5] Even Lloyd George might have been swayed by the obvious propaganda victory of liberating even a small part of 'little Belgium', from the clutches of the 'Uhlan baby killers'. Thus, on 4 May 1917, at a meeting in Paris, tentative agreement was given to allow Haig to mount his Flanders campaign, with French support, at the end of July. It is probable, however, that Haig envisaged it as a lot more substantial than did the other Allied generals or politicians!

After the victory at Messines, a buoyant Haig travelled to London for a series of meetings with the War Cabinet, during which he expected to be given complete support for his coming offensive. He was thus totally taken aback when he was met with a barrage of pessimism and doubt, predictably spearheaded by David Lloyd George. Nevertheless, at the 20 June meeting, at a critical moment for Haig's plans, Admiral Jellicoe, The First Sea Lord, dramatically informed the meeting that shipping losses were so heavy that unless Ostend and Zeebrugge were captured and the U-Boat offensive halted, Britain would lose the war. Although history does not prove the Admiral to be correct in his view, and even at the time some of the War Cabinet thought it an exaggeration, his pronouncement still hit the meeting like a bombshell. The Cabinet agreed to meet the following day, when its members had had time to assimilate all the possibilities. By the time the Cabinet met again, on 21 June, even the wily Lloyd George was wavering, because even he could not afford to ignore the possible implications. Therefore, despite a typical last minute attempt to subvert the Flanders attack and substitute it with an expedition on the Italian Front instead, he eventually but grudgingly agreed to let Haig carry on with preparations for the breakout from Ypres – providing that the French were able to support the offensive.

The plans were fairly straightforward. Plumer's Second Army, which had captured Messines Ridge, was to hold the right flank, and Rawlinson's Fourth Army was to hold the left flank on the coast by Nieuport. In the middle, Gough's Fifth Army was to assault the Germans on a line from Boesinghe in the north, across the Menin Road, to Zillebeke in the south. It was expected that it would take Gough's men about a week to capture the Passendale and Pilkem Ridges, and then as Plumer moved his troops forward, Rawlinson, in cooperation with French and Belgian forces under General Anthoine, would mount an amphibious attack on the coast, which would eventually meet up with Gough's advance. On 10 July, however, the Germans, having been well aware of the build-up of Rawlinson's troops, attacked in force, the positions on the east bank of the Yser opposite

The impossible attack conditions of the Third Battle of Ypres.

Nieuport and overwhelmed the troops there, killing or capturing two whole battalions. Although this was a blow to the ultimate success of the whole operation, Haig decided to go ahead with the main assault nevertheless. He had initially intended that this would commence on 25 July, but as the French would not be ready by then, he was forced to put forward the date by three days to 28 July. As a consequence, on 16 July, the preliminary bombardment began, but incredibly, even at this late stage, the War Cabinet had still not given Haig permission to start the offensive. This permission finally arrived only on the 21st, by which time the intense heat of mid-July had ominously begun to give way to cold and mist, not uncommon in Flanders at that time of year. On 27 July came the news that the French were still not in position for the attack and that they needed more time. Haig reluctantly agreed to a further postponement, which meant that zero hour would be 3.50 am on 31 July 1917.

One thing was certain, however, even if the Germans did not know when the attack would start, after Messines and the activity near the coast, they knew it was coming, and the proof of this knowledge was in the sudden growth of concrete fortifications, gleaming white over the whole of the German defensive line on the Flanders Plain. General Sixt von Arnim, who commanded the German forces in the Salient, had decided that his forces were definitely not going to be pushed off the high ground, and as it was impossible to fortify the mud, he had settled for near impregnable concrete positions, feverishly built in the front-line during the summer of 1917, each one able to give covering fire to the next. Ever innovative, and masters of military science, the Germans had decided to adopt a new defensive strategy. This consisted of lightly defending the front-line, to leave the main defensive force in the rear, where it was fairly safe from the inevitable bombardment of its front-line trenches. The German expectation was that the enemy would waste the impetus of his assault on the concrete fortifications, whereupon the main defensive force, specially trained in counter-attack skills, could be rushed forward and supported by its own artillery, catch the attacking forces in the open, where they could be annihilated. If the knowledge that the Germans were ready for the attack was not bad enough, by the evening of 30 July 1917, the weather had begun to break.

The Battle of Pilkem Ridge

II Corps' area of attack on 31 July 1917 was in the centre of the Vth Army front and covered the plateau which lay across the Menin Road. The 30th Division was to attack in the centre of II Corps' area, with the 8th Division on its left and the 24th Division on its right. Because of the battering that the Division had sustained on the Somme and at Arras, it was suggested at GHQ, that it be replaced with a fresher division, but this was not implemented in time for the battle,[6] and instead a brigade from the 18th Division was allotted to it. For the initial assault on 31 July, the 90th Brigade was to attack on the left, the 21st Brigade on the right, and the 89th Brigade was to be in reserve. All the objectives on 31 July were marked

Even moving up the line without enemy fire was a near impossibility.

out on the carefully prepared maps as coloured lines. The first objective was the Blue Line, the second objective was the Black Line and the third objective was the Green Line. The 18th Battalion of the 21st Brigade was given the task, on the left of the Brigade attack, of taking the Black Line when the Blue Line had fallen. This Black Line was a section of trenches which ran across the Menin Road, and contained Inverness Copse and the larger portion of Dumbarton Wood and Lakes. Once this had fallen, it was intended that the 17th and 20th Battalions, from the 89th Brigade, should move through the Black Line and capture the Green Line, further east, towards the village of Gheluvelt. The 19th Battalion was to remain in Divisional Reserve until needed.

On 24 July the 18th Battalion moved from Canal Reserve Camp at Dikkebus towards the front, and reached Chateau Segard on 30 July, where as much rest as possible was given to the men. At 8.00 pm the Battalion left the Chateau for the line and by 12.30 am on the 31st, it was in position for the attack, on the western edge of Sanctuary Wood. Meanwhile, the 17th and 20th Battalions had also left their training and reserve positions for Chateau Segard, and moved into assembly positions at Maple Copse in the early hours of the 31st, ready to carry out their supporting role. The Division thus occupied trenches from Sanctuary Wood to Observatory Ridge. Zero hour was set for 3.50 and at this precise time, all along the front, covered by the guns of the artillery, the Vth Army moved forward into battle. Generally, the initial advance was a great success, but unfortunately this success was not totally shared by the 30th Division.

The 18th Battalion had been ordered to follow up the 2nd Battalion, The Wiltshire Regiment, who were supposed to capture their portion of the Blue Line. However, the 21st Brigade's departure was delayed by heavy German shelling of its dugout entrances, and in the darkness and confusion of Sanctuary Wood, the battalions detailed to take both the Blue and Black Lines became intermingled. As a result, although No.1 and No.3 Companies of the 18th Battalion emerged from the tangle heading in the correct direction, No.2 and No.4 Companies were unknowingly swung around to the left and moved into the attack area of the 90th Brigade. This naturally weakened the advance, and as the Blue Line had not fallen on schedule, as soon as the men of No.1 and 3 Companies left the limited protection of the eastern edge of the Sanctuary Wood they found themselves engaged by the Germans from a fortified position known as Stirling Castle, which was part of the original Blue Line. They nevertheless advanced, but were held up on the right of the attack in a trench called Jar Row, where the Germans in one strongpoint put up a fierce resistance with machine-gun fire and stick grenades. Despite a concerted attack by some men of No.1 Company and a mixed bag of others from the Wiltshires, the Manchesters and the Royal Scots Fusiliers, the party was eventually forced to withdraw, having sustained heavy casualties. Although the two Companies of the 18th Battalion had not even been able to get through the first objective – the Blue Line, they had managed to advance in the correct direction and shown great endurance and fortitude. This was undoubtedly because of the leadership of Captain R P Heywood, MC, who had previously distinguished himself at Ligny-Thilloy in October 1916. He continued to encourage the advance until he was eventually hit and wounded by machine-gun fire, after which the advance came to a standstill. For his gallantry that day, he was later awarded a Bar to his Military Cross.

Meanwhile, another party of men from No.1 and 3 Companies was being held up by another strong-point further to the left, until it was eventually stormed and captured in an assault led by Second-Lieutenant W J Graham. Graham was unfortunately killed by machine-gun fire shortly afterwards in the newly captured strongpoint, and his men claimed later that the bullets that killed him were fired by a British tank. They stated that despite waving an artillery flag[7] at the tank to indicate that they were British troops, the tank crew either did not see it or ignored it, and fired just the same. The Battalion Commander, Lieutenant-Colonel W R Pinwill, was inclined to disbelieve their account, as he reasoned that machine-gun fire was coming from all directions at that stage of the fighting, but the fact remains that they were a lot closer to the action than he was, and tank crews in 1917 had very limited visibility! Not far away from Graham's strongpoint there was also a very broad belt of largely uncut wire which proved a deathtrap for all those attempting to get through it. Although some men found two gaps in it, the Germans must have known exactly where they were, because machine-gun fire cut down anyone who tried to get through.

Lieutenant-Colonel Fraser-Tytler's County Palatine Artillery unit, further back from Sanctuary Wood, had a somewhat more successful start to the battle.

The curtain rose at 3.50 am, accompanied by the usual racket, reek of HE, and rocket display by the Hun. On the whole our area was fairly peaceful, and Wilshin arrived with gun limbers at 5.45, having made an easy advance. We were due to cease firing and advance at 6.10 am, so at that hour, the backs of the gun-pits having been previously pulled down, we started the heavy labour of man-handling the guns out of their sunk pits. Luckily at that very moment a Highlander came down the track escorting twenty-five prisoners. I called to him to go to the cookhouse and have some tea, and to hand over his rifle and the prisoners to my tender mercies. The Hun was sending over some 8-inch shell, and when I ordered the prisoners to man the drag-ropes they started to argue that they ought not to be made to do it, but the arguments only lasted thirty seconds; the well-known sound (almost 'Esperanto') of a rifle bolt going to full cock and a few well-chosen words of abuse learnt on a Pomeranian barrack square, quickly got them to work, and meanwhile our gunners were safely under cover and able to have breakfast. Lieutenant-Colonel N Fraser-Tytler, DSO, TD, CLIst (County Palatine) Brigade, Royal Field Artillery.

A group of some of the 18th Battalion's Transport Section pictured somewhere on the Western Front in 1917. They are: standing 17241 Sgt W E Adams, 21913 Pte J Stockley, seated 21912 Pte J Rosbotham and 21904 L/Cpl J Routledge. Note the ammunition bandoliers and spurs worn by mounted soldiers.

Meanwhile, No.2 and 4 Companies, who should have followed behind No.1 and 3 Companies, were by this time, heading north-eastwards, towards the Menin Road and became mixed up with other battalions who were meant to be heading in that direction. Some of the Pals passed to the west of Stirling Castle and made for a position known as Clapham Junction, but all cohesion was soon lost, and eventually all but one of the experienced officers were killed or wounded by shell fire, or machine guns firing from Stirling Castle, or positions across the Menin Road. Despite this, some men did manage to cross the road at Clapham Junction, but no one at Battalion Headquarters, at a position known as Crab Crawl Tunnel, knew their whereabouts, and consequently they could not be supported and were eventually forced to withdraw.

Communications in a battle were always a problem, but the advance through Sanctuary Wood had made any confusion that already existed complete, and the Colonel and his Headquarters staff could only sit and hope that all was well, as they had virtually no news of the Battalion. Any kind of communication that might provide information was tried.

Passendale, oh gosh, that was terrible, mud up to your knees, sticky everywhere! The Headquarters was at Sanctuary Wood, and these French places sometimes had a brick cavern, you went down with a little ladder, would it be for keeping fodder? It was bricked anyhow. It was empty, and that was our headquarters, so we had to go down the ladder, with signals down below and Company Headquarters above. He only had to drop a bomb down there, and we would have all gone, but he

173

seemed to miss us somehow. It was impossible to keep a telephone line going, you couldn't do it, as soon as it was put down, it was smashed, so they sent a dog up, a trained dog. And of course you weren't supposed to feed the dog but the chaps were giving it bits of bacon, and we couldn't get the damn dog to go away, we had to hit it with sticks! We tried all kinds of things you know, because you couldn't keep the wires intact, – there was shelling all the time – the place was just one mass of shell holes. **16224 Private T H Brown, 18th Battalion KLR.**

Eventually, Second-Lieutenant A F Merry, MC, was sent from Battalion Headquarters to find out if the Blue Line had fallen, and exactly what had happened to the Battalion advance. He returned at 5.13 am, and reported that the Battalion had already moved forward into the German lines, presumably at Jar Row, and as the Intelligence Officer went forward with a small party to investigate further, he found that fighting was still going on in Bodmin Copse, which was a small wood south of Stirling Castle, on the line of advance. His party then came under fire from Stirling Castle and withdrew, which decided Lieutenant-Colonel Pinwill that it would be unwise to move Headquarters forward at that stage. Stirling Castle eventually fell to other troops of the 21st Brigade at about 6.00 am. At just after 8.00 am two runners were sent forward to try to establish contact with any of the Companies, and as they did not return, Second-Lieutenant Merry was once more sent forward to try to make contact and to order that the attack be continued. At 1.30 pm one of the original runners, 24451 Private G Reynolds, returned to report that the Battalion was in front of Stirling Castle, but all the Companies were mixed together. This was as far as the Battalion was able to advance that day, and by 9.00 pm, the forward position was held by Second-Lieutenant Futvoye, 25726 Company Sergeant-Major J R Crosby and about fifty other ranks. Far from being the 18th Battalion's original objective – the Black Line, this position was just short of the Blue Line, which was supposed to have been crossed early in the attack!

The two Pals Battalions of the 89th Brigade, the 17th and the 20th, had moved into assembly positions east of Maple Copse, between 5.00 am and 5.20 am, ready to advance through to the Green Line. At 7.00 am these positions were heavily shelled by the enemy which caused casualties to both Battalions. At 7.50 am they began to advance, the 17th Battalion on the left and the 20th Battalion on the right, and eventually found that the Blue Line had not fallen. As troops of the 21st and 90th Brigades were still pinned down in front of them, further attempts at advance were pointless. The Commanding Officer of the 17th Battalion, Lieutenant-Colonel T S Rendall then ordered his men forward on a north-easterly bearing, to reinforce troops whom he assumed were ahead of the Battalion in the Black Line. It was only when his men were out in the open on a ridge in front of Stirling Castle that it was realised that there were no British troops ahead of them at all. By this time the Germans had spotted their advance, and they began to come under shell and machine-gun fire. Despite this they managed to push on ahead until their left flank, led by Captain G G Rylands of 'C' Company, was just touching Clapham Junction. There, they dug in to wait the inevitable bombardment, which hit them soon after, killing Lieutenant F R Dimond, Second-Lieutenant E N Goldspink and thirty five other ranks, and wounding Second-Lieutenant C Bassingham, Second-Lieutenant L E L York, and one hundred and twenty men. Many of the casualties amongst the other ranks were senior NCOs, and as such were virtually irreplaceable. By 11.00 am the 20th Battalion, on the right of the 17th, had also realised that the advance ahead was blocked, and its men began to dig in beyond the old German support trench in a rough line which ran south from Stirling Castle. At about this time, an officer from the Tank Corps confirmed that Inverness Copse was still in German hands, as he

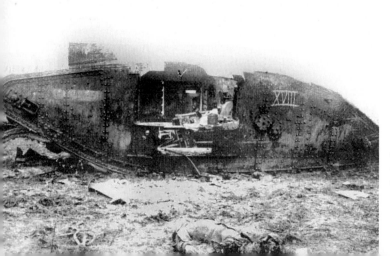

A destroyed British tank with the remains of one of the crew beside it.

had just been repulsed on the edge of the wood by machine gunners in concrete fortifications and had had his tank set on fire and destroyed. This convinced the 20th Battalion officers in the forward positions that they had advanced as far as they were able to go.

Nevertheless, at 11.06 am orders were received for the Battalion to attack at 10.20 am! Obviously it was impossible to comply with these orders and a further order to attack was received in the afternoon which was similarly rejected by the Commanding Officer, Major C N Watson, because of the impossibility of success whilst hostile machine guns were still active. Watson had temporary command of the Battalion whilst Lieutenant-Colonel J Douglas was home on leave. Watson's communication to Brigade Headquarters stated: 'front-line now roughly along road south of Clapham Junction . . . Enemy machine guns very active and any movement in forward positions is at once stopped by machine-gun fire . . . without careful artillery preparation, further attacks will be useless. I am therefore holding on to present line.' Thus, at nightfall, like the 17th and 18th Battalions, the 20th had only managed to get as far as a position just short of the 30th Division's first objective, the original Blue Line.

The 19th Battalion, in Divisional Reserve, had been called forward in the early morning of 31 July and reached Maple Copse in the late afternoon. At first it was detailed to continue with the morning's attack, but this was later cancelled, and it eventually relieved a battalion of the 53rd Brigade in the newly captured line, sustaining some casualties from shell fire in the process. By the morning of 1 August heavy rain had started to fall, which made it impossible to dig in any further, and still subjected to heavy enemy artillery fire, all the troops in the front-line became thoroughly miserable. The 18th Battalion was relieved on the evening of 2 August and made its way painfully back to Chateau Segard, which it reached the following morning, and then went into rest at Eecke. The 17th, 19th and 20th Battalions were relieved on the night of 3/4 August, and also initially made for Chateau Segard, and then rest at Brandhoek.

During its time in the line, the 18th Battalion, on the left of the 21st Brigade assault had lost seven officers and seventy-six men killed or died of wounds, and seven officers and one hundred and seventy-seven men wounded. The 17th Battalion lost three officers and seventy-nine men killed or died of wounds, and four officers and one hundred and ninety-eight men wounded, and the 20th Battalion lost one officer, Second-Lieutenant G G Nickel and forty-two men killed or died of wounds, and six officers and one hundred and forty-nine men wounded. The 19th Battalion, which was not actively engaged in the assault, lost twenty-six men killed or died of wounds, and four officers and one hundred and one men wounded.

Although they had not managed to achieve their objectives, they could hardly be blamed, as their success depended on the success of those in front of them. Once they joined the action, they performed as creditably as anyone could have expected, the 17th Battalion actually being the most advanced Battalion in the Division at one stage, despite the fact that it was supposed to have a supporting role. The 30th Division's task was undoubtedly the most difficult in the II Corps, because of the nature of the ground to be attacked, and the fact remains that the whole of the II Corps attack was not so successful as those either side of it, although in all places the German front and support line had been breached. This lack of success was almost certainly because of the central position of the II Corps attack area and the carefully prepared German fortifications which straddled the Menin Road.

Shot at dawn

The Battle of Pilkem Ridge was over, but the battle to break out of the Ypres Salient would continue for three more weary murderous months, and even then, only the tip of the Passendale Ridge would be reached. The 30th Division would take no further major part in the fighting, but would continue to move in and out of the line and suffer the wearing and inevitable casualties of trench manning and trench fighting. Most of the month of August was spent in training and reserve at Merris and Dranoutre, and in the third week of the month the 18th Battalion took over trenches in the Wytschaete area, where it remained until 30 August. The 89th Brigade and the other three Battalions moved back into the front-line at Hollebeke at the beginning of September and spent most of the month in and out of the trenches. On 5 September 1917, however, whilst the 17th Battalion was in support at Hollebeke, one small group from the Battalion, probably consisting of an officer and ten men, was in the rear at Kemmel, where a most unpleasant duty was performed – the execution of a comrade – 52929 Private J Smith, of 'C' Company. The story of this tragic episode was brought to the attention of the author in a most unusual way.

During the course of the research for this book, the author several times met a veteran of the 17th

24643 Pte R Blundell of the 17th Battalion photographed on the Western Front in 1917 beside a Bavarian pickelhaube. Blundell played an unwilling part in the execution of a comrade.

Battalion, 24643 Private Richard Blundell, who originally came from the Everton district of Liverpool. Although he would always chat generally about his Great War experiences, of which he was justifiably proud, he would never consent to a full interview. Sadly, Richard Blundell died in February 1989, and a few months later, his son, Mr William Blundell, contacted the author in the hope that he might be able to verify the details of a story that had disturbed him since his father's death.

Richard Blundell did not die quickly, but lapsed in and out of consciousness for some time before the end, sometimes murmering in a semi-delirium. One phrase which seemed important to him, and which he repeated time and time again, was 'What a way to get leave, what a way to get leave,' and the places Hellfire Corner and Kemmel were also mentioned a lot. In one of his lucid moments of consciousness, his son asked him what it all meant, and obviously anxious to talk about it before he died, the former soldier related his story. It seemed that sometime in 1917, whilst in the Ypres Salient, he had been chosen as part of a firing squad to execute a soldier from his Battalion who had been sentenced to death for cowardice. Private Blundell knew the soldier well. The sentence was duly carried out, but after the volley had been fired, it was discovered that the condemned man was still alive. By tradition, in such an event it was always the duty of the officer in charge of the firing squad to carry out the 'coup de grace' with his revolver, and kill the unfortunate victim. However, on this occasion, the officer in charge was unable to go through with it and giving Private Blundell his service revolver, he ordered him to kill the man. Blundell carried out the order. Then, as a 'reward' for carrying out this most unpleasant duty, he was granted ten days' immediate leave to the UK. After his father's eventual death, William Blundell naturally wanted to know if there was any way of verifying the story, as he was quite sure that it was true, being, as it was, in the nature of a death-bed confession.

By consulting a ledger in the Imperial War Museum library, the author was able to discover that the only time that Private Blundell was granted ten days' leave was in September 1917, the leave commencing on 5 September. A check through 'Soldiers Died' for the Pals Battalions, however, revealed that there were no deaths at all, in any circumstances, in any of the four Battalions, on this date. Executions were not always officially listed, however, and the author knew that the book 'Shot at Dawn'[8] was in the process of preparation by the authors Julian Putkowski and Julian Sykes. A 'phone call to the publishers solved the problem. There was only one Liverpool Pal executed during the Great War, 52929 Private J Smith of the 17th Battalion, who came from Bolton in Lancashire, and who was executed on 5 September 1917, at Kemmel, near Ypres. A simple check of all the cemetery registers in the immediate area of Mount Kemmel soon found the grave of Private Smith, and confirmed the date of his death. It was a straightforward process to establish further details of the service life of this unfortunate soldier.

James Smith was born in 1891, and was the son of James William and Elizabeth Smith of Bolton. In 1910, he joined the local county regiment The Lancashire Fusiliers, and became 2022 Private J Smith and on the outbreak of war was serving with the 1st Battalion in Karachi, India. The Battalion returned home early in 1915, but on 16 March 1915, it was dispatched to the Middle East to take part in the Gallipoli landings. On 25 April 1915, the Battalion won immortal fame, and six Victoria Cross awards, when it stormed the Turkish positions at 'W' Beach, Cape Helles, later to be called 'Lancashire Landing' in honour of the heroism of the soldiers who fought and died there. The Battalion stayed on the Gallipoli Peninsula until the final evacuation in January 1916, and eventually arrived in France in March 1916. At some stage,

Private Smith was transferred to the 15th Battalion of his regiment, and then, on 26 June 1917, was transferred to the 17th Battalion The King's (Liverpool Regiment) with the rank of Lance-Corporal.

Whether or not Smith thought that he had 'done his bit' in the war and should not have been sent to the front again can never now be established, but his time with the 17th Battalion was not a happy one. On 11 October 1916, the day before the Battle of Flers, he was buried by a German shell whilst in the front-line in front of Ligny-Thilloy, and when dug out was taken to 140 Field Ambulance with a bruised shoulder. He returned to duty a fortnight later, but not long after, committed a breach of military discipline, probably absence, for which, at a Field General Court Martial held on 29 December 1916, he was sentenced to ninety days Field Punishment No. 1 and ordered to forfeit one Good Conduct Badge. As we have already discussed, it is unlikely that he would have served his sentence in the front-line but in any case, it obviously did not have the desired deterrent effect on his conduct, for six months later, whilst serving with the Battalion in the Salient, he went absent again. At another Field General Court Martial convened on 15 July, he received another ninety days FP1, and was ordered to forfeit a second Good Conduct Badge, which, considering his record, must be considered as a light sentence! Perhaps there were extenuating circumstances. However, this punishment was obviously no more effective than the previous one, for some time later, almost certainly during the Battle of Pilkem Ridge, he refused to carry out an order and then deserted. He was tried on 22 August 1917 for desertion and disobedience, and once his guilt had been established, there could really be only one sentence – death. He was executed at 5.51 am on 5 September, and his body was buried in Kemmel Chateau Military Cemetery, in grave M. 25, where it lies today.

It is impossible after all these years to form an objective judgement of the whole subject of military executions, because there are far too many factors that would need to be considered, about which we simply do not have sufficient knowledge. No doubt there were cases during the Great War where obvious and irrevocable injustices occurred. It would seem, however, from the known details of his case, that Private Smith was given a lot more chances than those accorded to many soldiers executed during the war. Most of the Pals veterans interviewed shared the opinion that in cases of shell shock or similar incapacity it was totally unjust to execute a man, – but they had little sympathy for those who deliberately ran away. Their view was simple; if a man who deserted was merely put in prison, even for life, at least he was alive, and out of the firing line, and did not have to face all the terrors of trench life still being experienced by his comrades. Thus, what else could the authorities do but execute such a man? Equally, all interviewed shared a distaste for having to shoot a comrade, even one for whom all respect had been lost, and all were glad that they did not have to do it. Unhappily, Private Blundell was not in this fortunate position, and as all the other facts concerning Smith's execution tally exactly with his death-bed testimony, there is no reason at all to disbelieve that it was he who actually fired the fatal gunshot.

Painfully, it is probable that Smith's parents knew the circumstances of his death, although from what source can only be conjectured. At the beginning of the war some relatives were informed of the execution of a loved one, and the reasons for it, but by September 1917, the authorities had adopted a kinder attitude and usually reported the death as 'killed in action'. Despite a thorough search of the local press in the Bolton area, however, there is no mention of Smith's death at all, and as it was usual for the relatives to submit such information for publication, one can only presume the reason that this did not happen. Similarly, Bolton has quite a detailed Book of Remembrance for those men and women from the area who lost their lives in the Great War and Private Smith is not included in that either. Presumably, like most municipalities, Bolton relied upon relatives to supply the names to be included in the lists of commemorated dead. Smith's grave inscription simply states: *Gone but not forgotten*. Although some names of executed men were included in the HMSO publication 'Soldiers Died in the Great War', as we have seen, Private Smith's name was not one of them. He is, however, included in the revised list of Pals who died in the war in Appendix II of this book.

One thing is certain, the 'pour encourager les autres'[9] aspect of all judicial punishments was lost on the men of the Pals at the time of Private Smith's death, because apart from the men of the firing squad, they simply did not know about it. All those interviewed since, from all four Battalions, were unaware that any Pal had been executed at all. Even Private Blundell kept the affair secret until just before his own death. One other soldier who played a minor part in Smith's execution certainly did not realise that one of the men whom he had helped to dispatch was a Liverpool Pal. 25561 Private H Redhead of the 20th Battalion was attached at the time to a Royal Engineers dump near Poperinghe, where he was in charge of stores, including those necessary for military executions. In this account, he refers to the bungled execution of one soldier, and it is interesting to speculate whether or not this was Private Smith's execution, but with the true facts slightly distorted as the story was passed from mouth to mouth.

An interesting point that you would hardly believe was that when an attack was coming off, the joiners used to make execution chairs. They always had cases of cowardice, and people hiding and not going over, and they had these chairs made which I had in my shed. They were like a stool, so far from the ground, and they used to tie their legs to the legs, and at the back used to protrude an arm about six inches, each side, and they used to tie their arms each side. And with that chair, you sent about ten yards of green canvas, they wouldn't bury them in a blanket – ordinary soldiers that got killed, if they could, they would bury them in a blanket – but with a chap who was shot like that, they used to give him ten yards of green canvas. Also, they had a white disk, enamelled white, with a pin at the back, which they put in the breast button hole of the chap who was going to be shot.

The people who did it never knew what they were going to do, they'd be out on patrol or something, and they were suddenly taken to a hut, and handed a rifle each, and cartridges, and without any warning, they were confronted by this fellow in the chair, and were told to shoot him. They were given the order to aim and then fire, and I heard of one case where they said 'Ooh, we're not going to do that', and instead of killing the chap, they badly wounded him in the legs and the arms, and the APM, – the Army Provost Marshal, had to go up and finish him off with his revolver.

I was never on a party, thank goodness to do that sort of thing, and I never saw it done, but I issued the chairs with the canvas and the disk to the regiments that had these people. Over the time I was with the REs[10] I issued about five, I suppose. In the Orders, they used to publish these things, in order to warn people not to do it. I never had any sympathy with deserters, I used to look on them as enemies. **25561 Private H Redhead, 20th Battalion KLR.**

Whether or not Smith's execution was published, it is probable that given the esprit de corps that was the very essence of the Pals Battalions, those that did know would have been so ashamed of the event that they would have done their best to keep it secret. They seemed to have succeeded, at least for nearly seventy-five years.

Autumn and winter 1917

The remainder of September was fairly uneventful for the rest of the Pals Battalions, except for two trench raids made by the 19th Battalion on 20 September 1917, which, elsewhere on the Salient, was the opening day of the phase of the offensive later referred to as The Battle of the Menin Road. These raids were made for two purposes. The first was an attempt to confuse the enemy as to the intensity and direction of the main attack, and the second was to try to capture two blockhouses known as 'The Twins', which commanded the 19th Battalion's trench front, and thus was able to dominate all its movement. The raiding party, consisting of Captain C Laird and twenty-five other ranks left the British front line at 6.00 am and moved into No Man's Land. However, it was soon spotted, and machine guns opened fire from the blockhouses. Despite a most determined effort to carry the objectives, the situation was hopeless from the start, and Laird and six other ranks were killed and fourteen more soldiers were wounded.

On 24 September, a draft of sixteen officers and two hundred and ninety men from the Lancashire Hussars, an old yeomanry unit, was posted to the 18th Battalion in the Torreken Farm area of Wytschaete, to be amalgamated with the 2nd City Battalion. The High Command had finally accepted that it was most unlikely that there would be a sudden breakthrough that could be exploited by cavalry, and as a result it was wasteful of manpower to keep trained soldiers waiting idly behind the lines. Thus, many cavalry/yeomanry units were dismounted and their troops sent up the line to fight as infantry, or to be attached to or absorbed by infantry units.

'In 1917 we'd been doing little odd jobs such as road mending and things like that, and looking

after German prisoners, and everybody was getting a little fed up. Anyhow, in 1917, they took our horses from us, and the younger men of the regiment were sent to the infantry. The older men over thirty – and there was quite a lot – I don't know what happened to them, I never saw them again. We went for three months infantry training, and we were sent to the King's (Liverpool Regiment) – the 18th Battalion, a Pals Battalion, they were. 300087 **Private W R McLeish, 18th (Lancashire Hussars) Battalion KLR.**

Because of the effective seniority of the Lancashire Hussars, they did not lose their identity completely, and it was ordered that thereafter the amalgamated Battalion was to be known as the 18th (Lancashire Hussars) Battalion, The King's (Liverpool Regiment). Even the official abbreviation, 18th (LH) KLR, was a mouthful, and neither the originals of the Pals or the Hussars, both dissatisfied with the amalgamation anyway, were prepared to accept it willingly. The former Pals continued to refer to the Battalion as the 18th, and the

The cap badge of the Lancashire Hussars, who were amalgamated with the 18th Battalion on 24 September 1917.

former Hussars continued to refer to it as The Lancs. Hussars.[11] The latter were particularly irked at having to change their Red Rose of Lancashire cap badge for the Eagle and Child of the Pals, and a lot of leeway seems to have been allowed in this direction, with some former Hussars still wearing the rose badge in their caps well into 1918, whilst at the same time wearing ordinary 'KINGS' shoulder titles. It would also appear from contemporary photographs that after the war the original Hussars reverted to their former cap badge, although still officially part of the 18th (LH) Battalion until its disbandment. It was not long before some of the former cavalrymen, still not experienced in infantry skills, were tested in the line.

In the early hours of 1 October 1917, a force of about forty Germans made a raid on No.11 outpost, near Derry House, Wytschaete, which was manned by men of No.3 Company, some of whom were former cavalrymen. According to the Battalion War Diary, the NCO in charge, 57587 Lance-Corporal J Hughes opened up on them with a Lewis Gun, and the remainder of his party of six fired rifles and threw bombs at them, enabling the party to retire to a better position, from where they then repulsed another attack. For his 'skilful and determined resistance against odd', Hughes was awarded the Military Medal, as was Private McLeish, who had never been in action before. Many years later, however, McLeish related the true facts of the incident, which differ significantly from the War Diary account.

On 3 October 1917,[12] was put with a machine gun section out in No Man's Land, it must have been two hundred yards from our front line out in No Man's Land. I was a bomber, I had Mills Bombs, and the Germans attacked us at dawn. They put down a barrage on our front line and they came over and attacked us. We let them have some bombs, there was someone else throwing bombs as well as me, but I didn't know, I thought I was on my own! We killed about six Germans, and when things went quiet, I went to see who was throwing these bombs as well as me. It was the Corporal in charge of the post. I jumped into this shell hole, and he grabbed me and said ' Who are You' – I must have frightened him, and I told him, and said 'What's the next thing we do now Corporal?' and he said "Oh we can't do anything more now, the Germans have failed." The others had disappeared, you know, the machine gunner and the rest, they'd disappeared. There was one lad, "Oh Mother" he says, "they're coming", and he knocked me flying. I had my bayonet in the side of the trench, and he knocked me over, and it snapped the bayonet off. I think that lad upset the lot of us.

Anyhow, I got back, I found this here Corporal, and he'd been throwing bombs as well as me, and just as it was beginning to break daylight, he says 'Come on, we'll go back into the front line now, and report to the Colonel'. And we went in, and the officers treated us like blinkin' heroes, and we went up to the Officers' Mess, and one officer gave me a packet of Three Castles cigarettes, and we had coffee and some breakfast. They were quite delighted with us, and I never thought about a medal, or anything like that, I was only saving my own skin, and so was the Corporal. Then the Colonel[13] came and congratulated us, and a few days afterwards, the Colonel told us "You've been awarded the Military Medal, you and the Corporal".

The effects of British bombardment on a German concrete blockhouse.

We'd probably saved an attack on the front line. If there'd been nobody there, they would have gone into the front line. We stopped them with the Mills Bombs, but I had the 'wind up', you know, it was my first time in action, and I was only a kid, I was 22. The Colonel, when he told us that we'd been awarded the Military Medal, he took us into the Officers' Mess, and gave us a good stiff glass of whisky, and all the other officers were round us congratulating us. 300087 Private W R McLeish, 18th (Lancashire Hussars) Battalion KLR.

In the evening of the same day the Battalion Medical Officer, Captain W T Chaning-Pearce, MC, RAMC, who had won his Cross during the Battle of Pilkem Ridge, was killed in most unusual circumstances, when he seemed to mistake the direction of the British and German trenches, and was caught out in the open by machine-gun fire from a German blockhouse. Not long after this, on 14 October, General Williams must have decided that Lieutenant-Colonel T S Rendall, of the 17th Battalion did not after all, match up to the job of Battalion Commander and he was replaced by Major C N Watson, who had so ably commanded the 20th Battalion during the Battle of Pilkem Ridge, for which action he had been awarded the DSO. Watson was promoted Lieutenant-Colonel upon appointment to command.

For the rest of the year, the four Pals battalions were in and out of the line, without being committed to any major engagement. After the capture of the Passendale Ridge by Canadian troops on 10 November, after over three months of blood letting, the much-hoped-for break out to the coast was finally abandoned, and the offensive halted. The British did now at last command all three of the ridges around Ypres, and for the moment, the danger to the 'City of Fear' was alleviated. This was of little consequence, however, to the men who now had to endure another winter in the open, in the most appalling of conditions and who still faced daily, the ever present uncertainty of sudden and violent death.

Experiences of the Salient

Even though the Pals did not go 'over the top' again in the Salient after the Battle of Pilkem Ridge, their service in the area left deep and lasting impressions which were still vivid a lifetime later. Here are just a few accounts of some of their experiences.

Why on earth we have ever sat down under the insults which the French have continued to pour out on us as regards our weather, I cannot imagine. We really are a most foolish race. If we had only taken the trouble to come over to this country we should have found that their climate is ten times worse than ours was, and that they have just as much as us in the way of rain – further than that, when it does rain in their country the whole state of the ground becomes nothing less than a sea of mud. Ours is bad in some parts of England, but nothing like this. We were having at the time bad weather, and I dare say owing to the drainage having been upset, the whole place was completely water-logged. To move about, you had to stick to the duckboarded track, or else you practically got drowned or stuck in the mud. This was tiresome, because the Boche had a way of paying attention to these tracks, which of course, were clearly visible in his aeroplane photographs. At times he could make it most decidedly unpleasant. Brigadier-General F C Stanley, CMG, DSO.

I had to hot these shoes up to fit the horses' feet, whereas originally they put them on cold, and they were coming off hand over fist, they wouldn't keep on, so I had a little furnace. This day, I'm making a bit of custard, – even now, every time I make custard I think of this. I'd cut the bottom end off a petrol tin, and used it for a pan, and there was a gust of wind came along, it was a fine day, but a gust of wind came along. There was a captive balloon stationed nearby, and maybe they'd decided to bring it in or something, but it got away from them. There was only one man stood to – all the others let go, but that poor beggar dropped off and as I was making my custard, he dropped right next to me, I could have reached out and touched him. He was dead as a stone, he never moved a limb, he'd dropped from a considerable height, and that was the end of him. Not long after this, I was detailed off as NCO in charge, there was an officer with me, to take so many mules from each Battalion of the Brigade, and get them delivered up to the front-line, or as near as we could get. I had a mule with eight tins of water on it, and it had frozen that night, and the duckboard track was very slippery, so he lost his footing, and over he went into a shell hole, tins of water and mule and all. When they're in difficulty, mules, they don't just lie there, and he was kicking and everything, and we had to sit on his head to quieten him. If you sit on a horse's head for a while he'll generally give over kicking, and give in, once you let him rear his head up, he'll start to kick up and want to get up again. Anyway, we kept him down 'till we got all the water off, it was a tidy weight for a little mule, because they were only little things, but their ordinary load was two boxes of ammunition, and they were 70lbs apiece, and they'd march through the country with that regularly. But it was alright, in the end, he turned over, and we got him up again.

Coming home the same night – once your load was delivered, the order was to make the best way you could home, because they were shelling heavily all the time, they knew where the duckboard track was, and it was always getting shelled – this fellow unloaded his load and got away. They used to jump on the mule's back on the way home, it wasn't a proper saddle and it was a very rough ride. When he got home, this fellow just tied the mule to the line, he wasn't much interested in mules and he never looked at it, but that poor beggar had a shattered leg and he'd come all the way home on the stump! **21915 Lance-Corporal W Heyes, 18th Battalion KLR.**

'*Conditions were terrible, and if you slipped off the duckboards you could sink right in, right up to your neck. The Germans had built a series of concrete blockhouses, with machine-guns in the front, and I went forward to a captured one, on Pilkem Ridge, to try to get some line going. The snag was though, that the only entrance was the original back door, which faced the German lines. Nevertheless, I was so tired that I fell asleep, and a shell must have hit the top, because I woke up and felt the whole damn thing shaking, but they were made of solid concrete, and the shell never got through.* **16224 Private T H Brown, 18th Battalion KLR.**

'*I was at a large dump at Busseboom near Poperinghe, a very large place it was, and one of the Engineers had found a dud shell and took it to the workshops and tried to get the nose off it, and it went off and killed him, and blew the roof off. I was one of the party that went to bury him. We had a pontoon with a couple of mules and they galloped and galloped, and the corpse was nearly flung off the pontoon and we had to hold it on until we got to the graveyard where we buried him.*

This specially drawn 89th Brigade Christmas card was sent home to Merseyside from Belgium in 1917. The cap badges of the three Pals Battalions, and the 2nd Bedfords, which made up the Brigade at this time, can just be seen in the ornamental corners above the figure of Britannia.

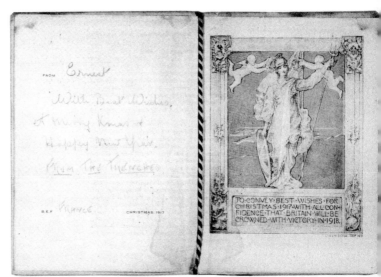

15485 Pte N E Etheridge of B Company, 17th Battalion, who came from Wavertree, near Liverpool, photographed in 1917. On his left arm, he displays, below the 'regular' pattern 'King's' shoulder title, a black 17th Battalion identification patch, (see page 37), the crossed flags trade badge of a signaller, an inverted 'V' good conduct chevron, and a wound stripe. He had been wounded in the right thigh on 29 January 1916, whilst the Battalion was at Maricourt on the Somme.

Anyway, we left Busseboom and went to another R E dump at Dikkebus, and when we got there, there was a broad gauge railway running right through, and there was a lot of camouflage where a gun had been. We got talking to some gunners, and they said they had had a new gun there, with a longer barrel and a longer range. The first shell they fired burst the gun, it just exploded. Then they said 'God, we've got another one coming in a few days', so we said 'Well when it does come, we're going to get out of the way.' Anyhow, the gun arrived and one morning we saw it levering up and down ready to fire. I had to go in the store, which was only 50 yards away from this gun, to get some nails for an order we'd had, and I got the nails, and as I was coming out, the gun went off.

I looked round, and it seemed to be all right, but there was smoke coming out of the end. When the officer looked down the barrel, all the rifling was torn out of the gun. The barrel was evidently too small for the shells, and the shell that it fired dropped somewhere in Scottish Wood,[14] behind our own lines. It must have cost at least £100,000 to make a gun like that, but anyway, it went back as scrap. **25561 Private H Redhead, 20th Battalion KLR.**

The Battalion had moved forward to the furthest point held, which was just beyond Hooge on the Menin Road, and we were told that next morning, at first light, Brigade transport had to move forward, because the rations had got to get through some way or other. But luckily it was one of those dull midsummer hazy wet days, with no visibility, except locally, and the horizon just tailed off into thin wet mist. So we all set off, wondering what the hell was going to happen to us, going up to the Salient in daylight. We came through the Lille Gate, and came past Shrapnel Corner, and the Engineers had built a log road across the fields and we were able to go along this with comparative ease, and no shelling at the time. There was the odd burst that came through, but apart from this, it was dead quiet. We came to the road just north of Zillebeke, to cross over the ruins of the village, and then we wound up to Sanctuary Wood. You didn't dare get off the track. Once I came off it, and went down to my knee, luckily I had one leg on firm ground, but it took me all my time to pull the other one out. Some people if they came off, were never seen again.

Close to the wood were shell holes full of water, and there sticking out of one of these was a yellow hand, it was sticking out of the top of the mud. And then there was the mask of a face, floating on top, – other people have mentioned that – but of course when you've got corpses in shell holes in water, anything can happen. This was in the late afternoon everything grey, and the blackened stumps of the wood standing out, and then we came to Hooge, and that was just brick-stained, but there was a little bit of trench still firm and undestroyed, and in it were two Germans.

One was just slumped at the bottom of the trench, in the normal position of someone who's had it, but the other one was sitting up, with his back against the trench, and some freak of artillery had taken apart the whole of the chest and the uniform, and taken it in one flap, and laid it over one side so that from his neck down to his middle stomach was laid out, just like an anatomy lesson. He was a nice looking chap too, in fact he reminded me of my father. That was interesting in many ways, but these were the freaks, but the other stuff, in the mud and the rest of it, they made your stomach turn. **17518 Corporal E G Williams, 19th Battalion KLR.**

Christmas 1917 was celebrated in style by all four Pals Battalions. As the funds of the Comforts Committee at home were running low, in mid-December The Optimists' were sent back to Liverpool, where they performed three shows for the public, which eventually raised £750. It must have taken a fair amount of 'fixing', even for a brigadier-general, to get the whole troupe sent home on leave at the same time, but the net result and justification was that each battalion was able to enjoy a good Christmas. For months the area around Ypres had been scoured for poultry and pigs for Christmas dinner, and once secured, these had been left at farms to be fattened until needed. The 18th Battalion was at rest at Chippawa Camp on Christmas Day, and celebrated with dinner, special entertainment and football. The 17th Battalion, also at Chippawa took a holiday on Christmas Day, and enjoyed Christmas dinner on Boxing Day, and a game of football against the 18th Battalion. The 19th out of the line at Ottawa Camp near Reninghelst, and the 20th Battalion at Swan Chateau, followed suit, with roast pork, plum pudding and free beer, on 26 December. The 19th Battalion entertainment troupe "The Duds' also put on a performance of the pantomime 'Aladdin' in the afternoon, which was attended by Brigadier-General Stanley. Some of the men of the 19th Battalion did suffer a festive disappointment, however.

An unknown private of the 19th Battalion in 1917. Although he is not wearing any shoulder title, on his upper arm, he displays a white 19th Battalion identification patch, which is wrongly positioned. It should have been sewn square-on to his shoulder. On his lower sleeve, he displays an inverted 'V' good conduct chevron.

Ottawa Camp was on top of a hill, where there was a windmill, which had been the focus of a spy scare with a Belgian supposedly passing information to the Germans by the setting of the sails. The windmill was still there, as the war had not got that far. The officers told us that in the evening, we could go into the nearby village of Reninghelst, but if we wanted to stay in camp, there would be milk and rum punch served under their supervision. I suppose about thirty to forty of us stayed, and three or four officers and the cooks brought up the dixies with what was supposed to be the rum punch in them, and one officer took a sip from a ladle, and immediately spat it out! Rum and whale oil[15] both came up the line in similar kinds of demijars, and the cooks had purloined the rum and substituted whale oil. I think there was one dixie with rum in it, but the cooks were roaring drunk in any case, and so all we got was whale oil and wasted milk. The troops that went into Reninghelst had this very steep hill to climb, on the way back, and furthermore, it had snowed and then frozen, and the duckboards and the hill itself were slippery, and for about two miles you could see people crawling on their hands and knees, because they couldn't stand up.
17518 Corporal E G Williams, 19th Battalion KLR.

Three of the four Pals Battalions returned to the front in time for New Year, only the 20th Battalion still being in the rear at Alberta Camp at Reninghelst, although it did provide some working parties for the line at Torr Top Tunnels and Railway Dugouts.

All along the Corps front, on the stroke of midnight, each artillery piece fired one round at the enemy, and each machine gun fired two belts of ammunition, in a show of force to celebrate

A railway ticket issued in 1917 to 25561 Pte H Redhead, of the 20th Battalion, for travel to and between Ypres, Poelcapelle and Comines. Note that the ticket is printed in both French and Flemish, (Yper. Poelcapel and Komen.)

the arrival of 1918. The Germans, however, did not retaliate. Once more, the Pals had survived a year of vicious fighting, and acquitted themselves with honour and distinction.

No one could have known on the first day of January 1918, that the New Year would bring them more wholesale slaughter, and then peace, on the Western Front at least.

However, their tenure of the Salient, for the moment, was coming to an end, as orders had been received just before the end of the old year, for the 30th Division to leave the Salient, to take over an area vacated by the French. Consequently, on 10 January the 18th Battalion entrained for the south, followed, the next day, by the other three Pals Battalions. No one was sorry to leave Ypres. By 29 January the move was complete, and the 30th Division relieved French troops opposite an old Vauban[16] fortress town called La Fere, south of St Quentin. Although only actually engaged in one major battle in the Ypres Salient, the four Pals Battalions had lost twenty-two officers and five hundred and eighteen men dead, and about three times that number wounded, during its seven months of duty in the accursed place!

NOTES

1. Confusion has always existed about the spelling and pronunciation of the town's name, and the British soldiery of the Great War was no less confused than anyone else. The inhabitants of the area are a mixture of Flemish, (people of Dutch origin), and Walloon, (people of French origin), and the place names of the region reflect both languages. Thus, the French name for the town is Ypres, whilst the Flemish name is Ieper. Soldiers of the Great War quite cheerfully mispronounced both of these in different ways, the commonest being 'Wipers', 'Eepray' and 'Yeeps'.

2. 17519 Private T S Williams was the brother of Corporal E G Williams, whose graphic and wonderfully poetical accounts of the war have been liberally used throughout this book. Originally from Formby, Liverpool, the brothers joined the 19th Battalion together in 1914, and both were captured on 22 March 1918, at Roupy, and survived captivity. T S Williams died in 1970, and E G Williams in 1989.

3. Although twenty-one mines were planted, only nineteen actually detonated. Of the other two, one exploded, apparently in a thunderstorm, on 17 July 1955, and the other is still down there somewhere, under the southern part of the ridge, its exact location now lost.

4. Stanley may not have known when he wrote about the mines, that there were, in fact, nineteen explosions. It is probable that from the 89th Brigade positions, only the two most northern explosions could be seen, as Private T S Williams also mentions just two sheets of flame.

5. The total number of British dead in the seventy-four days of the Falklands Campaign in 1982 was two hundred and fifty-five, and fewer than seventy of these were actually killed during land operations. The others were all killed in the air, at sea, on board ships hit by enemy action, or as the result of accidents.

6. Military Operations France and Belgium 1917. Volume II, page 152.

7. An artillery flag was usually made of red material, and was carried by advancing infantry units to place in the position of their furthest advance, so that FOOs or artillery spotter aircraft could note their progress, and not shell them by mistake. By necessity, it was a very haphazard arrangement, which often led to tragedy.

8. See bibliography.

9. The British Admiral Byng was executed on the quarterdeck of his ship in Portsmouth harbour on 14 March 1757, following his failure to engage more closely, the French fleet, under Admiral La Galissoneire, during the Battle of Minorca. French philosopher and playwrite Voltaire described the execution through one of the characters in his novel 'Candidc', written in 1759. 'Dans ce pays-ci (England) il est bon de tuer de temps en temps un amiral pour encourager les autres.' Roughly translated this means 'In this country it is thought well to kill an admiral from time to time, to encourage the others'. The phrase has become a cynical condemnation of the need to set a punitive example, ever since.

10. Private Redhead was attached to the Royal Engineers for about fifteen months, from after the Battle of Flers, to the German March offensive of 1918.

11. For the sake of clarity, the 2nd City Battalion will continue to be known as the 18th Battalion in the text, unless there is a reference to a former member of the Lancashire Hussars, in which case the amalgamated title will be used.

12. Private McLeish is obviously mistaken about the date.

13. This was Lieutenant-Colonel W R Pinwill, known as 'Puss in Boots' to his men, because of the riding boots he used to wear, even in the trenches.
14. Scottish Wood was a small wood on the Kemmel to Ypres road, at Elzenwalle, about a mile from Dikkebus.
15. Whale oil was issued in the front Line for soldiers to rub into their feet, to prevent them getting a condition known as 'trench foot'. This condition, which produced puffed flaky skin and swelling, which in extreme cases, could turn to gangrene, was brought about by long periods of immersion of the feet in water, or excessive damp. It was, in fact, a military offence to contract trench foot, as regular massaging with whale oil did prevent it.
16. Vauban was a brilliant Engineer General in the 17th Century French Army of Louis XIV. It was said of him that any town he besieged had to fall, and any town he fortified was impregnable.

Chapter Eleven

The others lay silent and prone in the trench, or around it ... the army order had been obeyed'

End of the 20th Battalion

Despite the fact that all the offensive moves of 1917 had been made by the Allies with more than even success, the dawning of 1918 brought no feeling of euphoria. The blood-letting on the Ypres Salient, with its limited victory, had taken away any real thought that the end of the war was near, and Sir Douglas Haig was still doubtful about the ability of the French Army to withstand an all-out assault by the Germans. On top of this, the situation in the east was uncertain. It was looking more and more likely that, after the Bolshevik Revolution of October 1917, the Russian Army would collapse and that would free some 400,000 Germans, to fight in the west. Haig was still in favour of continuing the offensive spirit in the west, to give the Germans no time to rest and reorganise, but for this he needed manpower, especially after his colossal losses at Passendale, which had left his divisions in the field short of about 70,000 men. It was true, however, that the Germans also had been hit very hard in the Salient; in fact they had been relentlessly ground down ever since the Battle of Verdun. Thus Haig and many of his staff believed that one great final blow would break them. He also argued that even if Russia did drop out of the war, her loss would eventually be more than compensated for by the arrival in France of the full force of the American Army.

Haig, however, was not in control of Allied or even British war plans. We have already seen that the British Prime Minister, David Lloyd George was not a supporter of Haig, and in fact openly tried to obstruct the Field-Marshal whenever he could. Although he could not always thwart Haig's military strategies and operations in the field, he could control, through the War Cabinet and the Army Council, the supply of men sent out to the Field Marshal at the front. This he had managed to do since the end of the Passendale battle, and this was the reason why Haig's divisions in the field were under strength. Quite simply, on Lloyd George's instructions, reinforcements to replace casualties were not being sent out from England. This might seem like an act of treason to modern readers, and it is just possible that the Prime Minister was acting out of humanitarian considerations. However, it is more likely that the wily politician, whilst wanting to be on the winning side, did not want to be tainted with the accusation of being a victorious butcher in any post-war political power struggle. It is an ironic tribute to his duplicity, but not to his honesty, that he managed to pass on this title to Haig himself, totally undeservedly, in the years that followed the war. In Lloyd George's defence, he probably thought that if Haig's assessment of German weakness was correct, then it was better to do nothing to incur huge casualties until the Americans arrived, after which total superiority of numbers over the enemy could definitely be achieved.

British battlefield dead.

Thus, at a meeting of a newly formed Supreme War Council at Versailles, in December 1917, Lloyd George agreed that the British Army should take over twenty-five miles of the front line from the French. From Haig's point of view, this stretched his already weakened forces still further, but from the point of view of Lloyd George, it not only showed him as anxious to act in accord with Allied needs, but it also made it most unlikely that Haig would be able to make any major strike against the Germans in the west. From the point of view of the Liverpool Pals, as we have already seen, it meant a move from the dreaded Salient and taking over the line in an area of unprecedented peace and quiet.

We went into our new piece of the line on the 29th of January, and the difference between that and our piece that we had left behind at Ypres was simply laughable. Here there was absolutely not a shell hole. It is true that all the villages had been destroyed, but this was done when the Boche evacuated this territory, and was brought about by mines and fires. Our front was enormous – in fact, the Divisional front was something between eleven and thirteen thousand yards; but we could afford to hold this line lightly when one sees that we were separated from the Boche by a 'No Man's Land' in most places of a width of not less than a mile, and the whole of this was taken up by a river and canal and the rest of it floods. In fact, except at one or two points, it was quite impossible for the Boche to get at us, added to which the French had put up masses and masses of wire. I have always found that whenever we are far away from the Boche, the wire is always excellent, for obvious reasons.

We had one of those places which the French call a Point de friction in front of us, in the shape of La Fere, which was an old Vauban fortress, but even here, we were separated from him by the river and canal, over both of which the bridges had been destroyed. The French whom we relieved, positively hated the idea of going out of the line. They had been there for six months, and had made themselves extremely comfortable. During the whole of their six months, the French Division had only had twenty-seven casualties, of which nine had been caused during a raid. No wonder that they were happy in their nice quiet spot!

Brigadier-General F C Stanley, CMG, DSO.

Haig was still short of manpower at the front, however, and his depleted and exhausted divisions were unable to carry out their duties with their customary efficiency. Even holding the line became an onerous task. Probably in an attempt to force Lloyd George's hand and get him to release the trained men still held in England, Haig had suggested after Passendale, in November 1917, that the number of divisions on the Western Front should be reduced, by the simple expedient of disbanding some, to make up the numbers in the divisions which would be retained. This would have left a shortfall of some twenty-five per cent of the Army in the field, and the resultant gap that this would have left would clearly not have been acceptable to the Allies. Not surprisingly, the War Office refused to accept this plan, but also still refused to send out the home reserves. Clearly something had to be done to avoid the exhaustion and possible collapse of those in the front line.

Thus, on 10 January 1918, the Supreme War Council at Versailles decided to 'solve' the problem by reducing the number of fighting battalions in each of the 47 Infantry Divisions on the Western Front from twelve to nine, and distributing the

The Americans arrive! Commander-in-Chief of the US Army, General Pershing, (left), and Commander-in-Chief of the US Navy, Admiral Sims (third left) are greeted at Euston Station, London, by Mr Walter Hines Page and (right) Lord Derby.

surplus men amongst those retained. Haig was ordered to effect this change as soon as possible, which would mean the disbandment of one hundred and fifteen battalions, the amalgamation of thirty-eight more into nineteen new ones, and the change of role of a further seven from infantry to pioneers. The War Office also ordered that no Regular, Guards or first line Territorial Force Battalions should be included in those disbanded, so the cuts would have to come entirely from units of the New Army. Thus, the newly constituted battalions would be up to strength, even if the number of men in each division was significantly reduced! In theory at least, this was not a catastrophe for Haig, as the Germans and the French had both adopted the same format for their infantry divisions some time earlier. However, the Germans had used their surplus trained men to form new fighting divisions, and the French had been able to reduce their front line commitment by bequeathing areas of it to the British.

Haig's men were now being asked to do the same job in the front line and take on extra territory as well! Although some vague promise was made that the disbanded battalions would be replaced eventually by American troops, this never materialised, if only because American Army policy was to be that American troops should be led only by American officers.

Thus, the 30th Division was ordered to get rid of three battalions, one from each Brigade. The 89th Brigade was ordered to disband the 20th Battalion and to disperse its officers and men amongst the three other Pals Battalions. The 20th was chosen simply on the grounds that it was the last Battalion of the original Pals Brigade to be formed. A further change that was also ordered was that the 18th Battalion should return to the 89th Brigade, from the 21st, to replace the 2nd Battalion, The Bedfordshire Regiment, which was ordered to be reassigned to the 90th Brigade. It will be remembered that the Bedfords had joined the 89th Brigade in December 1915 to replace the 18th Battalion when it was thought that New Army battalions on the Western Front might need some 'stiffening' by Regular Army troops. Although the Brigade had formed a very happy association with the Bedfords, there was naturally great joy over the fact that all the remaining Pals Battalions were once more to serve as part of Lord Derby's Brigade.

In early February 1918, the 30th Division was taken out of the line so that it could carry out the complicated battalion changes, and for the 20th Battalion it was naturally a very sad time. The Battalion had only received its orders to disband on 1 February, and on 5 February 1918, at a special parade held at Chauny, it was addressed for the last time by its last Commanding Officer, Lieutenant-Colonel J W H T Douglas,[1] by its Divisional Commander, Major-General W de L Williams, and then by its Brigade Commander, Brigadier-General Stanley. All praised the fighting record, efficiency, hard work and esprit de corps of the Battalion, and thanked the men for their loyalty and devotion to duty. On the following morning the drafts for the remaining three Pals Battalions marched off to join their new units, and on 8 February 1918 the 20th Battalion effectively ceased to exist. It had never been overwhelmed by the enemy and now it had finally been defeated by bureaucracy. At least, however, its men would still continue to wear the insignia of the Eagle and Child.

The defensive zones

On 21 February 1918, the newly constituted 89th Brigade went back into the line – this time opposite St Quentin, where the whole Divisional sector was overlooked from the lofty heights of the cathedral, most of which was still standing. The 17th Battalion was commanded by Lieutenant- Colonel C N Watson, and the 18th Battalion was commanded by Major G M Clayton, who had taken over from Lieutenant-Colonel W R Pinwill, after the latter had been appointed to the General Staff of the 32nd Division, on 12 January 1918. Both Pinwill and Clayton had been awarded the Distinguished Service Order in the New Year's Honours List, and Clayton was promoted to Lieutenant-Colonel, in April 1918. (The 18th Battalion officially joined the 89th Brigade on 11 February 1918.) The 19th Battalion was commanded by Lieutenant-Colonel J N Peck. It will be remembered that he had been in command of the 17th Battalion until sent home by General Williams in July 1917, but after the 19th Battalion's Commanding Officer, Lieutenant-Colonel G Rollo had been wounded by shellfire on 1 December 1917, Brigadier-General Stanley specifically requested that Peck be brought back from England to take over command of the 19th Battalion, and General Williams had acceded to his request.

By the end of February 1918, no one at the front was under any illusion that the enemy intended to attack some time in March. Logically, if the Germans were to win the war, the battle had to begin soon whilst they had superiority in numbers after the collapse of Russia, and before the American Army arrived in enough force to redress the numerical imbalance. The Allies knew the attack was coming, but did not know exactly where the blow would fall. The Fifth Army, under Sir Hubert Gough, of which the 30th Division was a part was the most badly stretched of all the Armies on the Western Front because it had only twelve Infantry and three Cavalry Divisions, to defend a frontage of forty-two miles. Moreover, the effective trench fighting strength of the three Cavalry Divisions was only that of one infantry brigade. The reason why the area around St Quentin was so poorly defended was that until almost the last moment before the Germans attacked, the British General Staff did not believe that the blow would fall on the Fifth Army Front. In any case, if the attack did come on the southern-most part of the British line, Haig had reasoned that it would be easier to contain it there, as there was a greater distance in which to manoeuvre between the front line and the sea coast. There was a distance of some ninety miles, for instance, between La Fere and the coast, and nothing of real military importance in between. Thus, it was militarily much more expedient for Haig to keep the main

US 'Doughboys' disembarking at Liverpool en-route for France.

defensive force of the British Army further north, to prevent the possibility of a Germans breakout to the coast. There, the distance between the front line and the sea was at most fifty miles. The dire consequences this might have brought, need not be imagined.

This was of little comfort to the men of the Fifth Army, however, overstretched to the extent that there was not even a continuous system of front, support and reserve line trenches, to which they had become so used during nearly four years of war. It was true that the nature of the countryside did not allow such defensive luxuries everywhere, but nevertheless, it was obvious to all that, if attack came, it would be impossible to hold any kind of front-line trench system for very long. Thus, instead, and probably taking a leaf from the Germans' Passendale book, the defence of the area was organised into a system of zones. The area known as the Forward Zone, or the Blue Line actually faced the enemy, but was to be lightly held, not necessarily with a continuous trench, but with a system of all-round fortified positions known as redoubts. It was hoped that the Germans would waste the main impetus of their attack on this zone, after which its surviving defenders could withdraw. The main defensive zone, the Battle Zone or the Red Line, was situated some two miles behind the Forward Zone, and was designed to be held at all costs. If the Germans managed to get through the Forward Zone, it was hoped that they could be contained, controlled and killed in the Battle Zone by troops who would have had time to be moved forward whilst the fighting was still taking place in the Forward Zone. Behind the Battle Zone was intended to be the Corps Zone, or the Brown Line, which was to be the final line of defence, should the need arise. This line was never completed, however, as the German attack took place before the defenders had time to construct it. In view of the fact that the Fifth Army line was so overstretched, Haig gave General Gough special permission to pull back, if the situation demanded it, to hold a new defensive line until British and French reserves could come to his aid. This fact was often conveniently forgotten by Gough's detractors after the battle was over.

It is doubtful, however, whether the men in the defensive zones would have derived any comfort from the fact, even had they known it, for they had been given their own instructions.

It was not long before the cushy time came to an end, and the alarm of the 'brasses'[2] as we used to call the Chiefs of Staff was communicated to the troops in the form of 'Instructions for Reception and Defeat of the Enemy.' Exhortations were made such as 'You will go into your positions and

will not retreat. You will be either killed or taken prisoner. There is to be no retreat, and anyone retreating will be fired on as enemy.' This came out in Battalion Orders one night! *Our time was occupied in and out of the line without much disturbance by Jerry, and we dug the most grandiose scheme of theoretical trench system that the planners could think up. Unfortunately. there were not enough men to do more than half the work required, and although the keeps and special strongpoints were excellently constructed, the rear lines of the Battle Zone proved to be death traps. Trenches three feet deep and six feet wide were dug, and, believe it or not, when the troops manned them they would then have to dig another three feet down in the middle, to provide themselves with a six foot trench, with a firing parapet on each side. That they would have the time or tools to do this was possible only in theory. The practical organisation of it was conspicuous by its absence, and that was not unnoticed by the troops themselves at the time.* **17518 Corporal E G Williams, 19th Battalion KLR.**

It is obvious, however, that it was not just the rank and file who had serious misgivings about the paucity of the defensive arrangements around the St Quentin sector.

Shoulder strap of a soldier from the German Reserve Infanterie-Regiment Nr.201. It was men from this unit that raided the 17th Battalion's positions on 5 March 1918.

As early as the beginning of March we were expecting him to make his attack. The 2nd March had been named as a likely day. On the whole the weather was not bad and did not hinder us unduly. We had bad spells but this country dried up very quickly.

It was a very anxious time indeed for the battalion holding the line. Our system was that one battalion was holding the whole Brigade front. It could not be considered anything else than a lightly held outpost line. They had a few posts out in front, about six in all, and each of these posts consisted of about six men. Behind this, we had a series of other posts, and again, behind these, a couple of strong points. This absorbed two companies of the battalion. Then there was one company which was detailed for counter- attacking purposes, and the fourth, and last company of the battalion was responsible for the garrison and up-keep of a redoubt called the Epine de Dallon. Here also was situated the Battalion Headquarters.

For any further defence, one had to go back to what was called the battle zone – a distance of some two or three miles back. This was a series of posts and strong points, which had been dug within the last month, and in our case was to be manned by the remaining two battalions at our disposal.

I might say that the Division had two Brigades holding each of them a sector, whilst the third Brigade was in Corps reserve. From this it will be seen, therefore, that if the enemy attacked in sufficient strength, there was no possibility of helping the battalion which was holding the line. In the first place, the distance to them was a great one, and in the second place, there were not the troops available to send to their assistance without depleting the already very thinly held defences of the battle zone. Therefore, the battalion in the forward zone knew that, if it was their fate to be in that zone when the Boche attacked, there was very little hope of support indeed for any of them. The officers and men were splendid, and their spirits were of the best. They were determined to give the enemy a real warm reception should he attack during their time, but, at the same time, it was anxious work, and we all knew it.

As time went on, it became more and more apparent to us that the enemy were going to attack on our piece. Day after day, we could see in their line, officers with maps, busy with their glasses, looking at our piece. We were a little bit relieved, however, by the fact that he did not appear to be registering any new guns, but this does not appear, with him, to be of the same necessity as it is

with us. He relies very much on the use of trench mortars, as was seen in a little incident which took place on our line at this time.

He suddenly, in the middle of the night, opened a very heavy trench mortar bombardment, and raided one of our posts. They fought most excellently, and it was not until the whole post, i.e. eight of them, had been wounded, that they succeeded, and were able to take away one of our men, who was wounded. On the other hand, we were able to take one of their men prisoner. It was unfortunate, however, that this man died within a few hours, because the identification had been particularly asked for by the Corps, and we had been straining every effort to get one for them. This is mentioned so as to show to what use they put their trench mortars. During a space of about three-quarters of an hour, they cut our wire in about twenty or thirty places, in addition to which they brought to bear a very heavy bombardment on the trenches and also on Company Headquarters.' **Brigadier-General F C Stanley, CMG, DSO.**

General Erich von Ludendorff.

The raid to which General Stanley refers, took place on 5 March 1918, on the 17th Battalion's positions. It was made by about a hundred and fifty of the enemy, who were in the British trenches for about ten minutes, before being repulsed. Although the prisoner Stanley mentions may have died, his unit was positively identified, as Reserve Infanterie-Regiment Nr.210, from Pomerania, a unit raised, like the Pals, especially for war service, and part of the German 45th Reserve Division. The raid was an obvious precursor to the main event, which would begin just over a fortnight later on the morning of 21 March. At this time, the 89th Brigade was in Corps Reserve behind the Battle Zone on the right sub-sector of the line near St Quentin. The 17th Battalion was at Villers St Christophe, the 18th Battalion was at Aubigny, and the 19th Battalion was at Dury.

Breakthrough

The German plan for breakthrough was quite simple. Masterminded by General Erich von Ludendorff, the intention was to strike a sudden and decisive blow at the Allies which would be so devastating that it would create a huge gap through which the German Army could strike out to the Channel coast with one massive thrust. Once the Germans controlled the Channel ports, it was hoped that the French would collapse, and the British and Americans, unable to land or evacuate troops, would have to sue for peace. For this attack, which would be the last chance the Germans would have of winning the war, with the Americans arriving in ever-increasing numbers, and their own war supplies running dangerously low, Ludendorff had planned four main simultaneous assaults, by four armies, from north to south. The first was to be made at Arras, the second at Croiselles and Bullecourt, the third at Bapaume and Peronne, and the fourth at St Quentin. Because of the enormity of the front and the scarcity of war materiel, all four attacks could not be made with the same pressure at the same time, and so the main impetus was to fall on the Fifth Army front, where the Germans knew the line was thinly held, and where they hoped to split the British and French forces and so exploit the gap. Having learned the lessons of 1917, the Germans also brilliantly changed their tactics to

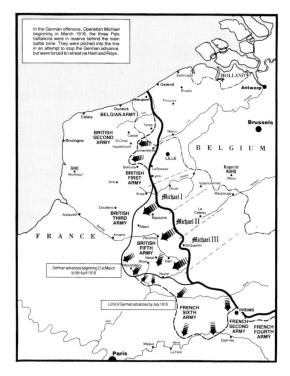

In the German offensive, *Operation Michael* beginning in March 1918, the three Pals battalions were in reserve behind the main battle zone. They were pitched into the line in an attempt to stop the German advance, but were forced to retreat via Ham and Roye.

German infantry advancing during the early stages of the March 1918 offensive.

achieve surprise and guarantee rapid success. For months they had trained special squads of 'stosstruppen' or storm troops, who, lightly armed with bombs and automatic weapons, and covered by devastating and rapid bombardments of the enemy lines, would swiftly punch a small but deep gap in the enemy positions. Once through the gap, the stosstruppen would fan out right and left to surround the enemy and wipe them out. They in turn would be followed through the gap by more troops and would relentlessly push forward to the next enemy position, where the operation would be repeated. The code name for the beginning of the March Offensive, sometimes called the Ludendorff Offensive was to be 'Michael', and as a result, this part of the fighting in 1918 is sometimes also referred to as the 'Michael' Offensive. This early method of 'blitzkrieg',[3] was to be in marked contrast to the way that attacks had been made so far in the war, largely unsuccessfully, when masses of troops in extended lines had attempted to punch wide gaps in the enemy's positions to achieve a breakthrough on a wide front.

The weather was certainly on the side of the Germans, for on 20 March 1918, it rained, with low cloud cover, which prevented air observation of the German lines. This enabled the German Army to move the stosstruppen forward into their assembly positions for the coming assault. Prior to this, the front line had been held by ordinary troops, whom it was not intended would be used in the attack. This was done for two reasons. Firstly, the stosstruppen would be completely fresh and unwearied by trench duties, and secondly, as the trench defenders knew nothing of the attack plans, the security risk was reduced, should any of them be captured in a trench raid. It is interesting to note, therefore, that the prisoner captured by the 17th Battalion on the night of 5 March, was an exception to this rule. His unit, a Storming Company of R.J.R. 210, had been specifically brought into the line just to make the trench raid, and his regiment was one of those specially trained to make the attack on 21 March![4]

By the early morning of 21 March the mild spring weather had combined with the light drizzle to form thick impenetrable fog, which further helped the Germans, at least at first, as it meant, incredibly, that the build up to their attack was still concealed from the British. Then, at exactly 04.30 am British time, a devastating concentration of artillery shells, a lethal mixture of high explosive and poison gas, began to fall, without any warning, on the British positions.

At 4.30 am I was sleeping very lightly, and heard a good deal of banging, so I was off downstairs like knife in very scanty attire, to the telephone. I rang up the Division and was answered by Brockholes, who said that the order to man battle stations had just come through. It was a pitch dark morning, and there was a thick mist. When it became light there was such a thick fog that one could not see twenty yards.' **Brigadier-General F C Stanley, CMG, DSO.**

Out at rest, as we then were, the usual inspection parades had taken place, with the usual stresses on the troops. Reminiscent of the bricks without straw of the Israelites in Egypt, we had to blanco[5] our webbing equipment without blanco. None of the local earthy substitutes was considered good enough, the weather was rough and wet, and the webbing equipment was hanging about around the barn billet, trying to dry for the next day's parade. Perhaps some psychic twinge motivated the senior corporal of the Company, that's myself, to get the equipment put together before night fell,

These four soldiers of the 18th Battalion's Transport Section pictured in 1918, all came from the little village of Bickerstaffe, near Ormskirk, Lancashire. They are: standing 17241 Sgt W E Adams, 21860 Pte J Berry, seated 2l9l5 L/Cpl W Heyes and 21912 Pte J Rosbotham. Note the farrier's horseshoe badge and the chevrons denoting two years' overseas service on Rosbotham's right sleeve, and the good conduct chevron on his left.

despite the grumbles, but there had been a pay parade and there was a Brigade canteen available, so all went to bed with full stomachs – luckily, as events turned out. My own supper, shared with my brother, was porridge oats, condensed milk and tinned strawberries.

The dark of early morning was broken by the crash of massed guns, which even some five or six miles behind the line was sufficient to penetrate the depths of sleep. Within twenty minutes, the Battalion was up and on the road in thick mist only some minutes before the long-range shells fell in the village behind them, as they marched away. By the time that the eerie march through the fog with the whistle of overhead long-range shells at frequent intervals, had ended, we found ourselves at the top of the hill looking down on the country just behind the Battle Zone, and a mile or so from its rear positions to the north of Roupy. **17518 Corporal E G Williams, 19th Battalion KLR.**

By 5.00 am, all three Pals Battalions had received the order to 'Man Battle Stations', and the 17th Battalion moved off for Beauvois, the 18th Battalion for Germaine, and the 19th Battalion for Vaux. Although all three battalions got there before their allotted time, no orders came through for them for the rest of the day, and no one knew what was happening up front. By midday the Germans had in fact, broken through to the Battle Zone, and by the mid-afternoon, the L'Epine de Dallon, which had originally been part of the 30th Division's defensive line, had fallen. Few troops who were in the front line when the attack began had survived, but most had sold their lives or their freedom dearly, which allowed those behind them to continue the fight the following day. It was only sheer providence that had found the 89th Brigade in Corps Reserve when the initial blow fell. When the situation became a little clearer on the evening of 21 March, the three Pals Battalions were split up and ordered to move forward to support other brigades. The 17th Battalion was to join the 184th Brigade of the 61st Division, which had been on the 30th Division's left flank, and at 10.30 pm it moved forward to the village of Atilly, two companies taking up positions in a railway cutting north-west of the village, and the other two in counter-attack positions near a copse known as Sword Wood. The 18th Battalion was assigned to the 90th Brigade of the 30th Division and proceeded to dig in south-west of Vaux, and the 19th Battalion was ordered to move forward from Germaine at 11.30 pm, and to mount a counter-attack on the newly captured German positions at Roupy the next day. 22 March 1918 would certainly be a day of destiny for the Pals which would rank alongside Montauban, Guillemont and Flers.

22 March 1918

The weather on the morning of 22 March was very similar to the previous morning – thick fog covered the battle front, which still favoured the attackers, who at least knew the direction of their advance, even if they did not know exactly what lay ahead.

As day broke, the 17th Battalion was in position at the village of Atilly awaiting a German attack that did not materialise. As a consequence, at 6.30 am 'D' Company was sent to find the 2/5th Battalion The Gloucestershire Regiment to help make a counter-attack on the Germans positions in the area of Holnon Wood. With the help of 'D' Company, the counter-attack was successful. As there did not seem to be much activity at Atilly, at 10.30 am the Battalion was ordered to move to a position known as Aviation Wood, immediately south of Fluquieres, to join the Headquarters of the 21st Brigade. Largely because of heavy shell fire, this move took four hours, and by the time the 17th battalion had got there the Headquarters and staff of the 21st Brigade had gone. About this time the enemy commenced a strong attack on Fluquieres itself, and the Battalion deployed around Aviation Wood in case the Germans broke through. Eventually, at 6.30 pm, orders were received that the 30th Division was to withdraw to the town of Ham, and the Battalion accordingly pulled back, reaching Ham at about 9.30 pm. Its casualties that day amounted to Second-Lieutenant Barnes and eight other ranks killed.

The 18th Battalion meanwhile, was established south-west of Vaux. At about 8.00 am No.1 Company, led by Captain J S Edwards, was ordered to proceed north to the village of Etreillers, to reinforce the 2nd Battalion The Royal Scots Fusiliers, which was dug in there. There they remained, engaging the enemy until late in the afternoon, when the order to withdraw was given. Shortly after No.1 Company left, Lieutenant-Colonel H S Poyntz, the acting GOC of the 90th Infantry Brigade called for a party to carry small arms ammunition and grenades to the 2nd Battalion, The Bedfordshire Regiment, which was helping to hold a fortified position known as Stevens Redoubt, on the northern edge of the Battle Zone. As a consequence, at 9.00 am, fifty men from No.4 Company, under the command of Second-Lieutenant J A Fisher, were despatched for this task, and when they arrived, they were retained by the officer commanding the redoubt, to help in its defence. At about 10.00 am, another carrying party, also from No.4 Company, was also sent forward, led by Second-Lieutenant H Derbyshire. By this time Poyntz had heard that the Germans had broken through on the left of the northern forward defences, and called for two companies from the 18th Battalion to make a counter-attack led by the Bedfords' Commanding Officer. As a result, at about 10.30 am No.2 Company under Captain J Lawson, and No.3 Company, under Captain F M Sheard MC were sent forward, but when they got there, they discovered that the situation was so serious, that a counter-attack was out of the question. The two companies were then retained for the defence of the Redoubt and fought on there with the main garrison until mid-afternoon.

The temporary Commanding Officer of the 18th Battalion at this time was Captain R P Villar, who had taken over when Lieutenant-Colonel G S Clayton had left for England on a month's leave, on 4 March 1918. When Poyntz called for the counter-attack to be made, Villar decided to go to Stevens Redoubt himself, to study the situation,

Back in Bapaume. This photograph shows German troops marching through Bapaume in March 1918, exactly one year after they had abandoned it during their withdrawal to the Hindenburg Line. The Allies recaptured it five months later, on 29 August.

and confer with its Commanding Officer. He was never seen again, and as his body was never found and identified, one can only assume that he must have been hit and killed by shellfire. By mid-afternoon it was obvious that Stevens Redoubt could not be held, and when both its flanks had given way, its commander ordered a withdrawal. Not all the 18th Battalion troops were able to retreat, however, and many were captured, especially from No.3 Company, which was evidently surrounded, as only about a dozen men from this Company managed to escape. Captain Sheard, who was taken prisoner by the Germans, later died in captivity from the wounds he had received during the action.

By late afternoon the situation had become critical, and after the Headquarters Company had destroyed all its signalling equipment and maps, the Battalion withdrew to Ham, arriving there at about 7.30 pm. As nothing further had been heard at all

German Uhlan cavalry advance past a captured British position near St Quentin on 22 March 1918. A dead British Lewis gunner, tunic and pockets rifled, lies below his gun.

from Captain Villar, the Adjutant, Captain F Lawless, assumed command of the Battalion, which then took up a defensive position just outside the town. Its losses for the day were quite heavy in terms of those wounded or taken prisoner, but only two officers and two other ranks are actually recorded as having been killed in action that day. They were Captain J Lawson, Captain R P Villar, 53741 Private R Driver and 300249 Private W H Farquhar.

The 19th Battalion, meanwhile, had the hardest task of the day, which was to counter-attack the German positions at the village of Roupy, most of which had fallen to the enemy, on the evening of 21 March. This attack was to be made at 01.15 am, on 22 March, from the edge of the walled cemetery outside the village, which meant that the men of the 19th Battalion had to approach in total darkness, with no real idea of exactly where the enemy was! Despite the night mists, there was enough light from the moon to make out some features, however, and the Battalion was eventually able to deploy in the original front trenches of the Battle Zone without having to engage the enemy. The subsequent events of the day were to leave no one alive or available to relate the true facts concerning the so-called counter-attack, and even the Official History is not accurate about what really happened,[6] but certainly by dawn the 19th Battalion was entrenched near the cemetery to await the inevitable German attacks which daylight would bring. Its Battalion Headquarters, shared with the 2nd Battalion, The Yorkshire Regiment, which had also been ordered to defend Roupy, was in a small fortified position situated near a group of houses south-west of the village and appropriately named 'Stanley Redoubt'.

Eventually, in the swirling fog and mists of the morning of 22 March, the long-expected assault began. The attacking infantry was the 5. Westfalisches Infantrie-Regiment Nr. 53, whose soldiers came from Cologne, and belonged to the German 50th Infantry Division. The Westphalians opened up the attack on defenders who were poorly dug in and poorly equipped for a long fight, yet were nevertheless determined not to be beaten. All morning the fighting continued spasmodically, with the main force of the attack being made in the early afternoon. Relentlessly and surely, the defence of the soldiers of the 3rd City Battalion, was worn down by the pressure and determination of the German attacks. Nevertheless they held out, despite the certain knowledge that they had neither hope of victory, nor hope of relief. At about 4.00 pm, the Commanding Officer, Lieutenant-Colonel J N Peck, realising that the situation was hopeless, gave the order to those who were able to hear it to withdraw to Stanley Redoubt, and just after this, he himself was wounded. Eventually, at about 4.30 pm, the inevitable happened, and the Germans finally broke through, having killed or captured the remnants of the defenders. Lieutenant-Colonel Peck and seven other officers were amongst those taken prisoner.

Also captured, with a bullet wound in his head, was the redoubtable Corporal E G Williams. His account of the action is reproduced here in full, not merely for its strength and intensity, but also because it is probably the only surviving account anywhere, of the last hours of the 19th Battalion at Roupy.

*17518 Corporal
E G Williams of
the 19th Battalion.*

The happenings of the 22nd of March remain in memory as disconnected episodes. The sense of time had vanished and all concentration wax focused on the immediate minutes. Only one's nearest companions in the trench made any impression – they and the enemv too, still enshrouded in the mist. The morning lightened and wore on. The first enemy attack on our front was repulsed by rifle and Lewis Gun fire – they came as shadowy forms, suddenly appearing in the faff close to the rifle sights, as we peered over the top of the trench. The silhouette of a man runs across the field of view – the target figure of a musketry course – 'Allow one yard in front for every hundred vards of distance and squeeze the trigger' – and up go his heels. Or one comes running straight on – 'Aim low and squeeze' – and down he comes, just dim shadows trying to kill us, especially the machine gun enfilading us from the Roupy Cemetery Walls, and the bullets hissing into the trench behind our backs as we pressed into our shallow cover.

Later there was commotion on the right flank, already enfilladed all morning by the machine- gun post in Roupy, a new young officer hurrying along the trench, shoulders hunched against the hissing bullets, his arm dripping blood. 'Lucky devil – he'll get away!" Looking back, I doubt it, for the infiltration patrols were already through on our right and behind us. Our right flank Platoon, No.3, held well in the hot spot, and no one got to us from that front – for a time at any rate.

Midday memory comes back of sunshine and clear vision, and in a lull, it is time to eat some of the bread ration, a third of a loaf, dry bread, but welcome, and with false optimism, I left some for tea. Our drink was cold water from the water bottle, and then turn again to the battle. Time had ceased to exist, only the warm sunshine, the clay of the trench and the yellow dead grasses. A few shells fall behind us. the first of the day to come to notice. There is a momentary black-out as a whizzbang[7] hits the parados[8] just behind, and I come to, to see the black smoke drifting away.

The sun is now behind us in the misty western sky, a round ball of orange red. The crackling sounds of musketry and machine gun come to ear, it is the fourth attack of the day. The Roupy machine gun no longer enfillades us, their comrades are breaking into the flank of our system with bombs. Somebody calls 'Jerry's in the Trench!' and a look down the wide traverse[9] shows a compact body of men some three hundred yards away down the open trench. A quick burst of rapid fire from the shoulder, and the hammer falls clicking on the empty breech, and a quick re-load allows a glance at the target – all gone. Snipers become active from the cemetery wall on our flank. The enemy are well through on our right flank, and our position in the wide shallow trench becomes a last stand, by half of No.4 Platoon, now in a trench facing this right flank. The other half of the Platoon has split up somewhere along the front trench, fighting it out on our left, very unequally with the enemy, who had broken in on that front.

No.3 Platoon has had it on our right, and gradually we also get it in the wide open shallow trench, some firing from a standing position, others more careful, kneeling behind the cover which the thin parapet affords. 'There is no retreat, you will stay in your positions until killed or captured!' – those were our orders. Ammunition is running short, in spite of the two extra bandoliers dished out last night, and Corporal Williams E G is shocked to find that he has now only five clips, that is twenty-five rounds, left, out of some three hundred or more rounds which he has fired in the preceeding hours. There is no reserve ammunition in the trench, and the bullet hail in the open is too thick to venture out to find some. The bullets are now ours, as well as theirs, for the enemy has been marked in our position.

Fragmentary memories come of falling men at one's elbow, of No.3 Platoon Sergeant, spread-eagled on the edge of a shell hole, a smile on his pale face, dead, and apparently untouched by blast, but caught by a bullet.[10] Then I heard a rattle of a fall close by, and someone said three or four yards back,'Poor old Dick's got it', and I looked round, and there was Dick Williams, with whom I had been on patrol, lying in a pool of blood, with the top of his head off.[11] I am just turning back to see what was in front of me, when about five million stars flashed, and I felt as if I'd been hit by a shell or an express train, with no pain, and I went off into a realm of indigo sky with millions of golden stars flashing about it. I remember thinking that I was killed, but I had a contented feeling

This German photograph shows the newly captured town of St Quentin in March 1918.

because the killing had been painless. This was the end of the last stand of 'A' Company[12] in the late afternoon of the 22nd of March 1918.

A patch of light appeared in the darkness in which I was swimming, and a pair of boots are seen at its edge. I am unable to move as I watch the slow dripping of blood from the end of my nose onto the red clay of the trench in the afternoon sunlight. I was slumped against the side of this shallow trench, creased across the forehead by a bullet which had left two neat holes in my tin hat, one to enter, the other to leave, having ploughed across the bone of my forehead with considerable shocking impact at close range. Shocked, I was just able to breathe, I was breathing shallowly, and there was the pair of boots on the edge of the visual angle. I knew that there was only one bloke with me in that bit of trench once Dick had got it. The feet at the edge of the window of light began to move away, but stopped when I made a kind of grunt, forced with willpower from the shocked paralysed casualty. 'So you're not dead', a voice says, and hands pull me upright and to my feet. My senses began to return, and I found a field dressing and he bound it around my head with the remark 'You're a lucky bugger!'

We retreated from the isolated forward position down the wide trench, no sniper fire now, as the patrols are through near us, but much machine gun fire from our guns. An officer and a few men rush round the traverse, and tell me to get away, which is not to my liking. Having lost rifle, and with returning senses, rage and aggressiveness induced by concussion makes me endeavour to get hold of other men's rifles, but I'm shooed away, and told to have sense, and get away. Out in the open, where the trench ends abruptly is too full of bullets to encourage just walking upright. Down on hands and knees, they are still too close for comfort, but stomach to the ground is fairly safe, and I crawled to what I thought was a known resistance point, behind our front lines.

A soldier in khaki, and without equipment, a Royal Engineer, gets up and holds up his hands in front of me, and with stunned surprise, I saw a lone German patrol walking forwards some yards away. There was nothing more to do but to get up and acknowledge capture by wave of hand, and loose off my equipment which was no longer of use for fighting. Out of one hundred men,[13] there were counted some eleven of us as POWs the next morning in St Quentin. The others lay silent and prone in the trench, or around it, except for, perhaps, two who went back wounded in the morning. The Army order had been obeyed!' **17518 Corporal E G Williams, 19th Battalion KLR.**

Incredibly, Corporal E G Williams' brother, Private T S Williams had also survived the trenches at Roupy and been captured, and there would be a rapturous reunion, quite by chance, the following morning in St Quentin. After spending the rest of the war in German prison camps, both would return to Liverpool in 1919, to be met with the rhetorical question from their father 'I hear you ran away?' Such was the power of the media and the way that the press treated Sir Hubert Gough and the stoical defenders of the Fifth Army!

After the forward defensive line had fallen, those men from the 19th Battalion who had managed to get back made another stand at Stanley Redoubt, but the ultimate result had to be the same. When the Redoubt finally fell at 5.15 pm the survivors fought their way through the encircling Germans and eventually reached Ham at about 2.00 am on 23 March. As well as those officers and men who were marched or carried off to St Quentin and captivity, another three officers and forty-seven men whose official date of death is given as 22 March 1918, would never leave the Battle Zone. The true figure for those killed at Roupy, however, is undoubtedly almost twice as high. The files and cemetery registers of the Commonwealth War Graves Commission list some thirty-nine men as having been killed in action on 30 March 1918, when the Battalion was actually safely on its way to St Valery sur Somme, and out of the area of fighting. Furthermore, apart from those whose bodies were not found and are commemorated on the Memorial to the Missing at Pozieres, all but two of these have burial sites at Savy British Cemetery, which is within a couple of miles of Roupy and contains most of the identified 22 March dead. Thus it would appear that, without definite information as to the exact date of death of these thirty-nine men, 30 March 1916, has been accepted as an arbitary one, and the true toll of 19th Battalion dead for the stand at Roupy is probably nearer a hundred.

Perhaps their stand had been futile from the start, but a magnificent stand it had been, and despite the criticism that would be heaped upon the men of the Fifth Army after the battle and the war was over, the stand at Roupy, and many others like it around St Quentin, probably cost the Germans the last chance they had of winning the war.

Retreat

Thus, by the early morning of 23 March 1918, the three Battalions of the 89th Brigade were in or around the defences of Ham. When the three Battalions had first been taken from him to help hold the line, Brigadier-General Stanley had been given the task of holding the town against the enemy, but as the military situation deteriorated, it was obvious that it could not be held for very long. Thus, as the remnants of the 19th Battalion were drifting into Ham, other units were already moving out, and the 89th Brigade Headquarters left at about 4.30 am. Volunteers were then called for to stay behind and try to delay the enemy as the Division began its retreat to the River Somme, and men from all three Battalions volunteered for this suicidal task. As the troops pulled out of Ham, vast amounts of stores were hastily abandoned, most of which would later fall into German hands.

When we got round the corner, there were some big hutments and they were full of stores, biscuits and chocolate and tinned food, and this officer wouldn't let us have anything, and we'd had nothing to eat for three days! I said 'Let's have some biscuits,' and he said 'You'll get Court Martialled for looting if you do!' So we walked past it. I heard later that Jerry got the lot! **16400 Private S R Steele, 18th Battalion KLR.**

Just outside Ham there was a very big canteen, run by the Army, and it was stocked with stuff, and of course the Germans hadn't had any food for some time, and the Colonel said, 'Destroy all the bottles of whisky.' But I think some of the chaps filled their bottles with it, 'and fill all you can with food'. I was eating chocolate for days, in case I got no food, I filled all the sandbags with choco-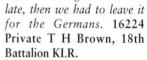*late, then we had to leave it for the Germans.* **16224 Private T H Brown, 18th Battalion KLR.**

The rations that we were supposed to have received to distribute to our men that day, the men who were in charge of them, the Army Service Corps, had left the whole lot on the roadside. So

British prisoners of war near St Quentin in March 1918.

as we went past, we could help ourselves to whatever we wanted, but we had no facilities to carry it. Our waggons were already heavily loaded, but I think we put an odd flitch or two on, but the Germans would get the rest the next day, or even later that day. **21915 Lance-Corporal W Heyes, 18th Battalion KLR.**

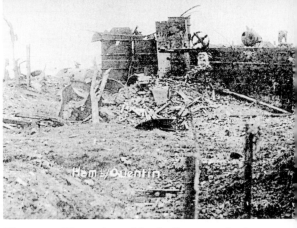

Eventually, the bridges over the Somme canal were blown, although not with total success.

I looked over the bridge, it was only a short bridge, over the canal, and I could see a man on the corner with a plunger. I went over to him and waited with him for a while, and eventually he said 'We'll have to blow it up', and he looked at his watch and he said 'Time's up', and he

The town of Ham, pictured by the Germans after its capture in March 1918.

pressed the plunger, and it just went 'puff', and he cursed like a trooper! So I went to see what had happened, and he said 'Don't go yet, it might be a delayed action one', so we waited a minute or so. and I said 'Shall I go and have a look now sir?' He said 'Yes, go and have a look'. I had a look, and it had blown the bridge up all right, but it was just leaningdown, hanging in the water, but you could still jump onto it from the bank, and troops could easily cross it. I went back and told him that troops could cross it quite easily, but artillery wouldn't be able to, but I'd seen them with pontoons anyway. So he packed up his things, and he got into his car and said 'Sorry I can't give you a lift, but it's against regulations.' So I said 'Best of luck, sir,' and he says 'You're the one that needs the luck!' So I started walking. **16400 Private S R Steele, 18th Battalion KLR.**

Ham officially fell to the Germans at 6.00 am, but German accounts state that the street fighting went on for another two hours after this, so it would seem that the remnants of the 89th Brigade fought on all this time to allow their comrades to escape.

Despite a daring counter-attack on 23 March by the 17th Battalion against the advancing Germans, which actually captured two machine guns, it was obvious that the front could not hold. As a result, gradually, over the next few days, the Brigade began to fall back, fighting actions with the advancing enemy all the time. It first moved back through Esmery Hallon, to Moyencourt and Libermont where it was almost surrounded and many men had to wade across the Canal du Nord to escape. Then by 25 March the French Army was moving up to give support and the Brigade was ordered to fall back on Omencourt and Solente, north of the Bois de Champien, to Roiglise, south-east of the town of Roye, which was reached at about midnight on 25/26 March. Despite the fact that the Brigade was virtually exhausted, having been without rest since 21 March, on the morning of the 26th it was ordered to take over some old 1914 French trenches on a defensive line from Bouchoir to Rouvroy, east of Roye. Roye had fallen to the Germans, by this time. On the following day, the line was strengthened by a minor withdrawal, and eventually on the 28th, after a German attack had been beaten off with determined artillery and small arms fire, the French 133rd Division took over the defensive positions and the Brigade was able to withdraw to Rouvrel, a village west of Moreuil and south-east of Amiens.

We got on to Ham, and just outside Ham was the first time that I saw an aeroplane, he came over us, hovering, and trying to see who we were. We gave a signal that we were British, and then we were nine days running back, until we got to Amiens. Nine days, and no one knew where we were. We used to dig in at night, and fall back again, retreat, retreat, retreat, because the flanks were giving way. Men didn't know where they were, we set stragglers posts, and we got back to just outside Amiens, and I think there was a river there, and I saw an officer with a revolver, telling our own men if they went back any further, he would shoot them. They'd been nine days running back, very little to eat, possibly, and then we were told to hold the line until reinforcements came up. The

At Rouvrel, on the morning of 29 March 1918, when it became possible to assess the approximate strength of the Brigade, Brigadier-General Stanley found that from all three Battalions, and including the combined transport sections and 89th Brigade Trench Mortar Battery, there were just under one thousand men fit for duty. Consequently, he decided to form them into one battalion, to be named the 89th Composite Battalion, with Lieutenant-Colonel G Rollo, recently returned to the front after his December 1917 wounding, as Commanding Officer, and Captain F Lawless, who had done such a fine job since Villar's disappearance, as Second-in- Command. It was assumed that this Battalion would be required to support the French, who were, at last, beginning to stem the German tide, but on 30 March the Composite Battalion was ordered to march to Saleux, where it eventually entrained for St Valery sur Somme. Once there, with the immediate danger over, the Composite Battalion was disbanded only two days after its formation. All three Battalions then began to reform their own strengths from those men who were left, returning stragglers and what new drafts were available.

Some of the signals section of the 18th Battalion pictured in 1918. They are: standing 16386 L/Cpl S Mole, 16400 Pte S R Steele, 16224 L/Cpl T H Brown, seated 16691 L/Cpl T Watson and Pte T Henderson. L/Cpl Watson is wrongly wearing his signaller's badge above the rank stripe on his right arm.

In the ten days from 21 March, twelve officers and two hundred and seven men from all three Battalions had been killed in action or died of wounds. On top of these figures for the dead must be added a total of some six hundred more who had been wounded or taken prisoner and would not serve with the Pals again. Many, of course, would die of their wounds in the weeks and months that followed. One very serious casualty, but without wounds, however, was Brigadier-General Stanley. Just after the Brigades's arrival at St Valery, General Williams ordered him to stand down from command of the Brigade. Officially, Williams had decided that he was too old for an active command in the field, but the timing behind such a decision does seem to be a little strange. If Stanley's age was not the reason for his dismissal, surely the performance of the Brigade after the German breakthrough could not have been the cause? Certainly, the Brigade had held its own extremely well since 21 March, and had performed its tasks honourably and stoically, incurring fewer casualties than the other two Brigades of the 30th Division. Also, Stanley was an experienced and reasonably successful Brigadier-General, and the British Army would need many more of these before the war was over.

Perhaps the 30th Division needed a scapegoat, or perhaps General Williams really did think that Stanley was too old. Whatever the true reason, which we may now never learn, one can only imagine how the Brigadier-General must have felt at relinquishing the command he had built up from nothing, and trained, hardened, protected, nurtured and led for over three and a half years. Only one statement made by him, hints that the real reason for his dismissal was something other than his advancing years.

It was a cruel blow, particularly as we all knew we had done really fine work during this retire-

ment, and that no troops could have done better. **Brigadier-General F C Stanley, CMG, DSO.**

If Stanley's dismissal was unfair, it would have been of little comfort to him to realise that he lasted longer in command than Sir Hubert Gough, who was also dismissed unfairly by Haig on the 28 March 1918. Thus, Gough became the Vth Army scapegoat for allowing the German breakthrough and seemingly the defensive miracles performed by some of his men were totally ignored![14] It would certainly have been more just if Lloyd George, or even Haig, had gone instead, for they were both certainly more blameworthy than Gough! Thus, although Gough's public reputation would be rehabilitated in later years, he and his Fifth Army had to take the immediate blame for what appeared, certainly at the end of March 1918, to be a debacle of major proportions. But it was really too early in the offensive for anyone to argue with this decision, and those who could have given the best argument of all, for both Gough and Stanley lay mute forever in the fields around St Quentin.

If the 89th Brigade thought that they had gone to St Valery for a well-earned rest they were sadly mistaken. By 4 April all three Battalions had entrained once more for their next destination, Elverdinghe, and the dreaded Ypres Salient once more, where, on 12 April, Brigadier-General R A M Currie took command of the 89th Brigade.

The Battle of the Lys

Just as the Pals returned to the Salient, Ludendorff disengaged the main force of his attack in the south, which had got nearly as far as Amiens, and on 9 April 1918, struck once again, on a line between Bethune and Ypres, with the intention of smashing the British once and for all, and taking the Channel ports. Within days his troops had reached the River Lys, and the normally taciturn Haig issued his famous order to his troops, *'Every position must be held to the last man . . . With our backs to the wall and believing in the justice of our cause, each one of us must fight on to the end.'* By this time, command of all the Allied Armies had passed to the French Marshal Ferdinand Foch, who fortunately had the remarkable ability to see all the fighting on the Western Front as one entity, and was thus able to command and unite all the troops at his disposal. Thus, Foch realised that if the Germans could be contained in Flanders, they would not have the reserves to maintain the offensive in that area for very long. As a result, although he sent some French troops to reinforce the British Army, he kept the main bulk of the French Army to defend the area around Reims, where he knew the next blow from the German Army would be bound to fall. This left Haig and his battle-weary troops virtually alone to withstand the German assault – and save Flanders and the Channel ports.

Marechal Foch.

The 89th Brigade had gone back into the front line on 7 April 1918, and on the 16th, because of German pressure outside Hazebrouck and Bailleul, the 18th and 19th Battalions were moved south by bus and attached to the 34th Division.

We were rushed up to Bailleul, and it was on fire. He started to attack through there, and we had to stem the attack, and this little village we came through,[15] it had been left entirely, the shops were full. Now our chaps hadn't had a change or a bath for a week, so they went into this ladies' place, and they were picking up knickers and chemises, and the Colonel said 'Go on boys, have a change while you can and get something clean on.' They helped themselves to the draper's shop.

There was a brewery there too, they took over a brewery, and the Colonel said that the men could have two or three glasses but no more – but Bailleul was in flames, I could see it at night time. Our signals headquarters was in a doctor's house, and they shouldn't have put us there, because it was a wine cellar. But it was silly drinking, because you had to have your wits about you, it wasn't worth while. **16224 Private T H Brown, 18tb Battalion**

The 17th Battalion followed the other two Battalions on 17 April, and then all three were sent into the line south-east of St Jans Cappel, a small town near Bailleul, to relieve troops of the 101st Brigade. The 17th Battalion was on the right, the 18th Battalion was on the left, and the 19th held in Brigade Reserve. Although there was no major infantry action during this time, patrols were sent out by both sides, which caused losses to both officers and men, and artillery duels were fierce. By this time the Germans had captured Bailleul, which, as Private Brown has remarked, was on fire and burning from British and French bombardments, and the German artillery was replying furiously, causing inevitable casualties to the Pals. On 20 April, Lieutenant-Colonel C N Watson DSO, commanding the 17th Battalion, was severely wounded by shell fire, losing one leg and having the other one badly shattered. He was evacuated to hospital, and eventually survived, and his place was taken by his Second-in-Command, Major J P Pitts, formerly of the Bedfords. He was eventually promoted Lieutenant-Colonel. On 21 April, with the immediate crisis over, the Brigade was relieved by French troops and left St Jans Cappel and rejoined the 30th Division at Busseboom on the Salient. It was not long before the Pals were in action again, however.

In the early morning of 25 April 1918, the Germans launched a furious assault on the French at Mount Kemmel and eventually captured it, and the small town of Dranoutre, the following day. For the first time in the war they were able to overlook the Salient from the Flanders Hills. At first, the Pals were mobilised to help the French, but when Kemmel fell, they were pulled back, and eventually sent up the line, in the early morning of 28 April to positions west of Voormezele, between Ridge Wood and Kruisstraathoek and the Brasserie on the road which connects Vierstraat with Ypres. The 17th Battalion was on the left, the 19th Battalion was in the centre, and the 18th Battalion was on the right. It was obvious that the Germans would soon try to attack the Scherpenberg, one of the Flanders Hills on their line of advance.

In the early afternoon of 28 April they mounted a furious assault on the unit to the left of the 17th Battalion, near Voormezele. As this unit began to fall back, the 17th Battalion also came into contact with the enemy, and for five hours, a front line hand-to-hand struggle took place. Eventually, Major Pitts ordered his reserve Company to make a counter-attack to clear the trenches, and this was scheduled to begin at 7.45 pm. However, the enemy obviously realised the intention, for at 7.35, a heavy bombardment fell on the British trenches, which lasted for some three hours, and halted any chance of a counter-attack. All the Battalion was able to do was to form a defensive flank to keep the Germans out. It was not a good omen! The following morning, at 3.00 am, the Germans opened their attack on the Scherpenberg with a terrific bombardment of high explosive, shrapnel and gas shells, which lasted for three hours. Most of the outpost lines all along the front fell at once and the 17th Battalion, seriously weakened from the previous night, soon fell prey to the enemy. Despite furious fighting, the Germans broke through on the left, and actually surrounded 'A' Company, whose men were overwhelmed. All were either killed or captured and the rest of the Battalion fell back until it could regroup.

The outpost lines of the 19th Battalion also fell in the initial assault, but the main body of the Battalion, expecting an attack after all the obvious signs of German offensive activity, stood firm, and the enemy was entirely repulsed along the whole of the Battalion frontage. Another attack was made at 8.00 am, but once more it was beaten back by the determined defenders. Throughout the day, the killing continued as the German survivors were

German troops sweeping through the streets of Bailleul on 15 April 1918, past hastily erected barricades.

picked off, as they tried to get back to their own lines. It was like Guillemont all over again – but this time in reverse! Lieutenant F E Sanders, who was in one of the outpost lines when the attack began, was later awarded a Military Cross for his action in bringing back thirty of his men.

The Germans also attacked the 18th Battalion positions, but this time the forward posts did not fall. The first assault, at 6.00 am, was directed on a listening post near Ridge Wood, held by men of No.4 Company. The post was initially captured and held by about twenty Germans, who began to move down the trench. Then, 16999 Company Sergeant Major G Button of No.1 Company leapt in amongst them, and single-handedly and with great gallantry drove them out with Mills bombs, killing and wounding some and putting the rest to flight. He was later awarded the Distinguished Conduct Medal for his action that morning.[16] After a strong defence along the rest of the Battalion's frontage, the Germans eventually withdrew. They continued their artillery barrage throughout the rest of the morning and afternoon, however, and at about 5 pm began to mass once more for another infantry attack. This was contained by rifle and Lewis Gun fire, however, until a British artillery barrage at 6 pm finally broke up all enemy activity.

By the end of the day the British line still held firm, the Germans had failed to take the Scherpenberg, and the Battle of the Lys was over. Although the Allied line was seriously bulged, Flanders was once more saved, as it had been in 1914 and 1915, largely by the stubborn defence of British infantrymen. Although Ludendorff would make one more assault on the French and the British at Soissons and Reims almost a month later, he would never again attack in great numbers in Ypres Salient. The Channel ports were safe for the British, and for the three million American servicemen who would eventually pour into France and tip the numerical balance back in favour of the Allies. On 1 May 1918, the three Battalions of the 89th Brigade were relieved in the line and moved back to Ouderdom and Busseboom. They would only fight one more action in the Salient, as their numbers were being ground down all the time, and replacements were just not available. The fighting for the Sherpenberg alone had cost the Brigade a total of over one hundred officers and men killed or died of wounds.

The end

On 2 May 1918, the Brigade moved to St Lawrence Camp at Brandhoek, where a head count revealed the fact that from all three battalions the total fit for active service was only twenty-seven officers and seven hundred and fifty other ranks. As a result, for the second time, one Battalion was formed from the three, this time to be called The 17th (Composite) Battalion, The King's (Liverpool Regiment). It was made up from the surviving members of all three Pals Battalions, and became a part of the newly formed 30th Composite Brigade, which in turn, was made up from the remnants of the 30th Division. The Battalion Commanding Officer was to be Lieutenant-Colonel G Rollo, DSO, who had just received the news that he had been awarded a Bar to his DSO, Major J P Pitts was to be Second-in-Command, with Major G M Clayton as Adjutant.

On the evening of 5 May, the Composite Battalion moved into the reserve trenches in the rear of a line from Ridge Wood to Klienvierstraat, and began to dig in. The 2nd Battalion, The Yorkshire Regiment, and the 2nd Battalion, The Bedfordshire Regiment, who made up the rest of the 30th Composite Brigade, were in front, the former on the left and the latter on the right. At 3.00 am on 8 May the German artillery put down an extremely intense barrage, and at 7.30 attacked all along the front,

Some more of the 18th Battalion's signal section in May 1918. They are: standing Pte Henderson, Pte Evans, Pte S R Steele, seated Pte 'Max' Pemberton, L/Cpl T Watson, and Pte Cheetham.

penetrating the Allied positions to a depth of some four to five hundred yards. It was decided that it would not be possible to mount a counter-attack until the evening, and as a preparation for this, the Allied artillery began to pound the newly captured German positions. It was intended that the counter-attack would be made by the French, and two battalions from the 19th Brigade of the 33rd Division. The 17th (Composite) Battalion was to stay in its reserve positions.

However, when the attack went forward, at 7.10 pm, Lieutenant-Colonel Rollo noticed that the two Battalions of the 19th Brigade had not moved forward to support the French, which left their flanks unprotected. Consequently, he moved 'A' and 'B' Companies of the Composite Battalion forward, (these were made up from troops of the former 17th and 18th Battalions, respectively), and with this support, the counter-attack was successful and most of the line was retaken. The following afternoon the Composite Battalion carried out a further attack, on a smaller scale, which succeeded in taking one officer and six men prisoners and capturing two machine guns.

Our last battle was really magnificent. The Boche had pushed back the other Brigade (Composite), and the French. The latter were to counter-attack in full daylight, with two battalions of the Brigade.[17] One of these was to go through my lines, but, somehow or other, it did not turn up. I had a feeling that something was going to happen, so I told my two forward companies (Henry's[18] and Williams') to be ready to assist with fire or men. When nobody came through to us, as we expected they would, over they went and did the job, and then, when I had got another two companies sent me, I thickened their line up, and we nibbled next afternoon, taking an officer and some men prisoners, and a couple of machine-guns.

We made a new line, and were able to hand over that night a complete show, with touch everywhere. It really was a pretty piece of work, most gallantly done with very tired troops, but showing extraordinary fine leadership on the part of the platoon commanders, who did the whole thing, including getting their men up to reinforce over 1,000 yards in daylight, in full view of the Boche on the ridge. **Lieutenant-Colonel G Rollo, DSO and Bar 19th Battalion KLR.**

The Composite Battalion came out of the line on the evening of 10 May and arrived back at St Lawrence Camp in the early morning of 11 May. It was disbanded that afternoon, its personnel going back to their original Battalions. In the nine days of its existence, it had lost four officers and forty-seven other ranks killed in action or died of wounds, but had done a magnificent job. It was the last time, however, that the Pals Battalions would serve alongside each other and was the end of Lord Derby's 89th Brigade.

On 13 May orders were received that the remnants of the

Western Front in 1918 showing extent of the German advance, and the Allied fight back to the final positions at the time of the Armistice.

three Battalions would be sent down to GHQ as training cadres, to train troops of the American Expeditionary Force, and all surplus fighting troops would be transferred to other battalions. These transferees left their Battalions for the Base on 14 and 15 May. On 16 May, the 17th Battalion moved to Meneslies to train the l/137th US Infantry Regiment, the 18th Battalion moved to Woincourt to train the 2/137th US Infantry Regiment and the 19th Battalion moved to Allenay to train the 3/137th US Infantry Regiment. On 19 June the Training Cadres of all three Battalions were transferred to the 66th Division, and then on 30 June, the 17th Battalion returned to England, transferred to the 75th Brigade, of the 25th Division. The delight of its members on returning to Blighty, would be short-lived, however, as fate would have other tasks for it before the decade was over.

On 8 August 1918, the 18th Battalion became part of the 197th Brigade, and then, on 13 August, it absorbed the 14th Battalion, The King's (Liverpool Regiment), which had recently returned from service in Salonika, and once more underwent a title change. It became The 18th (Lancashire Hussars Yeomanry), The King's (Liverpool Regiment). On 19 September, it joined the 199th Brigade. The Training Cadre of the 19th Battalion continued its task with the American Army, after its transfer to the 66th Division, until, on 31 July it was absorbed by the 14th Battalion KLR and effectively disbanded. Virtually all its personnel, however, would still serve as Pals when the 14th Battalion was itself absorbed by the 18th, a fortnight later. Nevertheless, it was a sad day for the few survivors of Knowsley, Grantham, Guillemont, Arras, the Salient and Roupy. Now there were only two Pals Battalions left to bear the badge of the Eagle and Child, although the newly reconstructed 30th Division, which no longer had any troops from Liverpool in it, still wore Lord Derby's crest as a Divisional insignia.

The 18th Battalion, meanwhile, had spent all of August and most of September in reorganisation and training. Lieutenant-Colonel G Rollo had been appointed Commanding Officer of the newly constituted Battalion, but on 29 September he left to take command of the 150th Brigade and his place was taken by Major R O Wynne, DSO. At about the same time, the Battalion moved forward to join the war again. Much had happened since it was last under fire. Ludendorff s offensives had finally ground to a halt and the German Army was itself 'on the run'. Haig, with foresight rarely credited to him after the war, had persuaded Foch that German morale had finally cracked, and that a victory was possible in 1918, if the Germans were relentlessly harried. Neither Foch nor the inexperienced, yet nevertheless forceful American Commander, General John J Pershing, had at first seen the possibility of victory until at least 1919, but Foch had allowed Haig to push forward and by October 1918, Haig's troops were once more up against the Hindenberg Line, whilst the Franco/American forces were attacking the enemy in the Meuse-Argonne area.

On 7 October 1918, the 18th Battalion arrived at the Hindenberg Line in the region of Bony and the following day moved into Divisional Reserve, to follow up the attacking troops who were relentlessly harrying the retreating enemy. By this time the war had become one of mobility and movement and the Battalion was constantly on the move. On the morning of 9 October, it advanced through L'Epinette, once troublesome machine guns had been dealt with by an armoured car, and by the following morning it was in position at Reumont, near Le Cateau. At 5.10 am, an attack was made towards Le Cateau which was also held up, at first by machine-gun fire and then by a battery of high-velocity field guns which were firing over open sights. Eventually, however, the Battalion was able to push forward to a position in a sunken road which crossed the Ruemont to Montay road, and by nightfall, was established in front of Le Cateau itself. The following afternoon, some men of 'B' Company surprised a party of some two hundred Germans in

30101 Pte T J Blincow of the 17th Battalion. On 3 September 1918, Blincow was part of a patrol which was suddenly fired upon by the Germans. Although most of the patrol members were hit, Blincow managed to dive into a shell hole, only to be captured by a German as he tried to climb out. Once in captivity, he discovered that the Prisoner of War Camp interpreter had once lived in Knotty Ash, near Liverpool, and when he discovered that Blincow had lived not far away in Huyton, he took him to a nearby barber's shop for a haircut and shave.

the main square of Le Cateau, and dispersed them with rifle and Lewis gun fire, killing about fifty. The same night the Battalion was relieved, and went into Billets at Reumont.

For the next month the advance continued, with the Battalion either following up the enemy or training for the next phase of the assault. Most men now felt sure that a victorious climax to the war was very close, and the idea of fatal casualties thus seemed even more tragic than at any other time during the past three years. On 3 November 1918, the Battalion moved back to Le Cateau, and on to Landrecies, where it expected to meet the enemy again, but Landrecies fell on the 5th, and the advance continued without making contact with the enemy. At 7.30 am, however, on 8 November, in what was to be its last action of the war, the Battalion was ordered to attack the enemy near Marbaix with support from the 25th and 100th Battalions of the Machine Gun Corps. The Pals met stiff resistance from the Germans, however, and although the attack was ultimately successful, Lieutenant H L Baker, MC, and fourteen other ranks were killed and three more soldiers would die of their wounds over the next few days. Perhaps their particular tragedy is that, apart from the fact that they almost made it to the Armistice, unlike the sacrifices made by their comrades in Prance and Flanders since December 1915, their deaths did not affect the outcome of the war in any way at all.

Three days later, on 11 November 1918, the Battalion was about to push forward from a position near a village called Clairfayts when at 10,59 am, Captain R West, the Staff Captain of the 199th Brigade, rode up with a message to the effect that all hostilities would cease at 11.00. Exactly one minute later the Western Front became silent.

> We had to turn out at 5 o'clock this morning to go to the Battalion, as they were going in action. The Battalion moved off at 7 o'clock just as if we were on a route march. We were the most forward infantry and really speaking were in the front line. Anyhow, we kept with the Battalion for about three miles until we came to a blown-up bridge that the infantry could cross but not the wagons. We took a by-road across the fields which took us a long way round through a wood.
>
> We had heard rumours in the morning that the war may be over at 11 o'clock, and the Battalion were told that they were not going to fight and if the enemy offered any resistance, they had to stop, but were not going to drop any mines. It was getting on for half-past ten as we were coming out of the wood, when we came across some Held guns that had been stopped, and told that the war was over. They were all shouting and telling us that the war was over and to take our tin hats off. When we caught the Battalion up, they were in billets and all the firing was over. **21913 Private J Stockley, Transport Section, 18th Battalion KLR.**

NOTES
1. Lieutenant-Colonel Douglas was a famous cricketer, playing for Essex and England, and having survived the war, he was tragically drowned, on 19 December 1930, after the craft he was in, was involved in a collision in the Baltic.
2. The 'Brasses' or 'Brass Hats', were so called because of the gold oakleaved braid worn on the peaks of their service caps. Every officer above the rank of Colonel was entitled to gold braid, and the more senior the rank, the more the braid that was worn. It was sometimes equally facetiously referred to by the troops as 'Scrambled Egg'.
3. The German word 'blitzkrieg' or lightning war, is more commonly applied to the devastatingly effective way in which the forces of Hitler achieved sudden and decisive victories in the early years of World War Two. Like the March 1918 offensive, it involved rapid and simultaneous attacks on the enemy, which pulverised his will to fight back, but in the Second War, using tanks, mobile infantry, artillery and overwhelming air support. The word 'blitz' has entered the English language to describe Hitler's bombing campaign of British cities.
4. This information comes from the War Diary of the 17th Battalion KLR, although the intelligence about the prisoner's unit identity is not totally accurate.
5. 'Blanco' is the traditional preparation in the British Army rubbed onto belts and webbing

equipment to clean them and make them look smarter. Although it came, and still comes today, in all shades of khaki, originally British Army belts and equipment were white, and the name 'Blanco' was a trade name, from the French word 'blanc' for white.

6. The Official History, page 200, states "in fact, at 1.15 am on the 22nd, some ground was regained by a counter-attack made by the 19/King's (89th Brigade) from the corps reserve, Lieutenant-Colonel J N Peck who led it falling at the close." In fact, far from "falling at the close", Lieutenant-Colonel Peck led the Battalion all day, until he was eventually captured in the late afternoon, when the Germans finally broke through. He spent the rest of the year in a Prisoner of War Camp, being repatriated in December 1918.

7. A 'whizzbang' was the name the British troops gave to any high velocity German shell, so called because its passage through the air was so swift, that unlike slower shells, which could be heard in flight, all that was heard from a 'whizzbang', was a sudden 'whizz', then a 'bang', as it exploded. Hated by the troops, the most common type was the 77mm, fired from a field gun.

8. In a well constructed trench, the defensive part in front of the occupant was the parapet, and the part behind, was the parados.

9. A normally constructed trench, viewed from above, was crenellated, like the battlements of a castle wall, to prevent an enemy being able to fire too far directly along it The forward parts of the trench crenellation were called firebays, and the backward parts were called traverses. In the case of the hastily dug Roupy trenches, there were was much greater length than was desirable in between each crenellation.

10. This was 17487, Lance Sergeant J T Topping, who came from Smithdown Road, in Liverpool, and was one of the original 1914 Pals.

11. This was 22896 Corporal R Williams, DCM, who came from Liverpool.

12. The change of Company designation from No.1 Company, No.2 Company etc., to 'A' Company, 'B' Company etc., seems to have taken place some time in November 1916, as in October 1916, the former is used, in the War Diary, whilst by the end of November 1916, and for the rest of the war, the latter style is used exclusively.

13. Corporal Williams is only counting the men from 'A' Company who had entered the Roupy trench system in the early hours of that day.

14. For a full account of Gough's career, and treatment after March 1918, see 'Goughie' by A H Farrar-Hockley.

15. This was probably the village of St Jans Cappel itself.

16. The citation for Button's award, published in a supplement to the 'London Gazette', on 9 September 1918 states, *For conspicuous gallantry and devotion to duty. During a violent attack by the enemy, a party of about twenty succeeded in entering one of our advance posts. This warrant officer immediately attacked them single handed with bombs, killing and wounding a number and recaptured the post. The enemy had penetrated at a most vital point and had he been given time to establish himself it would have taken a large force to eject him.* CSM Sutton came from Walton, a district of Liverpool.

17. This was the 19th Brigade of the 33rd Division, which Brigadier-General Stanley declined to name, when he published this letter written to him by Rollo, just after the action.

18. This was Captain N Henry, MC, who was killed during the counter-attack.

Chapter Twelve

'You didn't know which side they might come from, you didn't know where they were.'

Intervention

The remnants of the 17th Battalion were overjoyed when they first caught sight of Folkestone in the early afternoon of 30 June 1918. But some of them remembered leaving it in November 1915, when the Battalion had first set sail for France, and no doubt their memories dwelt on the events of the last three years and the comrades who would never see Folkestone or England again. With the war still in full progress on the Western Front, none of the Pals could reasonably have expected to have seen home for a long time, as leave had been cancelled when the Germans had broken through in March 1918, and the chances of making it home through a 'Blighty One'[1] were not very high. However, pleasant though it was to be back in England, the men were soon in for a shock they little expected. After a period of rest, re-inforcement and training at Aldershot and Clacton-on-Sea, news filtered through that they were about to go on active service again – in Russia!

The reasons for this northern move were, to say the least, complex. During the course of the war the

15790 CQMS J H Bracher, of the 17th Battalion who was awarded a Meritorious Service Medal for his devotion to duty in Russia.

Allies had provided vast amounts of military stores, mainly munitions, clothing and coal, with which the Tzarist forces could prosecute the war against the Central Powers. Most of these stores had been delivered to the north Russian ports of Murmansk on the Kola Peninsula and Archangel on the White Sea and through the sheer incompetence of the Russian forces, little of them had ever been distributed, let alone used. As a result, they remained in huge dumps on the quaysides. In the early spring of 1918, however, the military situation in the west was dramatically altered by the collapse of Russia and the impending Treaty of Brest-Litovsk between the Germans and the Bolshevik government. The Germans were in virtual control of nearby Finland, through their newly elected 'puppet' king, the former Frederick of Hesse, and at this time the Bolsheviks were quite happy to give away great tracts of Russian territory to secure peace. The reason for this was that Lenin, their leader, believed that all land would return to Bolshevik control very soon anyway, when the 'inevitable' workers revolution that had begun in Russia would spread throughout the whole of Europe. Thus, the Bolsheviks were on the point of accepting the Germans' proposal that they should cede the Kola Peninsula, on which Murmansk stands, to Finland. This would have given the German Army easy access to all the military stores, which if nothing else, would have seriously dented the effect that the Allied naval blockade was beginning to have on the German economy. Moreover, the stores had never been paid for by the Tzarist government, and the new Bolshevik government refused to honour the old debt, whilst at the same time claiming them as Bolshevik property.

Thus, notwithstanding the fact that there were upwards of two million tons of stores at Murmansk and Archangel that the British wanted to take back, it was possible that if the Germans had moved into northern Russia to capture the stores, they might have been able to operate U-Boats from Murmansk and break out from the Baltic into the North Sea. A further consideration was that in Siberia there was a force of some 50,000 Czecho-Slovak troops who had originally served with the Imperial Russian Army, and who were now anxious to get home. They had already been double-crossed by the Bolsheviks, and it was thought that if they could make their way to northern Russia, and more especially Archangel, they could swell the numbers of Allied refugees already cut off there by Bolshevik forces, and at the very least, form a barrier to the possibility of German expansion into the area. Furthermore, as Archangel was frozen in and cut off from the sea during the winter months, it became vital that both Murmansk and Archangel should be occupied by the Allies, for all these reasons, before winter set in. Thus, the purpose of what was later to be called 'Intervention', was not at first anti- Bolshevik, so much as anti-German and protectionist.

As a result, early in 1918, the Allies landed a force at Murmansk, which was at first welcomed by the local soviet, who invited the mainly British, French and American force to protect and maintain the area on its behalf, even handing over three former Imperial Russian Naval destroyers, to help in the task. Soon, however, the Bolsheviks, acting on orders from the interior, turned on the Allies, who replied promptly by occupying Murmansk, Archangel, and Onega, which was another port on an inlet of the White Sea. Soon thousands of Russians who were opposed to the Bolshevik cause joined the Allied force, although not all of them with total conviction, and by the time of the Armistice in Europe, the North Russian Relief Force, as it had became known, was firmly aligned against the Bolsheviks.

On 9 September 1918, the 75th Brigade, to which the 17th Battalion KLR was attached, was re-numbered the 236th Brigade and left the 25th Infantry Division, destined for other theatres of war. On 5 October 1918, the 17th Battalion, with a total of twelve officers and eight hundred and fifty other ranks, left Mytchett Camp at Aldershot, and entrained for Glasgow. Another twenty-two officers and fifty-six men would join them later. The following morning the main party embarked on the SS *Keemun* to sail for Russia. Twenty-four hours later, however, the Battalion disembarked, as the ship was not ready to sail. It then marched to Yorkhill Barracks, Glasgow, where it stayed until 10 October, when it embarked on the *Keemun* once more. This time the ship really was ready, and at noon she moved out of the dock and began her voyage north. Passing through the Hebrides and sailing by Iceland, she arrived at the port of Murmansk in the morning of 17 October and moored alongside a wharf there. On 20 October the Headquarters Company and 'B' and 'C' Companies transferred by tug to the troopship *Goentoer* under escort, for the voyage through the White Sea to Archangel. The following day, 'A' and 'D' Companies followed suit on board the troopship *Asturian*, all Companies having arrived at Archangel by 23 October. They would never actually serve together again as a complete Battalion, as each Company would be assigned to a different role in the northern interior.

We embarked in SS Keemun for Murmansk, (10th October) from Glasgow, arriving at Murmansk on 19th October.[2]

From there we moved to Archangel, arriving on 24th October[2], and from there to Ekonomiya. The Battalion were there made into two only fighting companies. I went with C Company up the Dwina[3] River to Beresnic[3]. From that day, on separation at Ekonomiya, our battalion never again reunited. **15755 Company Sergeant-Major E E Bryan, 17th Battalion KLR.**

We went into camp, and we got the shock of our lives, there were rumours flying about that we were going to India, and then we got kitted out for Russia, Russian rifle and everything – bayonet, but no scabbard. Then they locked us all up on a train and sent us up north. I can remember stop-ping at Preston – but no getting out – the doors were locked! Then we got to Glasgow, a Riding School in Glasgow, it was all red dust, we had to sleep on a red dusty floor for three or four days until the boat was ready – the Keemun, that was the boat.

We sailed from there on the 10th of October 1918, and we landed in Murmansk in about seven or eight days. We didn't get off the boat, we weren't allowed, we had to wait for three days, and then we moved out to the White Sea all the way to Bakharitza, which is similar to Liverpool and across the river to Birkenhead, its very very similar – a wide river, and there's Bakharitza, where all the sheds were, stores and everything, and Archangel across. We stayed in Bakharitza for a day or two before we moved to Archangel, and then we went back to Bakharitza, where the railway

started, the railway went south from there to Vologda, and that's where we started our adventures.
50752 **Private J Grogan, 17th Battalion KLR.**

In early November 'A', 'C' and 'D' Companies were billeted on board the 'Asturian', whilst 'B' Company was in Barracks in nearby Bakharitza. 'B' Company, led by Captain R G Smerdon, remained there on 4 November, when the rest went into barracks at Ekonomiya, a town north of the port. Two days later the Battalion suffered its first loss of the Russian campaign when 114148 Private G H Brown of 'C' Company died of pneumonia in the Barracks hospital. On 9 November 'B' Company left Bakharitza by rail, to join the Vologda Force at a place named Obozerskaya. By this time Archangel was frozen in, and Bolshevik activity was centered around the towns of Seletskoe and Tarassova. Seletskoe was about a hundred miles miles south of Archangel and Tarassova was about thirty miles further south still. On 14 November 'C' Company, consisting of two officers and two hundred other ranks, left Ekonomiya by rail to join the Dvina River Force at Bereznik, and on the 21st 'D' Company, consisting of two officers and one hundred and sixty other ranks left by rail to join the Vologda Force, but on temporary attachment to Seletskoe. Thus, with the exception of 'A' Company, which remained in Archangel, by early December the rest of the Battalion was in position, ready to repel any Bolshevik attempt to move north.

The military situation was by no means clear, however, and although there was not the constant threat of death or maiming that was ever-present on the Western Front, in Russia no one was ever sure who the enemy was, or where he was, and this brought a strange kind of nervous tension that the Pals had not experienced before. They were, in reality, fighting an early form of guerrilla warfare that the Russians would perfect during the winters of 1942 and 1943! In mid-December 1918, after putting under arrest the 1st Company of the Archangel Regiment for refusing to proceed to the front, 'B' Company was then ordered south by train to a position known as Verst 445, to relieve a company of the American 339th Infantry Regiment. As most of the front line was a wilderness anyway and there were few villages or settlements, outposts were simply labelled by the number of versts they were from the base. A verst was a Russian measurement, roughly equivalent to a kilometre, or about two-thirds of a mile. Thus Verst 445, on the edge of a forest, was roughly three hundred miles away from Archangel. The men of 'B' Company arrived in the post at 8.30 am, on 18 December and had just completed the relief by 11.00 when the enemy began to shell the front positions. The armoured train which had brought them then retaliated with shell-fire, and although the only casualty was a wounded Russian officer, the action was sufficient to cause disaffection amongst the Russians, and four of them deserted and eleven more were arrested under suspicion of desertion. As a consequence, the Russians were withdrawn from the front line. 'B' Company was relieved by French Colonial troops on Christmas Day, and went into reserve lines.

Meanwhile, 'D' Company-based at Seletskoe, had received information on 4 December 1918, that the Bolsheviks were preparing to make an attack on the town of Tarassova. As a consequence, they moved to the town on 5 December and 'stood to' to await the attack. This did not materialise, however, but a deserter from the Bolsheviks who surrendered to 'D' Company, revealed that the attack was in fact scheduled to take place at 7.00 am the following morning. With a daring plan, Captain E A Dickson, MC, decided to pre-empt the attack by ambushing the enemy in his own assembly positions. This meant that the Company, with a strength of two officers and seventy-six other ranks, and supported by thirty White Russians[4], had to circle the enemy positions at night in an unknown virgin forest without tracks and cover a distance of about eleven miles. Nevertheless, they managed to do it and by 6.25 am, found themselves about 4,000 yards behind the enemy position. They then noticed a blockhouse about a hundred yards away and they captured this without loss to themselves, killing

The SS Kermun of the Blue Funnel Line, which took the 17th Battalion to Russia, whilst in service as a troopship.

'Red' victims of the 'White' Russians.

seven Bolsheviks in the process and capturing four more. Soon after this, at 8.00 am, they spotted an enemy transport column on the road from Tarassova to Kochmas, and having attacked it, they were able to capture it intact. It consisted of eighteen limbers which contained two Maxim guns, small arms ammunition and provisions.

By this time, however, the Bolsheviks were alerted to the Company's presence and snipers in the forest began to cause problems. Nevertheless, the advance continued until the enemy's supply dump was reached, where a stiff resistance was encountered and the enemy lost many men before it fell. Twenty-two horses were captured by the Pals. By this time the main enemy position was only about a mile away and the opposition began to stiffen still further. Second-Lieutenant A Cousins was wounded at about 10.00 am and Captain Dickson tried to push home the assault as rapidly as possible. However, the enemy force was judged to be about six hundred strong and the Pals now came under quite fierce rifle and machine-gun fire. The White Russians were supposed to have attacked from the flanks at 6.30, and although they made a half hearted assault at about 7.00 am, they had not been seen since. Consequently, with ammunition running low and the Lewis guns jamming with compacted snow in their working parts, at 11.45 Dickson decided to withdraw. The party then killed most of the captured horses, wrecked the captured transport and retreated back into the forest with the two Maxim guns, seven horses and five prisoners. However, as it became impossible to move quickly with the machine guns, they were broken up and the bits thrown into the forest. The Company returned to Tarassova at about 4.00 pm.

Apart from Second-Lieutenant Cousins, four other ranks were wounded, 24938 Sergeant P M Greany MM, died from wounds received in the action, and 266299 Private C H Ainsworth, 33092 Private R Brown, 58354 Private J Houghton, 330381 Private A Owens and 114356 Private H J Turner were killed. Their bodies were never recovered and identified, and although their official date of death is given as 7 December 1918, this is a mistake, as the action took place on 6 December. All six are commemorated on the Memorial to the Missing of the North Russian Campaign at Archangel, and also on the Russian Memorial at Brookwood Military Cemetery, Surrey.[5] The War Diary states that one other rank was reported as missing in the action, and it is probable that this soldier turned up again alive and rejoined his unit, as no other casualties are listed for that day and the Bolsheviks were not in the habit of taking prisoners. Captain Dickson was later awarded the Russian Order of Saint Stanislav, Second Class with Swords, for his leadership during the action. Later, he decided to strengthen the defences around the Allied position by building blockhouses instead of relying on trenches and making regular patrols into local villages, and at the end of the month one section of 'D' Company made another attack, supported by machine-guns, on the Bolshevik positions on the Tarassova to Kochmas road. Once again, however, the troops of the Archangel Regiment refused to advance and the attack had to be aborted when the Pals were only about eighty yards from the objective. 114187 Private B R Sugden was killed in the raid and one other soldier wounded.

Company Quartermaster-Sergeant J Bracher kept a diary at the time, and his record confirms the events of the attack and outlines the Company's movement for the rest of the year and early 1919.

Dec 28th Zerkdvna attacked. Left Sredmekrenga by sleigh for Gora.
29th Gora. Issued Rations. Prepared and loaded convoy of 90 sleighs with shells, SAA,[6] rations, hay for stunt on Kotshmas. Proceeded with convoy at midnight to Terressova

30th Terressova i/c Convoy in advance on Kotshmas. Pie Sugden killed. Troops returned to Terressova. Convoy parked.

31st Established supply stores at Gooshsnea for Terressova district. Some New Year's Eve! - P – avec Sgt. Harrison and his cigars.

Jan 7th Xmas Day according to Greek Catholic Calendar. Bolshi Prisneck dashed around to district in fast sleigh. Visiting dance disturbed by Bolsheviks.

Jan 20th Terressova – Period of alarms and 'stand to's'. Left for Gora, rejoined Coy and proceeded to Sredmekrenga.

Jan 21st Left S/Mekrenga for Seletskoe. Seletskoe to on rest. Great sport on skis, some good Feb 4th nights in mess. Terressova and Gora lost.

4th Company rushed to Sredmekrenga, no sleighs for H.Q.

7th Rejoined Coy at Sredmekrenga.

12th Bolo[7] attacked 'Sred', some stunt, some 'wind up', some night.

26th Relieved by two companies of Yorks proceeded to Seletskoe (north end).

15790 Company Quartermaster-Sergeant J H Bracher, MSM, 17th Battalion KLR.

By the new year the weather conditions had begun to give as much trouble as the Bolsheviks, which made movement difficult in the vastness of the landscape.

In January 1919, we began a three-day trek on foot, (too cold to ride on droski[8] or sleigh) along forest trails towards Seletskoe, stopping two nights cramped in old blockhouses. It was so cold we could not thaw out our tinned rations. On this trek, we wore all the clothing we had, and the Balaclava helmets under our fur hats became frozen with breathing through them. Men not wearing Shackleton boots suffered frostbite and had to be returned to base.

On arrival in Seletskoe, we moved to Sredmakrenga, and into front-line blockhouses. Doing night sentry outside a blockhouse gave one a lonely and uncanny feeling, ever alert for the Bolsheviks whom we could not hear as they moved about in the perpetual snow of the silent forest. Compared to duty on the Flanders front this was another world. **50752 Private J Grogan, 17th Battalion KLR.**

As we have seen from CQMS Bracher's diary, Tarassova was captured by the Bolsheviks on 4 February 1919, and on the same day a detachment of twenty-three other ranks, led by Second-Lieutenant E A Stephenson from 'A' Company, arrived at Mejonovsky, near Seletskoe, from Ekonomiya, as reinforcements. On 7 February 1919, 'B' Company, including the reinforcements, took part in a combined attack with a French 'Coureurs des Bois'[9] Company and an American Trench Mortar unit on the Bolshevik forces in the town of Avda. Nearby Kodish was occupied at 8.20 am and two field guns were captured by the French and destroyed. However, enemy resistance around Avda itself was so strong, that the attack had to be called off and the Allied forces had to return to their original positions. During the course of the attack, however, Second-Lieutenant Stephenson, who was on attachment to the 17th Battalion from The Yorkshire Regiment, was wounded, and when the Allied force returned to

FIELD PASS. Permis militaire. Полевой пропускъ.

Loss of this pass to be reported immediately. / La perte de ce permis doit être signalée immédiatement. / Донести немедленно о потери сего пропуска

The Bearer / Le porteur / Предъявителю — is authorised to proceed to / est autorisé à se rendre à / разрѣшается ѣхать въ:

№ 1769

Signature of bearer / Signature du porteur / Подпись предъявителя

2/6/1919

(Signature of officer issuing pass).

A field service pass in English, French and Russian, issued to CQMS J H Bracher on 2 June 1919, to allow him to proceed on horseback to Seletskoe. (The Bracher family and the Liddle Collection).

'White' victims of the 'Red' Russians.

base, he was not amongst them. He must have died of his wounds or been killed by the Bolsheviks, for he was never seen again. 108580, Private W Graham, was also killed and four other ranks were wounded.

By this time the Bolshevik presence was growing stronger all the time, and two days later, on 9 February 1919, at 4.30 am they opened up a fierce and sudden artillery bombardment on both flanks of the Battalion's forward blockhouse positions, which immediately cut all communications with the rear. Soon they attacked with infantry, and although they were repulsed with considerable losses on the left flank, they eventually overran the right flank blockhouse after a fierce fight, and killed all its occupants, except one, 114144 Private J W Roberts, whom they captured. However, it was highly unusual for the Bolsheviks to take prisoners, unless it was to discover information, and they eventually killed him, his body being found later. The other dead were 391327 Private C E Corlett, 114116 Private J Kenworthy, 18619 Private C Maher, 29664 Private W McDonough, 50785 Lance Corporal C L Milton and 114191 Corporal A H Wright.

An attack was made on Kodish, and we moved into a deserted village, and after some skirmishing in the surrounding forest, we retired to our front-line posts before nightfall. Two days later the Bolshi made a reprisal raid on our right-wing outpost and wiped it out.

The right-hand post was called Church Post, on the bank of the River Empsa, and word came from our headquarters across the road that they couldn't get through on the 'phone, so a party went up, and I went with them. When we found them, they were dead. They'd all been dragged out and murdered.

They left behind some dead who were of the Mongol type. On this front, we found propaganda leaflets with the name of Phillips-Price, then Manchester Guardian correspondent in Moscow.
50752 Private J Grogan, 17th Battalion KLR.

During the rest of February men of the Battalion continued to garrison the forward positions and mount patrols, with varying results, and although contact was made with the enemy on several occasions, no casualties were sustained.

Most of March was spent in billets at Seletskoe and no significant action was joined again, apart from skirmishes between 'D' Company and the enemy on 1 May and 5 May 1919, at Bereznik, on the River Dvina. On both days the enemy mounted a strong attack on the Company's positions which was fought off with courage and determination. Forty-two prisoners were taken, but on 5 May 108666 Private J Murphy was killed, and four other ranks were wounded. Although Private Murphy has the distinction of being the last of the Pals to be killed in action, he was not the last to die on active service. This doubtful distinction belongs to 114100 Private B Murray who was drowned whilst swimming in the River Dvina on 28 June 1919. He is buried in Archangel Allied Military Cemetery.

By this time the Allied involvement in Russia was being scaled down and the British troops were the only ones left in the north in any kind of force, most of the other foreign 'interventionists' having gone home. Some of the 17th Battalion had also left for home in mid-June, as by this time, most of the military stores had been removed from Archangel and the rest had been allocated to White Russian troops, who

Former 17122 Pte W Gregory of the 18th Battalion, as a sergeant with the 241st Trench Mortar Battery, Royal Fusiliers in Russia. When he was discharged from the forces in 1918, and returned home, all his former comrades in the Pals had been killed, and so he rejoined the Army in an attempt to recapture the old spirit of camaraderie. He served in Russia from May until October 1919, but did not come across the 17th Battalion.

were trained in their use, so that they could take on the Bolsheviks themselves and cover the British withdrawal. This withdrawal slowed down in July and August, when a combined British/White Russian force took on and beat the Bolsheviks in the area of Troitsa, but the Pals were not involved in this fighting. During August the remaining men of the 17th Battalion were brought back from the front line to Archangel, and finally, on the last day of the month, the 2nd Battalion, The Highland Light Infantry took over all guards and duties from them and in the first week of September they embarked on the troopship *Kildonan Castle* for home. The following month, all stores having been removed from Murmansk, the port was evacuated by the British and 'The Intervention' was over.

In the eleven months that the Pals had served in Russia, they had earned two Military Crosses,[10] ten Military Medals, five Meritorious Service Medals and thirteen Russian decorations. One officer and twenty other ranks had perished. Although this is a paltry figure compared with a single day's average losses during trench-manning on the Western Front, it is a very poignant number when one considers that most of the men who were killed had already managed to survive the nightmare conditions of the war in France and Flanders. Also, unlike the sacrifices made on the Western Front, it is difficult to see how their deaths served any real purpose in the fight for Europe's freedom. However, perhaps it is not really fair to make comparisons between the two, when conditions and situations were totally different.

I can't compare Russia and France, they were so different. At the beginning, I would say that the Russian adventure was an adventure to someone with a Boy Scout mind like me, but France and the trenches, it was different. You were more alone in Russia than you were in the trenches. You weren't really alone in the trenches, you always had company at your side – even if it was only a sergeant. You were with others in the trenches or in the dugouts, but not in Russia. You weren't in the company of anybody in Russia. When you were doing a duty, you were alone, unless you were in a party going scrapping. Even going up there with your rifle cocked, it was a bit tense, because you didn't know which side they might come from, you didn't know where they were. **50752 Private J Grogan, 17th Battalion KLR.**

What had begun as a simple exercise to protect British interests had devolved into an anti-Bolshevik campaign which was to cause the Soviets to mistrust the Western Allies for the rest of the century. After all, as far as the Bolsheviks were concerned, they had just thrown off the shackles of Tzarist oppression and the Allies had then arrived in their country and tried to deny them their freedom – tenuous though this might have been!

A generation later, in another war, the Allies once again brought military stores to Murmansk and Archangel, and once again they were for the use of a Russian Army, which was once again fighting the Germans. Many of the Merchant and Royal Naval crews who brought these stores to the Kola Peninsula and the White Sea were from Liverpool, and doubtless some of them were the sons of Liverpool Pals. The reception they got from their hosts, despite the appalling rigours of their missions, and the vital aid they gave to the Soviet war effort, was one of sullenness, obstruction and outright hostility; so much so that many seamen considered that Britain was fighting the wrong enemy! Perhaps, in a perverse way, the harsh treatment meted out to the seamen in the early 1940s was an ironic tribute to the fighting skill and success of their fathers and their comrades who had fought so well in 1918 and 1919!

Some of the Transport Section of the 18th battalion at Huy, Belgium in 1919. Note the overseas service chevrons and that they are all wearing the Lancashire Hussars cap badge. They are: standing 21913 Pte J Stockley, 21912 Pte J Rosbotham, 16436 Pte E A Bryers, seated 21904 Cpl J Routledge, Pte 'Pickabit' Johnson and Pte J Beige.

Epilogue

When the last men of the 17th Battalion finally arrived home in September 1919, the war had been over for almost a year in Europe. Following the Armistice, Germany had been wracked by revolution and her representatives had eventually, though reluctantly, signed all the treaties which formally ended the hostilities on all the battle fronts. The Treaty of Versailles, which largely dealt with Europe, had been signed officially on 28 June 1919, exactly five years to the very day, after Archduke Franz Ferdinand of Austria had been assassinated at Sarajevo.

After the Armistice the 18th Battalion had been told that, in recognition of the 66th Division's excellent fighting record since 5 October 1918, it would be marching to the Rhine with the British occupation force. However, this did not materialise, and instead it marched to Huy, in Belgium, where it spent Christmas – the Battalion's one and only peacetime Christmas. By this time officers and men were being demobilised, almost by the day, and Battalion numbers were consequently reduced. Eventually, after having moved to Assesse and Natoye, the remnants, by now only at cadre strength, arrived at Antwerp on 10 May 1919, and five days later embarked on the SS *Sicilian* for England and home.

To all the returning Pals Merseyside had changed. It was not the same place they had left only a few short years yet several lifetimes ago. They had all longed to get home thousands and thousands of times since November 1915, and yet, once they were home, they could not settle. To them the real world was still one of trenches and danger and comradeship and death, and civilian life just did not seem what it had been. In fact it wasn't. The whole point of joining the Pals had been that men who worked and played games together could serve and fight together, and many of their pals would neither work nor play games ever again. During the war, the comradeship of pre-war life had given way to a much greater comradeship in the field and this was glaringly absent in civilian life. Some just could not adjust to it and rejoined the Army to try to recover what they had left behind. Former 17122 Private W Gregory of the 18th Battalion was one such. After recovering from the wounds he received at Flers in 1916, he volunteered for the RFC and was just about to gain his pilot's 'wings' when the Armistice was signed. Instead, he was 'demobbed', but on his return to Liverpool, he found that he knew nobody as all his friends had been killed. Missing the cameraderie of service life, he re-joined the Army and became a Trench Mortar Battery sergeant with the Royal Fusiliers and saw further active service in Russia. This eventually cured him of his need for Army life.

The inevitable recession which always follows any major war meant that it would be many years before some of the Pals ever worked again. For others, who were lucky enough to get their old jobs back, civilian life was so trivial compared with what they had seen and been through, and no one who had not been there, could possibly understand how they felt. Besides, the top jobs were now held by men who had been office boys when they had first joined up, or worse, held by cowardly incompetents who had refused to volunteer for the Army in 1914, and had somehow managed to avoid conscription. The 'land fit for heroes' that they had been promised had a somewhat hollow ring! Many Liverpool firms did not survive long after the war, particularly the small family firms, because after those who had run them during the

Part of a menu card celebrating the 89th Brigade 50th Anniversary Reunion Dinner in 1964. The cartoon was drawn 50 years earlier!

war retired or died there was no one left to take over. Besides, Liverpool's trade had depended on the sea, and after four years of war, many of the old customers had been forced to look elsewhere for suppliers of goods and simply did not return to their former Liverpool trading partners.

Great Britain was numbed and shocked in the early years after the Armistice, when the terrible price that was paid for peace gradually dawned on the grieving nation. There can have been few families in the land that didn't lose someone and few streets in any town that had not had the mourning curtains drawn for most months of the war. Just under 2,800 Liverpool Pals had perished during the conflict, or the rough equivalent of three complete fighting battalions. Although the level of decimation was nowhere near so great as in other Pals Battalions, the number of casualties was higher, because there had been four Liverpool City Battalions which had served for most of the war in almost continual action.

1 July 1916 was the blackest day of the war for some of the other northern Pals Battalions and, although, as we have seen, it was a day of triumph for the Liverpool Pals, nevertheless 1916 was their decisive year. After 1916 the character of the City Battalions began to change. Perhaps the early euphoria had been replaced with reality by this time, or perhaps it was merely that so many of the originals of 1914 had been killed or left the Battalions through wounds, that the original identity had been lost. Certainly, after 1916 many Pals who had not sought a Commission before decided to do so, and then continued to serve instead as officers. Many of them would have had the necessary qualities and qualifications needed to become officers at the start of the war, but the whole point of the Pals was that men could serve with their comrades. Even in the Liverpool City Battalions, where the relationship between officers and men was a lot closer than in many other units, most men still wanted to serve in the ranks. After the losses of 1916, however, so many men had lost comrades, that perhaps they thought they could better serve their country as officers. This is not to say that after 1916 the fighting strength and determination to win was at all diminished, however; it was merely different.

One thing is certain, however, from the original men of the four Battalions who left England for France in November 1915, over twenty per cent were dead by the end of 1919. For the 18th Battalion, which served the longest on the Western Front, and suffered most on 1st July 1916, this figure reaches twenty-five per cent. These percentages are incomplete, however, because they only include men who were actually serving with the Pals at their time of death, and do not include those who were transferred from the Pals and then killed with other units. If one then considers how many others would have been wounded during the war, it is probable that the total number of casualties amongst the 'Silver Badge Men' would have been somewhere in the region of seventy-five per cent. Some, of course, would recover from wounds and return to active duty, only to be wounded again, and perhaps again still. Others would be so badly maimed that they would be of no further use to the Army – or any other employer. In a perverse way, these figures illustrate how professional and successful the Pals were, for as we have seen with their service on the Somme, only successful units were put into action again and again, simply because they could be relied upon. However, there is no evidence at all that these crippling casualties, or the turnover of men that they brought, diluted the City Battalions' fighting spirit or dented their resolve to win. This fact is amply illustrated by the 19th Battalion's stand at Roupy in March 1918.

The last formal act involving the City Battalions was the laying up of their Colours in 1923. It has long been a tradition in the Army that Colours are laid up in the local church when they are no longer needed or the unit has been disbanded. Usually, in the case of local volunteer units, the church was a place of safe-keeping in case the Colours were ever needed again. For most of the Service Battalions, raised exclusively for Great War service, it was fitting that this tradition should continue once the war was over. The City Battalions, however, were a different case. Although they, too, were Service Battalions, they had, as we have already seen, been recruited almost exclusively in the first instance from Liverpool businesses and had never really lost this distinction. Thus, it was decided that the proper place for their Colours was not a church or cathedral, but Liverpool Town Hall. Thus, on 26 March 1923, after a parade in front of Lord Derby and the Lord Mayor of Liverpool on St George's Plateau, (the wheel had come full circle!), the Colours, escorted by former Pals, were marched to the Town Hall, behind the regimental band of the 1st Battalion of The King's (Liverpool Regiment), There they were handed over to the Lord Mayor for safe keeping by General Sir W H Mackinnon, Honorary Colonel of the Regiment. For many years, they hung from specially made holders, either side of the main Town Hall staircase, but the passage of time and the uneven heat from the gas lighting in the 1920s eventually caused them to decay. At the instigation of the author, and with the help of the Merseyside branch of the Western Front Association, they were taken down in the summer of 1990 and examined by professional restorers, and the cost of restoration estimated at £6,500. As the City Council say they have other priorities and cannot afford this price, it is hoped that the money to restore them to their original condition might be raised in other ways, outside the council.

As early as October 1918, an association called 'The 89th Club' had been formed, with Lord Derby as its Patron and General Stanley as its President. Its Vice-Presidents were equally well known to former Pals. They were Brigadier-General H W Cobham, CMG, DSO, Brigadier-General G Rollo, DSO and Bar, Colonel E F Gosset, Lieutenant- Colonel G S Clayton, DSO, and Lieutenant-Colonel C L Watson, DSO. The club, which was situated not far from Liverpool city centre, was meant to be a 'gentlemen's club' for former members of the Pals or from units who had been associated with them on active service. Its entrance fee was £1, with an annual subscription of £1 for 'Original Members', and £2 for'Associate Members', those who had not been in any of the four battalions before they had left for France in 1915. However, although the club's list of members for 1919 is impressive and contains a lot of familiar names, the club itself does not seem to have been a success and didn't survive beyond 1921. Also in 1918, the 89th Infantry Brigade Old Comrades' Fund was established, whose purpose was to give help to former members of the four battalions who had fallen upon distressed circumstances. Not only did it distribute money to deserving cases, but it found employment for those former Pals who were out of work, 'as a duty they all owed to the memory of those who did not come back.'[11]

More men of the Transport Section of the 18th Battalion at Huy Belgium, in 1919. They are: standing Pte Frank Davies, Pte Frank Hudson, seated Pte Billy Holland, Pte Jimmy Clare and Pte Jack Dauley. Note that Clare's cap displays the badge of the Lancashire Hussars.

It was natural that any band of men who had been raised through comradeship for war should tend to stick together in peace, and on 7 November 1925, exactly ten years after they had left England for France, the survivors of the 17th Battalion held their first reunion, with a special dinner in a Liverpool restaurant. It was not long before the other battalions followed suit, and reunion dinners soon became an annual event, only being interrupted by the six years of the Second World War. At first there were enough survivors still alive to have a separate reunion for each battalion, but as the years went by and numbers dropped, they became more and more difficult to organise. The

17th Battalion reunion was the first to be abandoned, in the 1950s, but the other three continued with them for about another ten years. However, in 1964 a special 89th Brigade 50th Anniversary reunion was held, which was such a success that thereafter, for a few years at least, combined reunion dinners were held, until the survivors became too old to attend. At least these dinners enabled old comrades to keep in touch, and kept the spirit of the Great War going whilst at the same time, they gave the survivors of the Pals the chance to recall former glories to people who actually wanted to listen.

It seems inconceivable that no one wanted to hear their experiences after the war, but after the euphoria of victory was over the whole nation began to lick its collective wounds and back away from the horrors its men had suffered. Until the war was over, many civilians had no real idea of what it had been like anyway, for the soldiers wanted to spare them the details, and rarely spoke of 'the front' when they came home on leave, but when the truth finally became known, many recoiled from it. The inevitable public reaction was to find scapegoats for what had happened, and the politicians were only too ready to supply them – the Generals. Thus, this faceless and nameless band of incompetents, 'the Generals', were suddenly saddled with the responsibility for all that had gone wrong in the war, and particularly the high casualties suffered. The truth was, of course, that the British High Command were no worse than the command staffs of any other army, and certainly a lot better than most; after all they did actually win the war. But they were no longer in public prominence, unlike the politicians, who were more than anxious to cover up their own shortcomings. There is no evidence at all, for instance, of criticism of their generals by front-line soldiers during the war and the affection with which General Stanley was regarded both during and after the war, exemplifies this. However, the evidence of the horrors of war was everywhere, if one wanted to look for it, even on street corners, where limbless ex-soldiers tried desperately to eke out a precarious living, and by some obscene irony, the soldiers who had fought and survived, often became objects of disdain. Perhaps the chief source of these detractors, acting out of repressed guilt, were the backsliders and avoiders who had stayed at home and never had to suffer, let alone win a war. Nevertheless, a climate was created for many years in which no one wanted to hear about the war, and especially not from the very ones who had fought it. This very reaction no doubt helped to fuel the appeasers of the 1930s, and one can only imagine the reactions of those who had lost loved ones in the 'war to end all wars', when the air raid sirens of September 1939 heralded a new conflict – against the same enemy.

The passage of time softened these ideas to an extent, however, but by then the veterans of the Second World War wanted to tell their stories, and once again, no one wanted to hear about the Somme and Passchendale, and the sacrifices of a different generation. By this time the view had changed somewhat to the idea that somehow the soldiers of the Great War were ignorant idiots who were duped, and deliberately sent to their deaths by uncaring stupid generals, who were never close enough to the front to see or care what had happened. Major anniversaries of the Great War which came round in the 1960s did eventually start to redress the balance, however, as people who had no vested interest in the Great War began to discover the true facts, although some of the 'lions led by donkeys'[12] views still persist today. Also, makers of documentaries, like most media personnel, do not want to hear of victories and triumphs, because defeats and disasters are much more sensational. Thus, everyone knows what happened to the

Accrington and Sheffield Pals on 1 July 1916, but who has ever heard of the unbelievable triumph of the Liverpool Pals on the same day until now? This in no way diminishes the sacrifices of the men of the 31st Division;[13] – it merely puts the record straight at last!

So were the Liverpool Pals better soldiers than most? Almost without exception, every veteran of the Pals interviewed, asked the question, 'Why do you want to talk to me? I didn't do anything special, I was just an ordinary soldier.' But equally so, all of them considered that the Liverpool Pals were something really special. All soldiers are taught to believe that their unit is the best one, for obvious reasons, but former Pals who had served in different units during the war still considered that the Liverpool Pals Battalions were the best. There is no doubt that their record is an impressive one, from the Somme to Passendale, and from St Quentin to Russia, and time and time again they proved that they were reliable in attack, and formidable in defence, so if they were better soldiers than most, why was this?

The Russia Memorial at Brookwood Military Cemetery, Surrey. The names of service personnel killed in Russia in two world wars are inscribed on this memorial, including men of the 17th Battalion. The memorial was erected in the early 1980s when the political situation in the Soviet Union made it seem unlikely that the Commonwealth War Graves Commission would be able to gain access to maintain existing memorials there. Those dead from the Battalion whose remains were not found and identified after the war, are also commemorated on the Memorial to the Missing of the North Russian campaign at Archangel.

The answer is quite simple. They were better educated and more intelligent as a whole than most units, and in any walk of life intelligence and education bring better results. Because of this, they were better led, and whilst trusting their officers, they were often able to see the reasoning behind a decision or an order, and if things went wrong, they didn't act like 'donkeys', but were able to seek an alternative. Also, soldiers often fight and sometimes die, not for an abstract, such as 'freedom', but for their comrades, with a determination not to let them down. Thus, although the Pals were motivated by love of country and defence of freedom, they were also much closer to their comrades than many other soldiers might have been. After all, their reasons for joining the Army were to fight alongside comrades, and thus the process of knitting the units together and promoting loyalty and comradeship was an easy one. Also, the mere fact that before the war most of them had worked together in the centre of Liverpool also meant that they came from similar backgrounds and had similar hopes and aspirations. Thus, they were welded together much more thoroughly, and at a much earlier stage of training, than many other units, and even after 1916, when the strain was diluted, the cameraderie still persisted, no doubt fostered by those who had survived from the originals. They did not just think they were the best, they knew it, and confidence in war, like education and intelligence, will always be a winner. Thus, even if one studies the home locations of the casualties in 1917 and 1918,[14] which show a significantly greater number who came from outside Merseyside, the fighting record of the Pals in those years shows no dilution at all of fighting spirit or military prowess. It could be argued that luck was with them, in their position in the line on 1 July 1916 for instance, and this is undoubtedly true, but there was nothing lucky about Guillemont thirty days later, and the scythe of death was no respecter of intelligence, education or confidence on that day, and yet decimated though they were, they recovered to fight again! Although they then knew what the war was really like, and that the task of defeating the Germans would take a lot longer than they had first thought, they did not feel duped, betrayed or let down; they just had a greater determination to get on with the job.

Like most of their generation, they were motivated by belief in God, love of country, a strong sense of justice and an equally strong desire to right the wrongs which they believed the Germans had perpetrated in France and Belgium. They were not supermen, but they were the finest that their generation could offer, and certainly amongst the best in the British Army. Today, very few are left, and in five years time; they will all probably be gone. It is doubtful if Britain will ever see their like again. Through their deeds and actions, they wrote their own history. This is their tribute!

NOTES
1. A 'Blighty One' was a wound which would necessitate the sending of a soldier back home to England for treatment and recovery. It was not usually serious enough to cause death, although it often caused mutilation, but it did allow the soldier a chance to get home. 'Blighty' was the soldiers' word for England, from the Hindi word 'bilayati', meaning anything foreign or European, picked up no doubt by the Regular Army in India.
2. CSM Bryan is wrong about the dates, which is hardly surprising, as he was recalling the events some forty years later, in 1961.
3. The spelling of Russian place names often differs from one soldier's account to another. This is because Russian place names were written in unfamiliar Cyrillic script, and thus the soldiers who recorded them would have written them down phonetically, when transposing them into English. As a result, there are bound to be differences in the English spellings.
4. Russians loyal to the Tzarist regime, or simply anti-Bolshevik were sometimes referred to as 'White Russians' as opposed to Communist 'Red' ones.
5. In the early 1980s, when the 'Cold War' had been in force for some time, it seemed unlikely that the Commonwealth War Graves Commission would be granted access to burial and memorial sites in the Soviet Union in the foreseeable future. As a result, all those servicemen who had perished in Russia during the Intervention, and also during the Second World War, were further commemorated on a specially built memorial in Brookwood Military Cemetery, Surrey.
6. This stands for Small Arms Ammunition, which are rounds for rifles and machine guns, as opposed to shells for field guns.
7. 'Bolo' was abbreviated slang for Bolshevik.
8. A droski or droshki is a low, four-wheeled, horse-drawn open carriage, quite commonly used in Russia, when the weather permits.
9. This phrase literally means 'Wood Runners', but was used in pioneering days, in the forests of Canada, to mean 'trappers', or 'woodsmen'. In this context, however, it obviously refers to special troops trained in forest and wood fighting.
10. Second-Lieutenant L N Worgan and Second-Lieutenant S G Dudley were both awarded the Military Cross for gallant conduct whilst serving with the 17th Battalion in Russia. Another officer, Captain F E Pritchard, who was nominally on the strength of the 17th Battalion, was also awarded the Military Cross, but little of his service in Russia was with the Pals, and the award was made when he was attached to another unit.
11. These are the words of Lieutenant-Colonel H L Brinson, speaking at the first reunion dinner of the 17th Battalion, on 7 November 1925.
12. The statement that the men of the British Army were 'lions led by donkeys' is first mentioned by General von Falkenhayn in his war memoirs, when recalling a conversation between General Ludendorff and General Max von Hoffman. Ludendorff has expressed the opinion that the English soldiers 'fight like lions', but Hoffman's reply is 'True. But don't we know that they are lions led by donkeys.' The statement has an ironic ring to it, when one considers the deliberate policies of attrition carried out by the German High Command, in relation to its own troops.
13. The Accrington Pals, the 11th (Service) Battalion of the East Lancashire Regiment, and the Sheffield Pals, the 12th (Service) Battalion of the York and Lancaster Regiment, were part of the 94th Brigade, 31st Division, on 1 July 1916, when they were all but wiped out at Serre.
14. See Appendix II.

Liverpool Pals Time Line
Chronology of Events

28/06/14 The Archduke Franz-Ferdinand, heir to the Austrian throne, is assassinated at Sarajevo.

04/08/14 Britain declares war on Germany.

05/08/14 Lord Kitchener, hero of Khartoum, accepts the position of Secretary of State for War.

07/08/14 Lord Kitchener makes his first appeal for 100,000 men to join the Army.

19/08/14 Lord Derby writes an open letter to the local press asking for volunteers to form a first Service Battalion of the King's Liverpool Regiment.

23/08/14 The new Service Battalion, the 11th Battalion, The King's, is fully manned and based at Seaforth barracks.

27/08/14 Lord Derby's idea to form 'Pals Battalions' is put forward in the Liverpool press, together with the announcement of a meeting to be held the next evening at the 5th Battalion's HQ, St Anne's Street.

28/08/14 Lord Kitchener calls for another 100,000 men. Lord Derby addresses a meeting at the drill hall, St Anne's Street, the hall being packed to capacity

31/08/14 A recruiting rally at *St* George's Hall raises a battalion of 1,050 recruits by 10 am. This to be the 17th (1st City) Battalion. Lord Derby tells thousands more waiting to enlist, to return on the following Wednesday 2 September.

02/09/14 Thousands more return to St George's Hall to enlist – The 18th (2nd City) Battalion is formed.

05/09/14 2,400 parade at the West Lancashire Riding School, Aigburth and form up to march into the City under the leadership of Lieutenant-Colonel The Hon F C Stanley DSO, Lord Derby's brother.

07/09/14 3,000 men have been enlisted and Lord Derby calls a temporary halt to the recruiting campaign after the formation of the 19th (3rd City) Battalion. At some time shortly after this date, Lieutenant-Colonel E H Trotter, ex-Grenadier Guards, is appointed to command the 18th (2nd City) Battalion, while Colonel E F Gosset who had served with the King's Liverpool, assumes command of the 19th (3rd City) Battalion.

Unidentified Pals Officer

14/09/14 The 1st City Battalion moves into the old watch factory at Prescot near Lord Derby's estate at Knowsley.

23/09/14 The 2nd City Battalion moves to the Hooton Park Race Course on the Wirral. The 3rd City Battalion trains at Sefton Park but as yet has no permanent barracks and so continues to be billeted 'at home'.

14/10/14 King George V sanctions the 'Eagle and Child' cap badge for the City Battalions. A solid silver version is manufactured, at Lord Derby's expense, and one is presented to each man. Some are personally presented by Lord Derby at ceremonies on his estate, while others are issued by the CO's of the Battalions.

**/11/14 The 3rd City Battalion moves into the camp built in the grounds of Lord Derby's estate at Knowsley. The 4th City Battalion is raised and the four battalions are formed into a Brigade (initially numbered 110th) under the command of Lieutenant-Colonel Stanley. The 4th City Battalion is quartered at the Tournament Hall in Knotty Ash. Lieutenant-Colonel B Fairfax, ex-Durham Light Infantry, is appointed to command the 1st City (17th King's) Battalion. Lieutenant-Colonel W Ashley is appointed to command the 4th City (20th King's) Battalion but later, because of ill-health, he is replaced by Lieutenant-Colonel H W Cobham.

03/12/14 The 2nd City (18th King's) Battalion moves to the hutted camp at Knowsley.

29/01/15 The 4th City (20th King's) Battalion moves to Knowsley and the four City (Pals) Battalions begin training as an infantry brigade.

/02/15 At the request of Lord Kitchener, Lord Derby raises two brigades for the Royal Field Artillery, which are later to serve in France and Flanders as the County Palatine Artillery, in the 30th Division.

20/03/15 Lord Kitchener reviews the (Pals) Battalions as they march past St George's Hall in Lime Street. Kitchener approves of what he sees and demands more of the same.

23/03/15 Lord Kitchener's prompting persuades Lord Derby to appeal for even more recruits for the Pals battalions.

30/04/15 The City Battalions leave Liverpool for Belton Park Camp, Grantham in Lincolnshire to train with other Brigades of the 37th Division. The Liverpool Pals become 89th Brigade, 30th Division, as a result of re-numbering by the Army Authorities.

Unidentified NCO

/05/15 All the infantry units of 30th Division are now assembled at Belton Park for training as a divisional unit.

27/08/15 The War Office finally adopts the 30th Division as a fully trained and equipped unit of the British Army.

31/08/15 Advance parties of the 89th Brigade depart Belton Park for Larkhill Camp on Salisbury Plain. The rest of the Brigade follows on at intervals.

07/09/15 The entire 89th Brigade is based at Larkhill Camp, Salisbury Plain. At some time shortly after this. Colonel E F Gosset relinquishes command of the 19th Battalion to be replaced by its 2 i/c Major L S Denham, who is promoted to the rank of Lieutenant-Colonel.

23/09/15 The whole of 89th Brigade takes part in a mock infantry assault at night, which ends in total chaos.

03/11/15 Lord Derby takes the place of the King, indisposed because of a riding accident, to inspect the 30th Division before its departure for France.

06/11/15 All Transport Sections of the four battalions leave Larkhill for Southampton and thence to Le Havre. Once landed, they pursue a long and slow journey to Pont Remy, and Ailly, near Abbville on the River Somme.

Unidentified Pals Officer

07/11/15 By 2.30pm all the Pals Battalions are embarked for France from Folkestone, to arrive later at Boulogne, and make their way to Ostrehove Rest Camp.

08/11/15 The battalions march back to Boulogne Central Station and entrain for Pont Remy, to be billeted as follows: The 17th Battalion at Bellancourt; the 18th Battalion at Monflieres; the 19th Battalion at Buigny-l'Abbe; the 20th Battalion at Pont Remy. Shortly afterwards they move to Vignacourt, north-east of Amiens and much nearer to the front line.

18/12/15 The 17th Battalion is sent to Englebelmer, Somme front. The 18th Battalion is sent to Hebuterne, Somme front. The 20th Battalion is sent to Bienvillers, Somme front. The 19th Battalion is split into two, one half being sent to La Haie and the other to Fonquevillers nearby, on the Somme front.

20/12/15 15206 Private E C Ecroyd (17th Battalion) becomes the first Pals casualty when shot through the chest He makes a complete recovery. 24620, Lance Corporal R Rezin is killed on night patrol in No Man's Land, the first of the Liverpool Pals to be killed in action. He is buried in Hebuterne Military Cemetery.

24/12/15 Three battalions, 17th, 19th and 20th come out of the line.

25/12/15 The 18th Battalion is transferred to 21st Brigade, but is still part of the 30th Division.

Souvenir from 89th Brigade
Liverpool Pals Grantham 1915.

06/01/16 The Pals Battalions take up position in the south of the Somme line, near Carnoy, as part of the 30th Division under the command of Major General W Fry.

16/03/16 All four battalions come out of the line and 30th Division as a whole spends the rest of March and most of April behind the lines.

30/04/16 The whole of the 30th Division returns to the line near Maricourt.

17/05/16 Major General J S M Shea takes command of the 30th Division.

25/05/16 The 17th, 19th and 20th Battalions come out of the line and move to Abbeville for specialist training for the 'Big Push'.

13/06/16 The 18th Battalion is released for training, for the 'Big Push', at Fourdrinoy and Vaux sur Somme.

24/06/16 Opening day of the British bombardment in preparation for the coming Battle of The Somme.

26/06/16 A German counter-barrage begins and inflicts quite serious losses on the 17th and 20th Battalions in the line near Maricourt.

Unidentified Private

27/06/16 The 18th Battalion returns to the front line taking over from the 2nd Battalion The Wiltshire Regiment at Maricourt.

28/06/16 Bad weather delays the start of the battle for 48 hours.

01/07/16 The first day of the Battle of The Somme. The 18th Battalion takes its objective, the Glatz Redoubt, incurring casualties of about 500 of all ranks. The 20th Battalion takes its objectives with casualties of 100 of all ranks. The 17th Battalion takes its objectives with losses of fewer than 17 of all ranks killed. The 19th Battalion is held in reserve at Maricourt Chateau. Montauban falls and all of 30th Division's objectives are achieved.

02/07/16 The 20th Battalion sends out two patrols to Bernafay Wood. These find it devastated and take prisoner 17 German soldiers found hiding among the ruined trenches and dugouts in the wood. The 18th Battalion is taken out of the line and moves back to its original assembly trenches at Talus Boise.

04/07/16 The 18th Battalion, together with the rest of the 21st Brigade, is relieved by the South African Brigade of 9th Division, and moves back to huts and tents at Bois de Tallies.

Unidentified NCO

05/07/16 The 9th Division takes over trenches from the 89th Brigade and the other Pals Battalions join the 18th Battalion at Bois de Tallies.

07/07/16 The 30th Division is sent back into the line to secure a defensive position from the Maltz Horn Ridge through the southern portion of Trones Wood.

08/07/16 At about 5.30pm. Colonel Trotter, Lieutenant-Colonel Smith (18th Manchesters), 2nd Lieutenant N A S Barnard and two other ranks are killed by a German shell in Train Alley Trench, while supervising the move forward of the 18th Battalion.

10/07/16 Companies of the 17th Battalion engage in operations to take the remaining tip of Trones Wood still held by the Germans.

11/07/16 The 17th Battalion is further engaged in operations to secure the wood. These are repulsed by German counter-attacks.

12/07/16 Lieutenant-Colonel Fairfax takes command of the fighting in Trones Wood, resulting later in the capture and holding of the wood. The 19th and 20th Battalions are relieved in the front line.

13/07/16 The 89th Brigade is relieved by the 55th Brigade and makes its way back to Bois de Tailles.

14/07/16 The 54th and 55th Brigades follow up the work of the Pals Battalions and finally clear Trones Wood of German occupation. The 89th Brigade moves from Bois de Tailles to Corbie and thence to its old camp at Vaux sur Somme.

15/07/16 Major General Shea addresses the 89th Brigade and warns the men that their rest will be short and they will soon return to action.

Unidentified Private

20/07/16 Lieutenant-Colonel L S Denham is replaced as Commander of the 19th Battalion by Major G Rollo of the 17th Battalion. Denham goes home to England. No explanation for this change of leadership is made.

23/07/16 The 21st Brigade makes an abortive attack on the fortified village of Guillemont, but the 18th Battalion of the King's is not involved, being designated as the carrying battalion for the assault.

29/07/16 The 17th, 19th and 20th Battalions move to take up their positions for the attack on Guillemont.

30/07/16 The attack on Guillemont fails and the three Pals Battalions losses, approaching 500 killed, make this a black day for Liverpool.

31/07/16 The 30th Division is withdrawn from the line and sent back to the Citadel Camp, south of Fricourt.

Unidentified Private

02/08/16 The Pals Battalions entrain for encampments around Abbeville to begin a period of rest and reorganisation.

03/08/16 The 30th Division temporarily becomes part of XVIIth Corps of 1st Army and moves north to the area around Givenchy.

08/08/16 Lieutenant-Colonel Fairfax returns to England, a casualty of the gassing he received on the night of 29 July before the attack on Guillemont. Fairfax is succeeded by Major J N Peck who is promoted Lieutenant-Colonel.

11/08/16 The 18th Battalion re-enters the line in the area around Givenchy.

26/08/16 The 17th, 19th and 20th Battalions re-enter the line, also in the area around Givenchy.

18/09/16 The 30th Division is ordered to join XVth Corps. The 18th Battalion entrains for Doullens and thence it takes up billets at Naours. The 17th, 19th and 20th Battalions entrain for Gezaincourt and thence take up billets at Vignacourt.

Unidentified Private

29/09/16 Major R K Morrison, the CO of the 19th Battalion, accused of drunkenness, is convicted after a Field General Court Martial; his sentence is 'to be dismissed the Service'.

04/10/16 Brigadier General F C Stanley and a group of commanding officers reconnoitre the Longueval-Flers area in front of the Transloy Ridges where the Pals Battalions are to re-enter the line.

11/10/16 The 89th Brigade is assembled in the Longueval-Flers area to take up its position for the coming assault on the Transloy Ridges.

12/10/16 The 89th Brigade takes part in the 'Battle of the Transloy Ridges'. The 17th Battalion is in the first wave, the 20th Battalion is in support and the 19th Battalion is Brigade reserve. The assault is not a success and the 17th Battalion is forced to consolidate a new position a mere 150 yards in front of the old front line. During the night and the early morning of the 13th it is relieved by men from the 19th and 20th Battalions.

Unidentified Private

224

15/10/16 The 21st Brigade is assembled in the Longueval-Flers area to take up its position for a further assault on the Transloy Ridges.

16/10/16 Major R K Morrison, dismissed the Service, returns to England.

18/10/16 The 21st Brigade takes part in the 'Battle of the Transloy Ridges'. The 18th Battalion is in the centre of the attack which is an abysmal failure.

22/10/16 All the Pals Battalions are withdrawn from the line by this date. Between October the 11th and the 22nd, 13 officers and 213 other ranks of the Pals have been killed in the 'Battle of the Transloy Ridges'.

31/10/16 The 30th Division moves north of the Somme to take over a section of the line opposite Bienvillers and Series.

18/11 /16 Official date of the end of the 'Battles of the Somme'.

06/12/16 David Lloyd George takes up office as Prime Minister.

31/12/16 During 1916, 64 officers and 1274 other ranks of the Pals Battalions have given their lives in the service of King and Country.

Unidentified Private

07/01/17 The 30th Division is relieved in the line and moves back to Halloy for a month's rest and training.

04/02/17 The 89th Brigade begins a move to the trenches around Agny, just south of Arras. The 17th Battalion is billeted in Arras. The 18th Battalion is split into two; one group is sent to Saulty to the 30th Divisional school, the other being sent to Beaumetz and eventually trench duty at Agny.

06/02/17 The 19th Battalion takes over part of the line at Agny. Subdivided, it alternates between the trenches and in training at the 30th Divisional school.

12/02/17 The 17th Battalion moves from Arras into the line at Agny.

22/02/17 German Army begins its withdrawal to the Hindenburg Line (Siegfried Stellung). This withdrawal continues through the rest of February and most of March. Accidental death of Private G Smith, 'C' Company, 17th Battalion, shot by Private W Houghton of the same company.

Unidentified Private

06/03/17 Lieutenant-Colonel J W H T Douglas of 2nd Battalion, The Bedfordshires assumes command of the 20th Battalion, The King's.

20/03/17 Major W L Campbell of 2nd Battalion, The Royal Scots Fusiliers assumes command of the 18th Battalion, taking over from Lieutenant-Colonel Pinwill.

09/04/17 Opening day of the 'Battle of Arras' The 18th, 19th and 20th Battalions are engaged in the opening assault with the 17th Battalion in reserve.

12/04/17 The 89th Brigade is relieved in the front line during the afternoon.

22/04/17 The 30th Division is ordered back into the line, opposite a strong part of the Hindenburg Line, west of the villages of Neuville Vitasse and St Martin sur Cojeul.

23/04/17 Second phase of the 'Battle of Arras'. The main assault is made by the 90th Brigade with the 21st Brigade in support and 89th Brigade in reserve. The 18th King's is in support as 'moppers up[1]. The attack fails and the battalions withdraw to their original iine.

24/04/17 The Germans vacate their trenches. These are occupied by the 18th Battalion which holds the position for four days.

26/04/17 Major W L Campbell relinquishes command of the 18th Battalion to Major M Owen of the Rifle Brigade. He is wounded on his first tour of inspection and Major G S Clayton assumes temporary command.

28/04/17 The 18th Battalion is relieved in the front line by the 6th Battalion, The Northamptonshires. The other three battalions are also relieved by other units on the same day.

Unidentified Private

29/04/17 The Pals Battalions leave the Hindenburg Line at Neuville Vitasse and march to Arras where they entrain for Petit Houvin, near Vaulx.

30/04/17 Major General Shea is dismissed from command of the 30th Division after making his feelings known about what he considers to be the unfair dismissal of Lieutenant-Colonel Weber, the GSO(1) of the 30th Division. Shea is replaced by Major General W de L Williams who remains with the 30th Division until the end of the war.

17/05/17 The fighting around Arras finally grinds to a halt and proves to be one of the costliest battles of the war in terms of days fought and numbers of casualties sustained.

20/05/17 The Pals Battalions leave Vaulx and begin the march north to the Ypres Salient.

22/05/17 Lord Derby, now Secretary of State for War, takes the salute of the 89th Brigade, near St Hilaire, on its march from Vaulx to Brandhoek.

Unidentified Private

28/05/17 The Pals Battalions arrive at Brandhoek and are billeted in different parts of the vast camp built and previously occupied by Canadian forces. 'A' and 'D' Companies of the 19th Battalion are sent into support trenches south of Hooge, holding the position for 24 hours.

07/06/17 The Battle of Messines. Nineteen mines are detonated beneath the ridge, facilitating its capture and making the outcome of the battle a great victory for the British. The 19th Battalion, not directly involved in the attack on the ridge, is again in the front line south of Hooge and the 17th Battalion is in the support line. Patrols of the 19th Battalion reconnoitre the German positions at Jackdaw Trench, north of the Messines Ridge. Several German counter-attacks are repulsed by the 19th Battalion.

Unidentified Private

10/06/17 The 17th Battalion is relieved in the support line and marches back to the Erie Camp. Brandhoek. The 19th Battalion is relieved in the line and marches back to the St Lawrence Camp, Brandhoek.

16/06/17 Brigadier General Stanley leaves for England to begin three months leave. He is temporarily replaced, first by Brigadier General W W Norman and then by Brigadier General W W Seymour.

05/07/17 Lieutenant-Colonel Rendall assumes command of the 17th Battalion, replacing Lieutenant-Colonel J N Peck who is sent home to England.

24/07/17 The 18th Battalion begins a move to the front line from the Canal Reserve Camp at Dikkebus.

30/07/17 The 18th Battalion arrives at Chateau Segard prior to the offensive the next day.

Unidentified Private

31/07/17 Opening day of the Third Battle of Ypres (Passendale). 30th Division occupies trenches from Sanctuary Wood to Observatory Ridge, in the centre of II Corps' area of attack, which covers the plateau across the Menin Road. The 17th and 20th Battalions move to their assembly positions at Maple Copse during the early hours of the morning. The 18th Battalion takes up its position on the western edge of Sanctuary Wood. The 19th Battalion held in Divisional Reserve is called up early in the morning, only to reach Maple Copse late in the afternoon. The Pals Battalions, for a variety of reasons beyond their control, are unable to achieve their objectives.

02/08/17 The 18th Battalion is relieved in the line and makes its way back to Chateau Segard, eventually going into rest at Eecke.

03/08/17 The 17th, 19th and 20th Battalions, relieved during the night and early the following morning, make their way back to Chateau Segard and thence into rest at Brandhoek. During their time in the line the Pals Battalions have lost 11 officers and 223 other ranks killed or died of wounds and 21 officers and 625 other ranks wounded.

Unidentified Private

/08/17 The battalions spend most of the remainder of the month training and in reserve at Merris and Dranoutre. The 18th Battalion eventually takes over trenches in the Wytschaete area, occupying the position until the end of the month.

22/08/17 No. 52929 Private J Smith of 'C' Company, 17th Battalion is tried by a Field General Court Martial on charges of desertion and disobedience. Found guilty, he is sentenced to death.

01/09/17 The 17th, 19th and 20th Battalions return to the front line and occupy trenches near Hollebeke.

05/09/17 No. 52929, Private J Smith of 'C[1] Company, 17th Battalion is shot by firing squad at Kemmel; the only soldier of the Liverpool Pals to be executed for a breach of military discipline.

24/09/17 16 officers and 290 men of the Lancashire Hussars are drafted to the 18th (2nd City) Battalion. The amalgamated battalion becomes officially known as the 18th (Lancashire Hussars) Battalion, The King's (Liverpool Regiment).

14/10/17 Lieutenant-Colonel Rendall is relieved of his command of the 17th Battalion. His place is taken by Major C N Watson, formerly of the 20th Battalion, who is promoted Lieutenant-Colonel upon appointment.

01/12/17 Inaugural Meeting of the Supreme Allied War Council at Versailles. Lieutenant-Colonel G Rollo is wounded by shell fire. Brigadier General Stanley specifically requests the return of Lieutenant-Colonel J N Peck to take Polio's place in command of the 19th Battalion. General Williams accedes to Stanley's request.

15/12/17 At a meeting of the Supreme Allied War Council at Versailles, Lloyd George formally agrees that the British Army should take over a 25 mile section of the line from the French; this consequent on a tacit agreement made at a conference at Boulogne on September 25th.

25/12/17 All four battalions now out of the line celebrate Christmas in style during the festive period.

/12/17 Orders are received for the 30th Division to leave the Salient.

31/12/17 The 17th, 18th and 19th Battalions are back in the front line for the New Year. The 20th Battalion is in the rear at Alberta Camp, Reninghelst The Pals Battalions have lost 22 officers and 518 other ranks killed during a seven month tour of duty in the Ypres Salient.

10/01/18 The 18th Battalion entrains for the south to take over an area vacated by the French.

11/01/18 The 17th, 19th and 20th Battalions follow the 18th battalion south. At a further meeting of the Supreme Allied War Council at Versailles it is decided to reduce the number of battalions in each of the 47 British Divisions from twelve to nine. Haig is ordered to effect the change as soon as possible. The cuts are to come from units of the New Army.

29/01/18 The move of the 30th Division is now complete and all four Pals Battalions are in a position opposite La Fre, near St Quentin.

08/02/18 The 30th Division leaves the line in order to carry out the battalion changes ordered by the Supreme Allied War Council. The 20th (4th City) Battalion effectively ceases to exist as a result of the reduction in the numbers of battalions in each division.

09/02/18 Drafts of the disbanded 20th Battalion move off to be distributed among the remaining three Pals Battalions.

21/02/18 The newly constituted 89th Brigade of 30th Division moves back into the line opposite St Quentin. The Pals Battalions' Commanding Officers at this time are as follows: The 17th Battalion; Lieutenant-Colonel C N Watson. The 18th Battalion; Major G M Clayton. The 19th Battalion; Lieutenant-Colonel J N Peck.

03/03/18 The Bolshevik Government signs the Treaty of Brest-Litovsk with the Central Powers.

21/03/18 Opening day of the German Offensive, code named 'Operation Michael[1], on the 5th Army front. The attack begins at 4.30 am in foggy weather which benefits the Germans. The three Pals Battalions of 89th Brigade are in Corps Reserve behind the 'Battle Zone'. At 5.00 am, they receive orders to 'man battle stations'. The 17th Battalion moves to Beauvois, the 18th Battalion moves to Germaine and the 19th Battalion moves to Vaux. Later in the day the 17th Battalion is ordered to join 184th Brigade, 61 st Division and moves to the village of Atilly. The 18th Battalion moves to 90th Brigade, 30th Division and digs in south west of Vaux. The 19th Battalion moves forward to prepare to counter attack the Germans at Roupy the next day.

22/03/18 The 17th Battalion deploys around Aviation Wood south of Fluquires in case of a German breakthrough. Later that night it withdraws to Ham, arriving at about 9.30 pm. The 18th Battalion after withdrawing from action around Stevens Redoubt also withdraws to Ham, arriving at about 7.30 pm. The 19th Battalion, defending the position at Roupy, suffers heavy casualties and is forced back, first to Stanley Redoubt and then the few survivors eventually retire, also to Ham.

23/03/18 The remnants of the three Pals Battalions are all in or around Ham but are eventually forced to retreat in the face of the continuing German onslaught.

25/03/18 During the night and the early morning of the 26th, the three Pals battalions fall back to Roiglise, south-east of Roye.

Unidentified
Private

26/03/18 The 89th Brigade is ordered to occupy some old 1914 French trenches on a defensive line from Bouchoir to Rouvroy, east of Roye.

28/03/18 After further German attacks are beaten oft, the French 133rd Division takes over the line and the 89th Brigade withdraws to Rouvrel, west of Moreuil and south-east of Amiens. General Sir Hubert Gough is dismissed from command of the 5th Army by Haig.

29/03/18 General Stanley finds that there are just under a thousand fit men left in the Brigade. These are combined to form the 89th (Composite) Battalion with Lieutenant-Colonel G N Rollo, recently returned after his recovery from wounds, in command.

30/03/18 The 89th (Composite) Battalion is ordered to march to Saleux and thence entrains for St Valery sur Somme. Here it is disbanded and the three Pals Battalions reform from survivors, stragglers and new drafts.

Unidentified
Private

01/04/18 General Williams orders Brigadier General Stanley to stand down from command on the grounds that he is too old for active command.

04/04/18 All three Pals Battalions entrain for Elverdinghe on the Ypres Salient.

07/04/18 The 89th Brigade moves into the front line in the vicinity of the towns of Hazebrouck and Bailleul.

09/04/18 The first day of a further German offensive. This known as The Battle of the Lys', on a front between Bethune and Ypres.

11/04/18 Haig issues his famous 'with our backs to the wall' order of the day as the German assault gains momentum.

16/04/18 The 18th and 19th Battalions are ordered south and attached to the 34th Division.

17/04/18 The 17th Battalion follows the other two battalions and the three are sent into the line south-east of St Jans Cappel, near Bailleul.

20/04/18 Lieutenant-Colonel C N Watson, commanding 17th Battalion is badly wounded in the legs by shell fire. He is replaced by the 2 I/C. Major J P Pitts, who is eventually promoted Lieutenant-Colonel.

Unidentified
Private

21/04/18 The 89th Brigade is relieved in the line by French troops and moves to rejoin the 30th Division at Busseboom on the Ypres Salient.

28/04/18 The Pals Battalions are sent into the line west of Voormezele, between Ridge Wood and Kruisstraathoek. Here, the 17th Battalion is involved in a five hour battle defending the approaches to the German objective, the Scherpenberg.

29/04/18 The Germans unleash a powerful assault on the Scherpenberg. Most of the outpost positions along the front fall quickly, but eventually the attack is repulsed along the whole of the front. The line stands firm and by the end of the day The Battle of the Lys' is over.

02/05/18 The 89th Brigade moves to the St Lawrence Camp, Brandhoek. From all three Pals Battalions the total personnel fit for service is 27 officers and 750 other ranks. A composite battalion is once again formed, becoming known as the 17th (Composite) Battalion, The King's (Liverpool Regiment), with Lieutenant-Colonel G N Rollo again in command.

Unidentified
Private

05/05/18 The 17th (Composite) Battalion moves into reserve trenches in rear of a line from Ridge Wood to Klienvierstraat.

08/05/18 The 30th Composite Brigade, of which the 17th (Composite) Battalion is a part, suffers an intense German artillery barrage and at 7.30 am, the Germans launch an attack along the whole brigade front penetrating to a depth of 400 to 500 yards in places. 'A' and 'B' Companies of the Battalion take part in a successful counter-attack.

09/05/18 The 17th (Composite) Battalion carries out a further, successful counter-attack.

10/05/18 The 17th (Composite) Battalion is relieved in the line and returns to the St Lawrence Camp early the next morning.

11/05/18 The 17th (Composite) Battalion is disbanded having lost four officers and 47 other ranks during its nine days existence. This is the last date on which the three Battalions serve together as part of Lord Derby's 89th Brigade.

13/05/18 Orders are received that the remnants of the three Pals Battalions are to be sent down to GHQ as training cadres to train troops of the American Expeditionary Force.

16/05/18 The 17th Battalion is sent to Meneslies to train the 1/137th US Infantry Regiment. The 18th Battalion is sent to Woincourt to train the 2/137th US Infantry Regiment. The 19th Battalion is sent to Allenay to train the 3/137th US Infantry Regiment.

19/06/18 All three training cadres are transferred to the 66th Division.

30/06/18 The 17th Battalion returns to England and is transferred to the 75th Brigade, 25th Division. After arriving at Folkestone the battalion moves first to Aldershot and then Clacton-on-Sea for rest, reinforcement and training.

31/07/18 The 19th Battalion Training Cadre is absorbed by the 14th Battalion, The King's (Liverpool Regiment) and is thus effectively disbanded. There are, therefore, only two surviving Pals Battalions.

08/08/18 The 18th Battalion becomes part of 197th Brigade.

13/08/18 The 18th Battalion is amalgamated with the 14th Battalion, The King's (Liverpool Regiment) recently returned from Salonika, which has already absorbed the 19th Battalion Training Cadre. The new battalion officially becomes the 18th (Lancashire Hussars Yeomanry), The King's (Liverpool Regiment).

09/09/18 The 75th Brigade, of which the 17th Battalion is a part, is renumbered 236th Brigade and leaves the 25th Infantry Division.

29/09/18 The 18th Battalion commander, Lieutenant-Colonel G N Rollo, leaves to take command of 150th Brigade and is replaced by Major R O Wynne.

05/10/18 The 17th Battalion leaves Mytchett Camp at Aldershot and entrains for Glasgow.

07/10/18 The 18th Battalion, having moved rapidly forward in the new 'war of movement', arrives at the Hindenberg line in the region of Bony.

09/10/18 The 18th Battalion advances through L'Epinette in pursuit of the retreating German Army and by nightfall is in position in front of Le Gateau. Here the battalion is relieved in the line and moves into billets at Reumont.

10/10/18 The 17th Battalion embarks on the SS *Keemun* bound for Northern Russia.

17/10/18 The SS *Keemun* arrives at Murmansk.

20/10/18 Headquarters Company and 'B' and 'C' Companies of the 17th Battalion transfer to the troop ship *Goentoer* and sail through the White Sea to Archangel.

21/10/18 'A' and 'D' Companies of the 17th Battalion sail to Archangel on the troop ship *Asturian*.

23/10/18 The whole of the 17th Battalion is based at Archangel but this is the last date on which all the companies serve together as a full battalion.

01/11/18 'A', 'C' and 'D' Companies of the 17th Battalion are billeted on the *Asturian* with 'B' Company in Barracks at nearby Bakharitza.

03/11/18 The 18th Battalion moves back to Le Gateau expecting to engage the enemy. However, no contact is made.

04/11/18 'A', 'C' and 'D' Companies of the 17th

Battalion move into Barracks at Ekonomiya. 'B' Company remains at Bakharitza.

Unidentified Private

06/11/18 The 17th Battalion suffers its first casualty of the North Russian Campaign when 114148 Private G H Brown of 'C' Company dies from pneumonia in the Barracks hospital.

08/11/18 The 18th Battalion attacks the Germans near Marbaix in what is its last engagement of the War.

09/11/18 'B' Company, 17th Battalion moves by rail to Obozerskaya to join the Vologda Force.

11/11/18 The Armistice is signed. Hostilities cease at 11.00 am. The 18th Battalion, in position near a village named Clairfayts, hears the good news; The War is over.

14/11/18 'C' Company, 17th Battalion moves by rail to Bereznik to join the Dvina River Force.

21/11/18 'D' Company, 17th Battalion moves by rail to join the Vologda Force but is placed on temporary attachment at Seletskoe, 100 miles south of Archangel.

01/12/18 With 'A' Company still in Archangel, 'B', 'C' and 'D' Companies, 17th Battalion are in position ready to repel any attempts by the Bolsheviks to move north.

05/12/18 'D' Company, 17th Battalion, having been warned of a Bolshevik attack, moves to the town of Tarassova to await the attack which does not materialise.

06/12/18 Captain E A Dickson MC decides to pre-empt the Bolshevik attack by ambushing the enemy in its assembly positions. After some success 'D' Company withdraws on Tarassova. Five men are killed during the action and one dies of wounds.

18/12/18 'B' Company, 17th Battalion arrives at Verst 445 to relieve a company of the American 339th Infantry Regiment. The company comes under enemy shell fire two and a half hours later. The armoured train which transported them retaliates. There are no casualties incurred by the Company.

25/12/18 'B' Company, 17th Battalion is relieved in the line by some French Colonial Troops and moves into reserve lines.

30/12/18 A further attack on the Bolsheviks at Tarassova has to be aborted just 80 yards from the enemy position when the Archangel Regiment refuses to attack in support of 'D' Company.

04/02/19 Tarassova is captured by the Bolsheviks.

07/02/19 'B' Company, 17th Battalion, *a* French 'Coureurs de Bois' Company and an American Trench Mortar Unit make a combined attack on the Bolsheviks at Avda. Enemy resistance is strong and the attack is called off.

09/02/19 The Bolsheviks make a fierce attack on 'B' Company's position and during the action the Company loses seven men killed. The attack, however, is eventually repulsed.

01/05/19 The Bolsheviks attack 'D' Company's position at Bereznik on the River Dvina. The attack is repulsed.

05/05/19 A further attack on 'D' Company's position by the Bolsheviks is once more repulsed.

15/05/19 The remnants of the 18th Battalion, now only at cadre strength, leave Antwerp for England on the SS *Sicilian* after spending time at Huy, Assesse and Natoye in Belgium, following the Armistice.

16/06/19 Some of the men of the 17th Battalion leave North Russia for home.

28/06/19 The Treaty of Versailles is officially signed, exactly five years to the day that Archduke Franz Ferdinand was assassinated at Sarajevo.

31/08/19 Having already been withdrawn to Archangel, the remnants of the 17th Battalion are relieved by the 2nd Battalion, The Highland Light Infantry.

02/09/19 The remainder of the 17th Battalion leaves Archangel for home on the troop ship *Kildonan Castle*.

26/03/23 The Colours of the four City Battalions are laid up in Liverpool Town Hall, having been handed to the Lord Mayor by General Sir W H Mackinnon, Honorary Colonel of The King's (Liverpool Regiment).

1914/19: *During the course of the Great War and the 17th Battalion's service in North Russia, just under 2,800 men from the City Battalions have perished.*

Appendix II

List of Liverpool Pals who died on active service during the Great War

This list has been compiled from a number of different sources, but mainly from the records held by The Commonwealth War Graves Commission, which is the most reliable source for details of Great War casualties, and covers the official time period accepted for Great War deaths, i.e. 4th August 1914 to 31st August 1921. Places of birth of casualties are shown within county boundaries that were in use during the same period. Every appropriate cemetery or memorial register, and the Commission's ledger relating to deaths in the King's (Liverpool Regiment), has been consulted in the compilation. Occasionally, the author has discovered a mistake made in the original records, or overwhelming proof that the original records are not accurate, in which case the liberty of altering the Commission's version has been taken. Inevitably, however, this list can never be complete in all details, but it is certainly the most accurate one ever compiled to date. A number of men enlisted under false names, and where these are known, they are shown in brackets alongside the true name.

To allow the reader to see which battles and engagements were the most costly for each battalion, the author has taken the unusual step of setting out the list in date order, rather than alphabetically. This way, the full impact of the course of the war on the Liverpool Pals may be seen at a glance.

No.	Name	Place of Birth	Rank	Bn.	Cause & Date of Death		Burial/Memorial Place
16578	Hoban. P.	Liverpool.	Pte.	18th	Died	23 12 14	Anfield Cem, Liverpool.
17050	Ellis. E.	Liverpool.	Pte.	18th	Died	05 01 15	Toxteth Park Cem, Liverpool.
24573	Hayworth. W.	Liverpool.	Pte.	18th	Died	24 01 15	Everton Cem, Liverpool.
21976	Jackson. M.F.	Ballakaneen, I.O.M.	Pte.	17th	Died	24 01 15	Toxteth Park Cem, Liverpool.
23800	Prue. G.	Birkenhead, Cheshire.	Pte.	20th	Died	06 03 15	Kirkdale Cem, Liverpool.
16840	Rowlands. T.O.	Bootle, Liverpool.	Pte.	18th	Died	30 05 15	Kirkdale Cem, Liverpool.
24649	Dodd. A.A.	Ellesmere Port, Cheshire	Pte.	18th	Died	02 06 15	Christchurch CY, Ellesmere Port.
21522	Hull. G.T.	Preston, Lancs.	Sgt.	19th	Died	06 06 15	Leyland Cem, Lancs.
	Ryder. C.E.	Liverpool.	Lieut QM.	17th	Died.	06 11 15	Tidworth Mil Cem, Hants.
24620	Rezin. R.	Mitcham, Surrey.	L/Cpl.	18th	K.I.A	20 12 15	Hebuterne Mil Cem.
24482	Greenhalgh. T.P.	Bradford, Yorks.	Pte.	20th	K.I.A	07 01 16	Cerisy-Gailly Mil Cem.
16862	Pearson. A.J.	Anglesey, N.Wales.	Pte.	18th	D.O.W	11 01 16	Carnoy Military Cem.
16351	Barker. J.	Liverpool.	Pte.	18th	K.I.A	3 01 16	Bronfay Farm Mil Cem.
17468	Stirrup. W.	Liverpool.	Pte.	19th	D.O.W	3 01 16	Carnoy Military Cem.
16050	Brownlie. L.C.	Bootle, Liverpool.	Pte.	17th	Died	4 01 16	Cerisy-Gailly Mil Cem.
15887	Guthrie. M.	Birkenhead.	Pte.	17th	K.I.A	4 01 16	Corbie CC.
24934	Harvey. C.E.	Seacombe, Cheshire.	Pte.	17th	Died	4 01 16	Cerisy-Gailly Mil Cem.
15563	Roberts. R.	Beddgelert, Caernarvon.	Pte.	17th	Died.	4 01 16	Cerisy-Gailly Mil Cem.
25348	Grace. J.	Hunts Cross, Liverpool.	Pte.	19th	K.I.A	5 01 16	Carnoy Military Cem.
26647	Burgess. H.C.	Wallasey, Cheshire	Pte.	18th	K.I.A	17 01 16	Carnoy Military Cem.
16257	Mullock. J.	Egremont, Cheshire.	Pte.	18th	K.I.A	20 01 16	Carnoy Military Cem.
17322	Downie. J.A.H	Liverpool.	L/Cpl.	19th	K.I.A	22 01 16	Carnoy Military Cem.
15805	Gordon. J.S.	Liverpool.	Pte.	17th	K.I.A	22 01 16	Cerisy-Gailly Mil Cem.
15429	Trinick. J.H.	Salcombe, Devon.	L/Cpl.	17th	K.I.A	22 01 16	Cerisy-Gailly Mil Cem.
22758	Lehan. F.J.	Liverpool.	Pte.	20th	Died.	23 01 16	St.Ouen CC.
15903	Dyall. L.A.	Birkenhead, Cheshire.	Pte.	17th	K.I.A	24 01 16	Cerisy-Gailly Mil Cem.
16906	Gell. J.F.	Liscard, Cheshire	L/Sgt.	18th	K.I.A	26 01 16	Carnoy Military Cem.
16857	Murray. H.	Liverpool.	Sgt.	18th	K.I.A	26 01 16	Carnoy Military Cem.
21940	Ackerley. T.	Liverpool.	Pte.	20th	K.I.A	28 01 16	Cerisy-Gailly Mil Cem.
15518	Bryan. C.	Liverpool.	Pte.	17th	K.I.A	28 01 16	Cerisy-Gailly Mil Cem.

15528	Ellicott. E.	Wallasey, Cheshire	Cpl.	17th	K.I.A	28 01 16	Cerisy-Gailly Mil Cem.
16652	Evans. S.	Everton, Liverpool.	Pte.	18th	K.I.A	28 01 16	Carnoy Military Cem.
15476	Ferguson. J.	Kirkdale, Liverpool.	Pte.	17th	K.I.A	28 01 16	Cerisy-Gailly Mil Cem.
22710	Halewood. E.	Great Crosby, Liverpool.	Cpl.	20th	K.I.A	28 01 16	Cerisy-Gailly Mil Cem.
16784	Kitchen. J.H.	Liverpool.	Pte.	18th	K.I.A	28 01 16	Carnoy Military Cem.
16806	Eady. J.A.	Wavertree, Liverpool.	Pte.	18th	K.I.A	29 01 16	Carnoy Military Cem.
16808	Eaton. R.	Aintree, Liverpool.	Pte.	18th	K.I.A	29 01 16	Carnoy Military Cem.
24990	Cheeseman. G.	Liverpool.	Pte.	17th	D.O.W	30 01 16	Chipilly CC.
25751	Pearson. G.	Liverpool.	Pte.	18th	D.O.W	31 01 16	Corbie CC.
17496	Whitehead. W.F.	Liverpool.	Pte.	19th	D.O.W	04 02 16	Corbie CC.
15537	Parry. H.	Aberffran, Anglesey.	Pte.	17th	D.O.W	05 02 16	Corbie CC.
	Wainwright. H.C.	Liverpool.	Lieut.	17th	D.O.W	05 02 16	St.Pierre Cem, Amiens
16922	Miles. P.S.	Wallasey, Cheshire	Pte.	18th	D.O.W	06 02 16	Carnoy Military Cem.
15004	Beeston. J.B.	Liverpool.	CQMS.	17th	K.I.A	10 02 16	Suzanne CC Ext.
24978	Meaker. T.A.	Liverpool.	Pte.	17th	D.O.W	10 02 16	Chipilly CC.
15229	Yates. A.H.	New Brighton, Cheshire.	Pte.	17th	K.I.A	10 02 16	Suzanne CC Ext.
22334	Burgess. W.T.	Liverpool.	Pte.	20th	K.I.A	11 02 16	Cerisy-Gailly Mil Cem.
16846	Thomas. E.W.	Liverpool.	Pte.	18th	D.O.W	11 02 16	Corbie CC.
16879	Wood. T.S.	Tranmere, Cheshire.	Pte.	18th	K.I.A	13 02 16	Carnoy Military Cem.
16924	Melvin. J.E.	Liverpool.	Pte.	18th	D.O.W	14 02 16	Carnoy Military Cem.
15072	Byrne. J.G.	Liverpool.	Pte.	17th	K.I.A	16 02 16	Cerisy-Gailly Mil Cem.
22225	Pyper. E.C.	Kimberley, South Africa.	Pte.	20th	D.O.W	17 02 16	Chipilly CC.
15031	Bonser. C.	Egremont, Cheshire.	Pte.	17th	K.I.A	18 02 16	Corbie CC.
16726	Leach. T.H.	Rock Ferry, Cheshire	Pte.	18th	D.O.W	18 02 16	St Sever Cem, Rouen. *
24609	Nolan. D.	Seacombe, Cheshire	Pte.	18th	K.I.A	20 02 16	Carnoy Military Cem.
22298	Yardley. F.	Liverpool.	Pte.	20th	D.O.W	22 02 16	Corbie CC.
17466	Smith. G.A.	Garston, Liverpool.	Pte.	19th	K.I.A	23 02 16	Carnoy Military Cem.
16329	Price. L.	Liverpool.	Pte.	18th	K.I.A	26 02 16	Carnoy Military Cem.
21996	O'Sullivan. J.	Liverpool.	Pte.	17th	K.I.A	02 03 16	Cerisy-Gailly Mil Cem.
17433	Morgan. J.	Liverpool.	Pte.	19th	K.I.A	04 03 16	Cerisy-Gailly Mil Cem.
22995	Farrell. T.	Liverpool.	Pte.	20th	K.I.A	06 03 16	Cerisy-Gailly Mil Cem.
23889	Bennett. J.A.	Liverpool.	Pte.	20th	D.O.W	07 03 16	Corbie CC.
22022	Bowler. J.F.	Liverpool.	Pte.	20th	K.I.A	07 03 16	Cerisy-Gailly Mil Cem.
22120	Housden. C.H.	Liverpool.	Pte.	20th	K.I.A	07 03 16	Cerisy-Gailly Mil Cem.
30831	Craven. F.J.	Liverpool.	Pte.	19th	K.I.A	09 03 16	Cerisy-Gailly Mil Cem.
22524	Searson. C.	Liverpool.	Pte.	20th	D.O.W	09 03 16	Corbie CC.
15590	Heyes. S.B.	Wavertree, Liverpool.	Pte.	17th	Died	20 03 16	Mericourt CC.
44077	Smith. L.	Liverpool.	Pte.	17th	K.I.A	22 03 16	Pozieres Memorial.
31010	Wilkinson. A.	Preston, Lancs.	Pte.	20th	D.O.W	22 03 16	Corbie CC.
24231	Evans. F.	Liverpool.	Pte.	20th	Died.	24 03 16	St.Marie Cem, Le Havre.
17262	Ankers. H.	Liverpool.	Pte.	19th	D.O.W	29 03 16	St.Pierre Cem, Amiens
17194	Henshaw. A.	West Kirby, Cheshire	Pte.	18th	D.O.W	09 04 16	Abbeville C.C.
15260	Porter. W.	Formby, Lancs.	Pte.	17th	D.O.W	19 04 16	Toxteth Park Cem, Liverpool.
21843	Birch. H.L.	Liverpool.	Pte.	19th	K.I.A	24 04 16	Cerisy-Gailly Mil Cem.
17817	Eagles. E.	Liverpool.	Pte.	19th	K.I.A	24 04 16	Becourt Military Cem.
15959	Fogg. N.B.	Willaston, Cheshire.	L/Sgt.	17th	K.I.A	01 05 16	Cerisy-Gailly Mil Cem.
15847	Beacham. H.	Rock Ferry, Cheshire.	L/Cpl.	17th	K.I.A	03 05 16	Cerisy-Gailly Mil Cem.
22038	Chorley. W.	Whiston, Lancs.	L/Cpl.	20th	D.O.W	06 05 16	La Neuville CC Corbie
22854	Sutton. H.	Liverpool.	Pte.	20th	K.I.A	07 05 16	Cerisy-Gailly Mil Cem.
17115	Curran. J.	Liverpool.	Cpl.	18th	K.I.A	08 05 16	Carnoy Military Cem.
26603	Cole. G.H.	Liverpool.	Pte.	17th	K.I.A	11 05 16	Cerisy-Gailly Mil Cem.
15546	Evans. D.	Ormskirk, Lancs.	L/Cpl.	17th	K.I.A	11 05 16	Cerisy-Gailly Mil Cem.
17614	Hodgson. N.R.E.	New Ferry, Cheshire.	Pte.	20th	D.O.W	11 05 16	La Neuville CC Corbie
15775	Holland. C.A.	Egremont, Cheshire.	Pte.	17th	K.I.A	1 05 16	Cerisy-Gailly Mil Cem.
23656	Kain. C.J.	Birkenhead, Cheshire.	Pte.	17th	K.I.A	1 05 16	Cerisy-Gailly Mil Cem.
15814	Moore. A.J.	Liverpool.	Pte.	17th	K.I.A	1 05 16	Cerisy-Gailly Mil Cem.
22221	Purnell. R.	Widnes, Lancs.	Pte.	20th	D.O.W	1 05 16	Corbie CC Ext.
17786	Bell. C.M.	Liverpool.	Pte.	19th	K.I.A	2 05 16	Suzanne CC Ext.

17790	Butcher. G.N.	Liverpool.	Pte.	19th	D.O.W	12 05 16	Suzanne CC Ext.
17840	Hoos. R.	Liverpool.	L/Cpl.	19th	K.I.A	12 05 16	Suzanne CC Ext.
22500	Parr. F.E.	Liverpool.	Pte.	20th	D.O.W	12 05 16	Corbie CC Ext.
27530	Nickson. W.J.	Liverpool.	Pte.	19th	D.O.W	13 05 16	Bray Mil Cem.
15233	Philipps. E.S.	Liverpool.	Pte.	17th	K.I.A	15 05 16	Cerisy-Gailly Mil Cem.
16977	Lucas. P.	Liverpool.	Pte.	18th	K.I.A	18 05 16	Carnoy Military Cem.
17036	Abrahams. E.G.	Bootle, Liverpool.	Pte.	18th	D.O.W	22 05 16	La Neuville CC Corbie
16827	Lawrence. G.A.	Anfield, Liverpool.	Pte.	18th	K.I.A	28 05 16	Cerisy-Gailly Mil Cem.
16841	Rowlands. I.	Liverpool.	Pte.	18th	K.I.A	28 05 16	Cerisy-Gailly Mil Cem.
16531	Lucy. F.W.	Liverpool.	Pte.	18th	K.I.A	29 05 16	Cerisy-Gailly Mil Cem.
22805	Penny. R.	Bootle, Liverpool.	Pte.	20th	K.I.A	02 06 16	Cerisy-Gailly Mil Cem.
25530	Skinner. R.H.	Widnes, Lancs.	Pte.	19th	Died	04 06 16	Farnworth, Widnes, St.Lukes
17454	Rimmer. R.	Liverpool.	Pte.	19th	Died	13 06 16	Westburn Cem.Cambuslang.Scot
24584	Holme. J.	Walton, Liverpool.	Pte.	17th	K.I.A	17 06 16	Bray Mil Cem.
15128	Cubbin. C.	Liverpool.	Pte.	17th	K.I.A	20 06 16	Cerisy-Gailly Mil Cem.
15831	Neame. T.W.S.	Southport, Lancs.	Pte.	17th	K.I.A	20 06 16	Cerisy-Gailly Mil Cem.
30251	Woodley. H.	Liverpool.	Pte.	17th	K.I.A	20 06 16	Cerisy-Gailly Mil Cem.
24808	Lewis. A.J.	Liverpool.	Pte.	17th	D.O.W	22 06 16	La Neuville CC Corbie
15493	Stott A.E.	Liverpool.	Pte.	17th	D.O.W	23 06 16	Abbeville C.C.
17742	Wild. H.	Liverpool.	Pte.	19th	K.I.A	25 06 16	Cerisy-Gailly Mil Cem.
22304	Aitken. W.H.	Aintree, Liverpool.	Pte.	20th	K.I.A	26 06 16	Cerisy-Gailly Mil Cem.
31013	Askew. J.	Liverpool.	Pte.	20th	K.I.A	26 06 16	Cerisy-Gailly Mil Cem.
22617	Bennett. M.W.	Derby.	Pte.	20th	K.I.A	26 06 16	Cerisy-Gailly Mil Cem.
22931	Bowden. W.	Liverpool.	Pte.	20th	K.I.A	26 06 16	Cerisy-Gailly Mil Cem.
21465	Carroll. J.	Liverpool.	Pte.	19th	K.I.A	26 06 16	Cerisy-Gailly Mil Cem.
22958	Colligan. J.	Liverpool.	Pte.	20th	K.I.A	26 06 16	Cerisy-Gailly Mil Cem.
29712	Davis F	Liverpool.	Pte.	20th	K.I.A	26 06 16	Cerisy-Gailly Mil Cem.
22728	Howell. A.E.	Bootle, Liverpool.	Pte.	20th	K.I.A	26 06 16	Cerisy-Gailly Mil Cem.
22793	Norman. G.	Liverpool.	Pte.	20th	D.O.W	26 06 16	Cerisy-Gailly Mil Cem.
26568	Prince. P.	Liverpool.	Pte.	20th	K.I.A	26 06 16	Cerisy-Gailly Mil Cem.
22531	Simmonds. P.G.	Ramsbottom, Lancs.	Cpl.	20th	K.I.A	26 06 16	Cerisy-Gailly Mil Cem.
33230	Spence. R.M.	Bromborough, Ches.	Pte.	17th	K.I.A	26 06 16	Cerisy-Gailly Mil Cem.
31679	Watkinson. W.C.	Liverpool.	Pte.	20th	D.O.W	26 06 16	Cerisy-Gailly Mil Cem.
26097	Arkinstall. C.H.	Liverpool.	Pte.	17th	K.I.A	27 06 16	Cerisy-Gailly Mil Cem.
29749	Colligan. J.	Liverpool.	Pte.	17th	K.I.A	27 06 16	Bronfay Farm Mil Cem.
15906	Don. P.H.	Liverpool.	Pte.	17th	K.I.A	27 06 16	Cerisy-Gailly Mil Cem.
15449	Hufton. H.	Liverpool.	L/Cpl.	17th	K.I.A	27 06 16	Cerisy-Gailly Mil Cem.
21535	Hurry. P.W.	Liverpool.	Pte.	19th	K.I.A	27 06 16	Cerisy-Gailly Mil Cem.
15751	Keay. H.	Liscard, Cheshire.	Sgt.	17th	K.I.A	27 06 16	Cerisy-Gailly Mil Cem.
15404	King. J.E.	Chester.	Pte.	17th	K.I.A	27 06 16	Cerisy-Gailly Mil Cem.
31191	Lammie. J.H.	Birkenhead, Cheshire.	Pte.	17th	K.I.A	27 06 16	Cerisy-Gailly Mil Cem.
26090	Lang. M.	Liverpool.	Pte.	17th	K.I.A	27 06 16	Cerisy-Gailly Mil Cem.
26644	Lawton. S.	Liverpool.	L/Cpl.	17th	K.I.A	27 06 16	Cerisy-Gailly Mil Cem.
21554	Lewis. J.B.	Warrington, Cheshire.	Pte.	19th	K.I.A	27 06 16	Cerisy-Gailly Mil Cem.
29160	Marsh. W.	Seacombe, Cheshire.	Pte.	17th	K.I.A	27 06 16	Cerisy-Gailly Mil Cem.
24939	Nolan. R.	Heswell, Cheshire.	Pte.	17th	K.I.A	27 06 16	Cerisy-Gailly Mil Cem.
26106	Rea. R.M.	Belfast.	C.S.M.	17th	K.I.A	27 06 16	Cerisy-Gailly Mil Cem.
32806	Smith. W.T.	Liverpool.	Pte.	17th	K.I.A	27 06 16	Cerisy-Gailly Mil Cem.
15001	von Schwartz. C.H.	Southport, Lancs.	Pte.	17th	K.I.A	27 06 16	Cerisy-Gailly Mil Cem.
30243	Walker. B.	Liverpool.	Pte.	17th	K.I.A	27 06 16	Cerisy-Gailly Mil Cem.
15369	Whittaker. W.J.	Wavertree, Liverpool.	Pte.	17th	K.I.A	27 06 16	Cerisy-Gailly Mil Cem.
15472	Wynne. G.	Llanrwst, Denbigh.	L/Cpl.	17th	K.I.A	27 06 16	Cerisy-Gailly Mil Cem.
27334	Chadwick. G.O.	Liverpool.	Pte.	17th	K.I.A	28 06 16	Cerisy-Gailly Mil Cem.
16523	Hilton. J.	Bolton, Lancs.	Pte.	18th	K.I.A	28 06 16	Carnoy Military Cem.
	Jowett. W.H.	Grassendale, Liverpool.	2nd Lieut.	20th	D.O.W	28 06 16	La Neuville CC Corbie.
16388	Mythen. W.G.	Liverpool.	Pte.	18th	K.I.A	28 06 16	Carnoy Military Cem.
15271	Nimmo. K.P.	Liverpool.	Pte.	17th	K.I.A	28 06 16	Cerisy-Gailly Mil Cem.
29268	Worrall. T.	Liverpool.	Pte.	17th	K.I.A	28 06 16	Cerisy-Gailly Mil Cem.

21780	Griffiths. L.	Widnes, Lancs.	Pte.	19th	K.I.A	30 06 16	Cerisy-Gailly Mil Cem.
17635	Lipton. S.H.	Liverpool.	Pte.	19th	K.I.A	30 06 16	Cerisy-Gailly Mil Cem.
17706	Rymer. J.R.	Liverpool.	Pte.	19th	K.I.A	30 06 16	Cerisy-Gailly Mil Cem.
	Adam. A.de Bels M.C.	Hooton, Cheshire.	Capt.	18th	K.I.A	01 07 16	Thiepval Memorial.
16695	Adlington. F.	Liverpool.	Pte.	18th	K.I.A	01 07 16	Thiepval Memorial.
25705	Ainsworth. F.	Garston, Liverpool.	Pte.	18th	K.I.A	01 07 16	Thiepval Memorial.
22599	Allen. H.	Hale, Liverpool.	Pte.	20th	D.O.W	01 07 16	La Neuville Brit Cem, Corbie
16155	Allman. I.	Aughton, Lancs	Pte.	18th	K.I.A	01 07 16	Danzig Alley Cem, Mametz.
16350	Arrowsmith. J.F.	Sunderland	Pte.	18th	K.I.A	01 07 16	Danzig Alley Cem, Mametz.
29620	Ashcroft. C.	Liverpool.	Pte.	20th	K.I.A	01 07 16	Thiepval Memorial.
16792	Atherton. W.	Liverpool.	Pte.	18th	K.I.A	01 07 16	Thiepval Memorial.
16287	Bailey. C.	Liverpool.	Pte.	18th	K.I.A	01 07 16	Thiepval Memorial.
22314	Baker. B.	Walton, Liverpool.	Pte.	20th	K.I.A	01 07 16	Thiepval Memorial.
16951	Bales. C.W.	Liverpool.	Pte.	18th	K.I.A	01 07 16	Danzig Alley Cem, Mametz.
	Barnes. F.	Edge Hill, Liverpool.	2nd Lieut.	20th	D.O.W	01 07 16	Danzig Alley Cem, Mametz.
16435	Barnes. S.F.	West Kirby, Cheshire	Pte.	18th	K.I.A	01 07 16	Thiepval Memorial.
26050	Bellamy. A.G.A.	Liverpool.	Pte.	18th	K.I.A	01 07 16	Thiepval Memorial.
22015	Birch. T.A.	Liverpool.	Pte.	20th	K.I.A	01 07 16	Thiepval Memorial.
17169	Bird. W.H.	Liverpool.	Cpl.	18th	K.I.A	01 07 16	Danzig Alley Cem, Mametz.
16952	Blinkhorn. H.	Liverpool.	Pte.	18th	K.I.A	01 07 16	Thiepval Memorial.
22020	Boggild. D.	Liverpool.	Pte.	20th	K.I.A	01 07 16	Thiepval Memorial.
16441	Bolger. T.	Eastham, Cheshire.	Pte.	18th	K.I.A	01 07 16	Thiepval Memorial.
21450	Booth. S.	Wallasey, Cheshire.	Pte.	19th	K.I.A	01 07 16	Thiepval Memorial.
16639	Bradley. J.	Liverpool.	L/Cpl.	18th	K.I.A	01 07 16	Thiepval Memorial.
21941	Braham. P.	Liverpool.	Pte.	20th	K.I.A	01 07 16	Thiepval Memorial.
	Brockbank. C.N.	Waterloo, Liverpool.	Capt.	18th	K.I.A	01 07 16	Danzig Alley Cem, Mametz.
16768	Buckley. R.	Fleetwood, Lancs	Cpl.	18th	K.I.A	01 07 16	Thiepval Memorial.
22945	Byrne. J.	Widnes.	Pte.	20th	K.I.A	01 07 16	Thiepval Memorial.
16442	Carter. C.	Widnes, Lancs	Pte.	18th	K.I.A	01 07 16	Thiepval Memorial.
16227	Carter. E.A.	Tranmere, Birkenhead, Ches.	L/Cpl.	18th	K.I.A	01 07 16	Thiepval Memorial.
21479	Catherall. E.	Chester.	Pte.	19th	K.I.A	01 07 16	Thiepval Memorial.
16358	Cheshire. R.B.	Liverpool.	Pte.	18th	K.I.A	01 07 16	Danzig Alley Cem, Mametz.
16773	Clare. G.	Walton, Liverpool.	Cpl.	18th	K.I.A	01 07 16	Thiepval Memorial.
24456	Clarke. J.H.	Tuebrook, Liverpool.	Pte.	20th	K.I.A	01 07 16	Thiepval Memorial.
26656	Clemson. W.	Wavertree, Liverpool.	Pte.	18th	K.I.A	01 07 16	Thiepval Memorial.
17176	Cohen. J.	Liverpool.	Pte.	18th	K.I.A	01 07 16	Danzig Alley Cem, Mametz.
16359	Conlan. E.	Everton, Liverpool.	Pte.	18th	K.I.A	01 07 16	Danzig Alley Cem, Mametz.
16513	Cooper. E.R.	Birkenhead, Cheshire.	Pte.	18th	K.I.A	01 07 16	Danzig Alley Cem, Mametz.
17179	Copple. J.	Prescot, Lancs	Pte.	18th	K.I.A	01 07 16	Danzig Alley Cem, Mametz.
16960	Comah. WJ.	Liverpool.	Pte.	18th	K.I.A	01 07 16	Thiepval Memorial.
27376	Courtliff. E.	Liverpool.	Pte.	18th	K.I.A	01 07 16	Thiepval Memorial.
16164	Cowley. A.E.	Prenton, Cheshire.	Sgt.	18th	K.I.A	01 07 16	Danzig Alley Cem, Mametz.
16897	Davenport. E.	Liverpool.	Pte.	18th	D.O.W	01 07 16	Peronne Rd Cem, Maricourt.
	Dawson. G.M.	Hooton, Cheshire.	Lieut.	18th	K.I.A	01 07 16	Thiepval Memorial.
25302	Deane. H.	Liverpool.	Pte.	19th	K.I.A	01 07 16	Thiepval Memorial.
27524	Denson. H.S.	Liverpool.	Pte.	18th	K.I.A	01 07 16	Thiepval Memorial.
26076	Douglass. R.	Brookfield, Mass. USA	Pte.	18th	K.I.A	01 07 16	Thiepval Memorial.
16775	Edwards. E.V.	Margate, Kent.	Pte.	18th	K.I.A	01 07 16	Thiepval Memorial.
16172	Edwards. H.	Tranmere, Cheshire.	Pte.	18th	K.I.A	01 07 16	Thiepval Memorial.
32802	Edwards. J.	Liverpool.	Pte.	18th	K.I.A	01 07 16	Thiepval Memorial.
25328	Evans. T.C.	Crewe, Cheshire.	Pte.	19th	K.I.A	01 07 16	Thiepval Memorial.
27513	Evans. W.	Bootle, Liverpool.	Pte.	18th	K.I.A	01 07 16	Thiepval Memorial.
16714	Farrance. W.W.	Liverpool.	L/Cpl.	18th	K.I.A	01 07 16	Thiepval Memorial.
24716	Fergus. J.	Liverpool.	Pte.	18th	K.I.A	01 07 16	Thiepval Memorial.
	Fitzbrown. E.	Hunts Cross, Liverpool.	2nd Lieut.	18th	K.I.A	01 07 16	Thiepval Memorial.
16815	Gastrell. C.H.	Liverpool.	Cpl.	18th	K.I.A	01 07 16	Danzig Alley Cem, Mametz.
24583	Gerrard. P.	Litherland, Liverpool.	Pte.	18th	K.I.A	01 07 16	Thiepval Memorial.
16519	Gibbons. H.	Liverpool.	Pte.	18th	K.I.A	01 07 16	Thiepval Memorial.

22091	Gilks. A.W.	Oxhill, Warwicks.	Pte.	20th	D.O.W	01 07 16	Thiepval Memorial.
	Golds. G.B.		2nd Lieut.	18th	K.I.A	01 07 16	Danzig Alley Cem, Mametz.
25725	Grace. H.C.	Liverpool.	Pte.	18th	K.I.A	01 07 16	Thiepval Memorial.
29239	Griffiths. H.	Liverpool.	Pte.	17th	K.I.A	01 0" 16	Thiepval Memorial.
16574	Griffiths. R.H.	Liverpool.	Pte.	18th	K.I.A	01 07 16	Danzig Alley Cem, Mametz.
27342	Hagan. P.L	Liverpool.	Pte.	18th	K.I.A	01 07 16	Danzig Alley Cem, Mametz.
32890	Halliday. G.	Liverpool.	Pte.	18th	K.I.A	01 07 16	Thiepval Memorial.
26080	Hamer. G.A.	Dingle, Liverpool.	Pte.	18th	D.O.W	01 07 16	Peronne Rd Cem, Maricourt.
16521	Hartley. A.	Liverpool.	Pte.	18th	K.I.A	01 07 16	Peronne Rd Cem, Maricourt.
24737	Heal. W.H.	Ribbleton, Preston, Lancs.	Pte.	18th	K.I.A	01 07 16	Danzig Alley Cem, Mametz.
	Herdman. G.A.	Liverpool.	2nd Lieut.	18th	K.I.A	01 07 16	Thiepval Memorial.
16369	Hewitt. T.	Warrington, Cheshire.	Pte.	18th	K.I.A	01 07 16	Thiepval Memorial.
24811	Hodson. W.	Liverpool.	Pte.	17th	K.I.A	01 07 16	Thiepval Memorial.
21820	Horrocks. R.A.	Liverpool.	Pte.	19th	K.I.A	01 07 16	Peronne Rd Cem, Maricourt.
16659	Houghton. G.H.	Liverpool.	Pte.	18th	K.I.A	01 07 16	Thiepval Memorial.
25791	Howard. H.	Liverpool.	Pte.	18th	K.I.A	01 07 16	Thiepval Memorial.
16244	Hughes. H.V.	Port Sunlight, Ches	L/Sgt.	18th	K.I.A	01 07 16	Thiepval Memorial.
24580	Hughes. S.H.	Seacombe, Cheshire	Pte.	18th	K.I.A	01 07 16	Danzig Alley Cem, Mametz.
24724	Hull. T.G.	Everton, Liverpool.	Pte.	18th	K.I.A	01 07 16	Thiepval Memorial.
16373	Humphrey. P.A.	Maidstone, Kent.	Pte.	18th	K.I.A	01 07 16	Thiepval Memorial.
16970	Ingham. L.E.	Liverpool.	Pte.	18th	K.I.A	01 07 16	Thiepval Memorial.
21918	Jackson. B.A.	Liverpool.	Pte.	18th	K.I.A	01 07 16	Thiepval Memorial.
16186	Jackson. R.F.	Chester	L/Cpl.	18th	K.I.A	01 07 16	Danzig Alley Cem, Mametz.
16527	Jermy. A.W.	Liverpool.	L/Cpl.	18th	K.I.A	01 07 16	Thiepval Memorial.
33247	Johnson. S.	Runcorn, Cheshire.	Pte.	18th	K.I.A	01 07 16	Thiepval Memorial.
17159	Johnston. J.	Liverpool.	Pte.	18th	K.I.A	01 07 16	Thiepval Memorial.
16913	Johnston. J.R.	West Kirby, Cheshire	Pte.	18th	K.I.A	01 07 16	Thiepval Memorial.
16582	Jones. C.R.	Liverpool.	Cpl.	18th	K.I.A	01 07 16	Thiepval Memorial.
17240	Jones. E.	Liverpool.	Pte.	18th	K.I.A	01 07 16	Thiepval Memorial.
24621	Jones. F.J.	Birkenhead, Cheshire.	Pte.	18th	K.I.A	01 07 16	Danzig Alley Cem, Mametz.
17225	Jones. H.E.	Liverpool.	Pte.	18th	K.I.A	01 07 16	Thiepval Memorial.
17028	Jones. J.A.	Liverpool.	Pte.	18th	K.I.A	01 07 16	Danzig Alley Cem, Mametz.
29686	Jones. T.A.	Liverpool.	Pte.	18th	K.I.A	01 07 16	Thiepval Memorial.
16247	Jones. T.H.	Liverpool.	L/Cpl.	18th	K.I.A	01 07 16	Delville Wood Cem, Longueval.
16825	Joyce. E.	Co.Carlow. Ireland	L/Cpl.	18th	K.I.A	01 07 16	Danzig Alley Cem, Mametz.
16971	Kemp. D.W.	Meols, Cheshire	L/Cpl.	18th	K.I.A	01 07 16	Thiepval Memorial.
16584	Kirkwood. A.	Liverpool.	Sgt.	18th	K.I.A	01 07 16	Thiepval Memorial.
17199	Kitchen. P.J.	Liverpool.	Pte.	18th	K.I.A	01 07 16	Peronne Rd Cem, Maricourt.
	Laughlin. J.C.	Belfast.	2nd Lieut.	20th	K.I.A	01 07 16	Cerisy-Gailly Mil Cem.
34553	Leach. J.A.	Rock Ferry, Cheshire.	Pte.	18th	K.I.A	01 07 16	Danzig Alley Cem, Mametz.
27597	Leath. A.S.	Shifnal, Salop	L/Cpl.	18th	K.I.A	01 07 16	Thiepval Memorial.
16590	Leatherbarrow. E.	Leeds, Yorks	Pte.	18th	K.I.A	01 07 16	Thiepval Memorial.
24860	Leech. A.	Liverpool.	Pte.	18th	K.I.A	01 07 16	Danzig Alley Cem, Mametz.
24623	Leeson. H.	Liverpool.	L/Cpl.	18th	K.I.A	01 07 16	Danzig Alley Cem, Mametz.
16614	Leigh. A.	Liverpool.	Pte.	18th	K.I.A	01 07 16	Thiepval Memorial.
16380	Lewis. F.S.	Liverpool.	Pte.	18th	K.I.A	01 07 16	Danzig Alley Cem, Mametz.
16465	Limb. A.	Runcorn, Cheshire	Pte.	18th	K.I.A	01 07 16	Thiepval Memorial.
17415	Under. C.P.	Daluth, Minnesota, USA.	Pte.	19th	D.O.W	01 07 16	Peronne Rd Cem, Maricourt.
17247	Littlewood. C.J.	Hove, Sussex	Pte.	18th	K.I.A	01 07 16	Thiepval Memorial.
24652	Lock. W.J.	Ellesmere Port, Cheshire	Pte.	18th	K.I.A	01 07 16	Thiepval Memorial.
25780	Lomax. J.	St.Helens, Lanes	L/Cpl.	18th	K.I.A	01 07 16	Thiepval Memorial.
22475	Maclean. A.	Inverness.	L/Cpl.	20th	K.I.A	01 07 16	Thiepval Memorial.
16917	Malley. L.	Birkenhead, Cheshire.	L/Cpl.	18th	K.I.A	01 07 16	Thiepval Memorial.
15850	Manifold. J.A.	Liscard, Cheshire.	Pte.	18th	K.I.A	01 07 16	Thiepval Memorial.
24589	Marsh. T.A.	Runcorn, Cheshire	Pte.	18th	K.I.A	01 07 16	Thiepval Memorial.
29118	May P	Liverpool.	Pte.	18th	K.I.A	01 07 16	Danzig Alley Cem, Mametz.
16597	McAllister. A.F.	Liverpool.	Pte.	18th	K.I.A	01 07 16	Thiepval Memorial.
22168	McCormick. J.	Liverpool.	L/Cpl.	20th	K.I.A	01 07 16	Thiepval Memorial.

26145	McCoy. F.	Fazackerley, Liverpool.	Pte.	17th	K.I.A	01 07 16	Thiepval Memorial.
16665	McGuinness. G.	Egremont, Cheshire.	Pte.	18th	K.I.A	01 07 16	Thiepval Memorial.
24201	McNabb. J.W.	Liverpool.	Pte.	19th	K.I.A	01 07 16	Peronne Rd Cem, Maricourt.
	Merry. R.V. MC & 2 Bars	Liscard, Cheshire.	2nd Lieut.	18th	K.I.A	01 07 16	Danzig Alley Cem, Mametz.
16318	Millar. W.G.	Liverpool.	Pte.	18th	K.I.A	01 07 16	Danzig Alley Cem, Mametz.
16320	Miller. F.E.	Liverpool.	Pte.	18th	K.I.A	01 07 16	Danzig Alley Cem, Mametz.
16920	Milne. J.	New Brighton, Cheshire	Sgt.	18th	K.I.A	01 07 16	Thiepval Memorial.
16540	Moore. J.	Liverpool.	Pte.	18th	K.I.A	01 07 16	Thiepval Memorial.
16192	Morrell. G.	Liverpool.	Pte.	18th	K.I.A	01 07 16	Danzig Alley Cem, Mametz.
21989	Morris. E.	Wrexham.	Pte.	18th	K.I.A	01 07 16	Thiepval Memorial.
24226	Moseley. J.R.	Liverpool.	Pte.	19th	K.I.A	01 07 16	Peronne Rd Cem, Maricourt.
30261	Mounsey. T.	Ellesmere Port, Cheshire.	Pte.	18th	K.I.A	01 07 16	Danzig Alley Cem, Mametz.
17136	Murray. J.	Liverpool.	Pte.	18th	K.I.A	01 07 16	Peronne Rd Cem, Maricourt.
17081	Newall. H.J.	Liscard, Cheshire	L/Cpl.	18th	K.I.A	01 07 16	Danzig Alley Cem, Mametz.
24383	Ney. D.	Liverpool.	Pte.	20th	K.I.A	01 07 16	Peronne Rd Cem, Maricourt.
16671	North. H.	Liverpool.	Pte.	18th	K.I.A	01 07 16	Thiepval Memorial.
24636	O'Brien. P.	Liverpool.	Pte.	18th	K.I.A	01 07 16	Thiepval Memorial.
16198	Parry. HP.	Port Sunlight, Cheshire.	Pte.	18th	K.I.A	01 07 16	Thiepval Memorial.
16332	Pearson. A.	Liverpool.	Cpl.	18th	K.I.A	01 07 16	Thiepval Memorial.
22211	Pearson. J.	Blundellsands, Liverpool	Pte.	20th	K.I.A	01 07 16	Thiepval Memorial.
15765	Pearson. J.R.	Northallerton, Yorks.	Pte.	17th	D.O.W	01 07 16	St Sever Cem, Rouen.
27510	Pendleton. J.	Liverpool.	Pte.	18th	K.I.A	01 07 16	Thiepval Memorial.
22810	Phillips. A.	Liscard, Cheshire.	CQMS.	20th	K.I.A	01 07 16	Thiepval Memorial.
16604	Pickersgill. T.S.	Liverpool.	Pte.	18th	K.I.A	01 07 16	Thiepval Memorial.
17206	Pierce. J.L.	Liscard, Cheshire.	Pte.	18th	K.I.A	01 07 16	Thiepval Memorial.
16333	Pitcher. J.S.	Liverpool.	Pte.	18th	K.I.A	01 07 16	Thiepval Memorial.
16330	Potter. S.	Egremont, Cheshire.	Pte.	18th	K.I.A	01 07 16	Thiepval Memorial.
16737	Powell. G.	Handsworth, Birmingham.	Pte.	18th	K.I.A	01 07 16	Thiepval Memorial.
30184	Prescott. H.	Maghull, Liverpool.	Pte.	18th	K.I.A	01 07 16	Danzig Alley Cem, Mametz.
16331	Price. H.S.	Seacombe, Cheshire.	L/Sgt.	18th	K.I.A	01 07 16	Thiepval Memorial.
17928	Pringle. LH.	Seacombe, Ches.	L/Cpl.	19th	K.I.A	01 07 16	Peronne Rd Cem, Maricourt.
16863	Pritchard. J.	Liverpool.	Pte.	18th	K.I.A	01 07 16	Thiepval Memorial.
17442	Proctor. O.	Liverpool.	Pte.	19th	K.I.A	01 07 16	Thiepval Memorial.
16742	Range. J.H.	Liverpool.	Pte.	18th	K.I.A	01 07 16	Thiepval Memorial.
16268	Rawlinson. J.C.	Liverpool.	Pte.	18th	K.I.A	01 07 16	Danzig Alley Cem, Mametz.
21619	Rawsthorne. W.F.	Liverpool.	Pte.	19th	K.I.A	01 07 16	Peronne Rd Cem, Maricourt.
30259	Richmond. T.	Liverpool.	Pte.	18th	K.I.A	01 07 16	Thiepval Memorial.
15470	Riding. J.S.	Liverpool.	Pte.	17th	D.O.W	01 07 16	La Neuville CC Corbie
16934	Roberts. D.S.	Birkenhead, Cheshire.	Pte.	18th	K.I.A	01 07 16	Thiepval Memorial.
16336	Roberts. N.F.	Liverpool.	Pte.	18th	K.I.A	01 07 16	Thiepval Memorial.
16549	Roberts. O.W.	Chertsey, Surrey.	L/Cpl.	18th	K.I.A	01 07 16	Thiepval Memorial.
17144	Roberts. R.J.	Liverpool.	L/Cpl.	18th	K.I.A	01 07 16	Danzig Alley Cem, Mametz.
25756	Roberts. W.	Liverpool.	Pte.	18th	K.I.A	01 07 16	Danzig Alley Cem, Mametz.
16479	Robinson. W.	Eastham, Cheshire.	Pte.	18th	K.I.A	01 07 16	Thiepval Memorial.
16680	Rose. G.	Liverpool.	Cpl.	18th	K.I.A	01 07 16	Thiepval Memorial.
26654	Roskell. R.	Ambleside, Westmorland.	Pte.	18th	K.I.A	01 07 16	Thiepval Memorial.
27585	Rylands. J.	Liverpool.	Pte.	18th	K.I.A	01 07 16	Danzig Alley Cem, Mametz.
17162	Sanders. T.	Liverpool.	Pte.	18th	K.I.A	01 07 16	Thiepval Memorial.
16997	Scarff. G.L.	Liverpool.	Pte.	18th	K.I.A	01 07 16	Danzig Alley Cem, Mametz.
16209	Scarlett. D.	Liverpool.	Pte.	18th	K.I.A	01 07 16	Danzig Alley Cem, Mametz.
24615	Scholes. J.C.E.	Torquay, Devon.	L/Cpl.	18th	K.I.A	01 07 16	Peronne Rd Cem, Maricourt.
16396	Schulke. T. (Shaw)	Liverpool.	Pte.	18th	K.I.A	01 07 16	Thiepval Memorial
16613	Seanor. F.A.	Liverpool.	Pte.	18th	K.I.A	01 07 16	Thiepval Memorial.
16870	Seddon. G.R.	Fazackerley, Liverpool.	Pte.	18th	K.I.A	01 07 16	Thiepval Memorial.
16397	Seiffert. F.F.	Liverpool.	L/Cpl.	18th	K.I.A	01 07 16	Thiepval Memorial.
23827	Sharp. W.H.	Birkenhead, Cheshire.	Pte.	20th	K.I.A	01 07 16	Thiepval Memorial.
16396	Shaw. T.	Ilford, London.	Pte.	18th	K.I.A	01 07 16	Thiepval Memorial.
17090	Short. W.L.	Liverpool.	Pte.	18th	K.I.A	01 07 16	Thiepval Memorial.

17722	Shortall. W.	Liverpool.	Pte.	19th	K.I.A	01 07 16	Thiepval Memorial.
16998	Showell. C.H.	Cressington, Liverpool.	Pte.	18th	K.I.A	01 07 16	Danzig Alley Cem, Mametz.
16212	Silver. FT.	Saltney, Cheshire.	Cpl.	18th	K.I.A	01 07 16	Danzig Alley Cem, Mametz.
24594	Simpson. M. M.M.	Liverpool.	Pte.	18th	K.I.A	01 07 16	Peronne Rd Cem, Maricourt.
22539	Sixsmith. R.	Upholland, Wigan, Lancs.	Pte.	20th	K.I.A	01 07 16	Thiepval Memorial.
24494	Smaller. J.W.	Kirkdale, Liverpool.	Pte.	20th	K.I.A	01 07 16	Thiepval Memorial.
22259	Smith. G.	York.	Pte.	20th	D.O.W	01 07 16	Thiepval Memorial.
16423	Smith. J.H.	Liverpool.	Pte.	18th	K.I.A	01 07 16	Thiepval Memorial.
21899	Smith. J.W.	West Lavington, Wilts.	Pte.	18th	K.I.A	01 07 16	Thiepval Memorial.
21789	Smith. S.H.	Liverpool.	Pte.	19th	K.I.A	01 07 16	Cerisy-Gailly Mil Cem.
16205	Sparrow. H.	Hoylake, Cheshire.	Pte.	18th	K.I.A	01 07 16	Thiepval Memorial.
16210	Stanley. E.G.	Birkenhead, Cheshire.	Sgt.	18th	D.O.W	01 07 16	Thiepval Memorial.
25703	Stanway. W.A.	Ellesmere Port, Cheshire.	Pte.	18th	K.I.A	01 07 16	Thiepval Memorial.
16490	Taylor. F.J.	Eastham, Cheshire.	Pte.	18th	K.I.A	01 07 16	Thiepval Memorial.
16402	Thomas. J.P.	Llandaff, Glam.	L/Cpl.	18th	K.I.A	01 07 16	Thiepval Memorial.
16492	Thompson. G.H.	Liverpool.	Cpl.	18th	K.I.A	01 07 16	Thiepval Memorial. «
16788	Todd. A.R.	Liverpool.	Pte.	18th	K.I.A	01 07 16	Danzig Alley Cem, Mametz.
16278	Tunstall. W.E.	Wallasey, Cheshire.	Pte.	18th	K.I.A	01 07 16	Danzig Alley Cem, Mametz.
17005	Turner. W.	Bootle, Liverpool.	L/Cpl.	18th	K.I.A	01 07 16	Thiepval Memorial.
21871	Wake. G.T.	Liverpool.	Pte.	18th	K.I.A	01 07 16	Thiepval Memorial.
26696	Walker. G.	Liverpool.	Pte.	18th	K.I.A	01 07 16	Thiepval Memorial.
24670	Waring. E.	Runcorn, Cheshire.	Pte.	18th	K.I.A	01 07 16	Thiepval Memorial.
16494	Waters. R.	West Kirby, Cheshire.	Pte.	18th	K.I.A	01 07 16	Thiepval Memorial.
30738	Webb. G.	Kirkdale, Liverpool.	Pte.	19th	K.I.A	01 07 16	Thiepval Memorial.
22882	Webber. W.D.	Liverpool.	Pte.	20th	D.O.W	01 07 16	Peronne Rd Cem, Maricourt.
17097	West. G.	Manchester.	Pte.	18th	K.I.A	01 07 16	Thiepval Memorial.
16343	Westmorland. F.J.	Liverpool.	Pte.	18th	K.I.A	01 07 16	Danzig Alley Cem, Mametz.
17014	Whitfield. R.	Poulton, Cheshire.	Pte.	18th	K.I.A	01 07 16	Danzig Alley Cem, Mametz.
23996	Whyte. W.H.	Runcorn, Cheshire.	Pte.	20th	D.O.W	01 07 16	La Neuville Brit Cem, Corbie
17229	Williams. F.C.	Seaforth, Liverpool.	Pte.	18th	K.I.A	01 07 16	Thiepval Memorial.
22895	Williams. G.	Liverpool.	Pte.	20th	K.I.A	01 07 16	Serre Rd No 2 Cem, B-Hamel.
17034	Williams. J.S.	Liverpool.	L/Cpl.	18th	K.I.A	01 07 16	Thiepval Memorial.
17251	Williams. R.K.	Liverpool.	L/Cpl.	18th	K.I.A	01 07 16	Danzig Alley Cem, Mametz.
16420	Wilson. J.	Liverpool.	Pte.	18th	K.I.A	01 07 16	Thiepval Memorial.
17100	Wilson. J.	Liverpool.	Pte.	18th	K.I.A	01 07 16	Danzig Alley Cem, Mametz.
24992	Wilson. W.	Liverpool.	Pte.	18th	K.I.A	01 07 16	Thiepval Memorial.
17223	Wood. J.	Liverpool.	L/Cpl.	18th	K.I.A	01 07 16	Danzig Alley Cem, Mametz.
24386	Wood. W.	Liverpool.	Pte.	19th	K.I.A	01 07 16	Cerisy-Gailly Mil Cem.
27502	Woods. A.R.	New Brighton, Cheshire.	Pte.	18th	K.I.A	01 07 16	Thiepval Memorial.
29117	Woodward. R.	Liverpool.	Pte.	18th	D.O.W	01 07 16	Peronne Rd Cem, Maricourt.
16558	Woolmer. S.	Bolton, Lancs.	L/Cpl.	18th	K.I.A	01 07 16	Thiepval Memorial.
30195	Wright. W.	Liverpool.	Pte.	18th	K.I.A	01 07 16	Danzig Alley Cem, Mametz.
16431	Yates. H.	Liverpool.	Pte.	18th	K.I.A	01 07 16	Danzig Alley Cem, Mametz.
11057	Zedickson. H.	Liverpool.	Pte.	18th	K.I.A	01 07 16	Danzig Alley Cem, Mametz.
15287	Arnold. S.	Liverpool.	Sgt.	17th	K.I.A	02 07 16	Thiepval Memorial.
27569	Bray. H.S.	Walton, Liverpool.	Pte.	17th	K.I.A	01/02 07 16	Thiepval Memorial.
16888	Byrne. W.	Birkenhead, Cheshire.	Pte.	18th	D.O.W	02 07 16	Dive Copse B.C. Sailly-le-Sec.
16119	Crawford. H.	Liverpool.	Pte.	17th	D.O.W	02 07 16	La Neuville Brit Cem, Corbie
21952	Gilroy. C.	Liverpool.	Pte.	17th	K.I.A	01/02 07 16	Thiepval Memorial.
15114	Harris. S.F.	Liverpool.	Pte.	17th	K.I.A	01/02 07 16	Thiepval Memorial.
31161	Holroyd. W.G.	Liverpool.	Pte.	17th	K.I.A	02 07 16	Peronne Rd Cem, Maricourt.
17228	Jones C.D.	Liscard, Cheshire.	Pte.	18th	D.O.W	02 07 16	Daours CC Ext.
24976	Jones. W.E.	Liverpool.	Pte.	17th	K.I.A	02 07 16	Thiepval Memorial.
15410	Kelsall A.J.	Blundellsands, Liverpool.	Pte.	17th	K.I.A	02 07 16	Thiepval Memorial.
16190	Laid. H.	London	L/Cpl.	18th	D.O.W	02 07 16	Heilly Station Cem.Mericourt
15062	Lee. J.C.	Roby, Liverpool.	Pte.	17th	K.I.A	01/02 07 16	Thiepval Memorial.
24974	McLauchlen. J.	Liverpool.	Pte.	17th	K.I.A	01/02 07 16	Thiepval Memorial.
15840	Passmore. C.C.	Ditton, Widnes, Lancs.	Pte.	17th	K.I.A	02 07 16	Danzig Alley Cem, Mametz.

	Scott DH.	Birkenhead, Cheshire.	Lieut.	17th	D.O.W 02 07 16	Heath Cem, Harbonnieres.
16147	Sharpe. R.D.	Liverpool.	Pte.	17th	K.I.A 01/02 07 16	Thiepval Memorial.
26024	Spencer T.R.	Seaforth, Liverpool.	L/Cpl.	18th	D.O.W 02 07 16	Daours CC Ext.
23660	Thelwall. E.	Tarporley, Cheshire.	Pte.	17th	K.I.A 01/02 07 16	Thiepval Memorial.
16031	Thomas. G.W.	Wallasey, Cheshire.	L/Cpl.	17th	K.I.A 01/02 07 16	Thiepval Memorial.
15412	Tucker. P.J.	Liverpool.	Pte.	17th	K.I.A 01/02 07 16	Thiepval Memorial.
	Walker. T.R.	Eastham, Cheshire.	2nd Lieut.	18th	D.O.W 02 07 16	Peronne Rd Cem, Maricourt.
	Withy B	West Hartlepool, Durham.	Lieut.	18th	D.O.W 02 07 16	La Neuville CC Corbie.
21845	Bateman. N	Liverpool.	Pte.	19th	D.O.W 03 07 16	Dive Copse B.C. Sailly-le-Sec.
24692	Bell. J.	Bootle, Liverpool.	Pte.	17th	K.I.A 03 07 16	Thiepval Memorial.
22936	Bradshaw. N.	Liverpool.	Pte.	20th	K.I.A 03 07 16	Thiepval Memorial.
32817	Brown. J.H.	Liverpool.	Pte.	17th	K.I.A 03 07 16	Thiepval Memorial.
22637	Carroll. J.J.	Liverpool.	Pte.	20th	K.I.A 03 07 16	Thiepval Memorial.
29648	Cave. P.	Croston, Lancs.	Pte.	17th	K.I.A 03 07 16	Thiepval Memorial.
22956	Colvin. G.	Liverpool.	Pte.	20th	K.I.A 03 07 16	Daours CC Ext.
22655	Cosgriff. B.	Liverpool.	Pte.	20th	D.O.W 03 07 16	Heilly Station Cem.Mericourt
23931	Dilworth. J.	Seacombe, Cheshire.	Pte.	20th	K.I.A 03 07 16	Thiepval Memorial.
15540	Downey. H.F.	Liverpool.	Pte.	17th	K.I.A 03 07 16	Thiepval Memorial.
16899	Ellis. T.W.	Bootle, Liverpool.	Pte.	17th	K.I.A 03 07 16	Peronne Rd Cem, Maricourt.
22085	Garton. S.V.	Liverpool.	Pte.	20th	D.O.W 03 07 16	Daours CC Ext.
16178	Green. H.	New Ferry, Cheshire	L/Cpl.	18th	D.O.W 03 07 16	Heilly Station Cem.Mericourt
24876	Halsall. W.F.	Liverpool.	Pte.	17th	K.I.A 03 07 16	Thiepval Memorial.
16085	Hivey. C.H.	Liverpool.	Pte.	17th	K.I.A 03 07 16	Thiepval Memorial.
22731	Hunt. N.W.	Bootle, Liverpool.	Cpl.	20th	K.I.A 03 07 16	Thiepval Memorial.
34393	Jones. R.	Liverpool.	Pte.	20th	K.I.A 03 07 16	Thiepval Memorial.
33193	Kirby. J.	Liverpool.	L/Cpl.	17th	K.I.A 03 07 16	Thiepval Memorial.
22177	Morris. H.	Liverpool.	Pte.	20th	K.I.A 03 07 16	Thiepval Memorial.
23877	Murphy. J.J.	Ballyboy, Co.Tyrone.	Pte.	20th	K.I.A 03 07 16	Thiepval Memorial.
22217	Pinkerton. W.L.	Birkenhead, Cheshire.	Pte.	20th	K.I.A 03 07 16	Thiepval Memorial.
29135	Schofield. F.G.	Liverpool.	Pte.	20th	K.I.A 03 07 16	Thiepval Memorial.
15808	Shone. E.E.	Liverpool.	Pte.	17th	K.I.A 03 07 16	Peronne Rd Cem, Maricourt.
23845	Todd. H.J.	Rock Ferry, Cheshire.	Pte.	20th	K.I.A 03 07 16	Thiepval Memorial.
15707	Waldron. W.J.	Liverpool.	Pte.	17th	K.I.A 03 07 16	Thiepval Memorial.
22897	Williams. D.	Beddgelert, N.Wales.	Pte.	20th	K.I.A 03 07 16	Thiepval Memorial.
24572	Wing. W.	Liverpool.	Pte.	17th	K.I.A 03 07 16	Thiepval Memorial.
15103	Wright. A.W.	Liverpool.	Pte.	17th	K.I.A 03 07 16	Thiepval Memorial.
17191	Grundy. R.	Liscard, Cheshire.	Pte.	18th	D.O.W 04 07 16	Daours CC Ext.
22123	Hughes. J.	Litherland, Liverpool.	Pte.	20th	K.I.A 04 07 16	Thiepval Memorial.
16976	Lindon. F.	Liverpool.	Cpl.	18th	D.O.W 04 07 16	St.Pierre Cem, Amiens
	Davies. LR.	Waterloo, Liverpool.	2nd Lieut.	18th	D.O.W 05 07 16	La Neuville CC Corbie.
17392	Jones. T.H.	Montgomery.	Pte.	19th	D.O.W 05 07 16	Daours CC Ext.
16730	Murphy. E.	Liverpool.	Pte.	18th	D.O.W 05 07 16	St Sever Cem, Rouen.
15309	Cochrane. G.	Liverpool.	Pte.	17th	D.O.W 06 07 16	St Sever Cem, Rouen.
	Barnard. N.A.S.	Liverpool.	2nd Lieut.	18th	K.I.A 08 07 16	Peronne Rd Cem, Maricourt.
16585	Kirkwood. J.	Liverpool.	C.S.M.	18th	K.I.A 08 07 16	Thiepval Memorial.
16485	Scott. S.	Eastham, Cheshire	L/Cpl.	18th	K.I.A 08 07 16	Thiepval Memorial.
16617	Tabron. T.	Seaforth, Liverpool.	Pte.	18th	K.I.A 08 07 16	Peronne Rd Cem, Maricourt.
23846	Toombs. J.H.	Omeath, Co.Louth.	Pte.	20th	D.O.W 08 07 16	Etaples Mil Cem
	Trotter. E.H. D.S.O.	London.	Lt.Col.	18th	K.I.A 08 07 16	Peronne Rd Cem, Maricourt.
17488	Tunstall. G.	Liscard, Cheshire.	Pte.	19th	D.O.W 08 07 16	Abbeville C.C.
17253	Williams. D.R.	Liverpool.	L/Cpl.	18th	D.O.W 08 07 16	Abbeville C.C. „
17044	Boosey. C.A.	Liverpool.	Pte.	18th	K.I.A 09 07 16	Peronne Rd Cem, Maricourt.
24657	Jones. N.E.	Liverpool.	Pte.	18th	D.O.W 09 07 16	St Sever Cem, Rouen.
22012	Beattie. T.B.	Liverpool.	Pte.	20th	K.I.A 10 07 16	Thiepval Memorial.
15962	Caldicott. W.	Ellesmere Port, Cheshire	Pte.	17th	K.I.A 10 07 16	Thiepval Memorial.
22051	Davies. J.	Birkenhead, Cheshire.	Pte.	20th	K.I.A 10 07 16	Bernafay Wood, Brit Cem.
24927	Elliott. H.	Liverpool.	Pte.	17th	K.I.A 10 07 16	Thiepval Memorial.
15164	Freestone. H.	Bromborough, Cheshire	L/Cpl.	17th	K.I.A 10 07 16	Thiepval Memorial.

15646	Grantham. R.E.	Liverpool.	Pte.	17th	K.I.A	10 07 16	Thiepval Memorial.
15855	Harrald. H.J.	Liverpool.	Pte.	17th	K.I.A	10 07 16	Bernafay Wood, Brit Cem.
	Higgins. G.F.	Rock Ferry, Cheshire.	Major.	17th	K.I.A	10 07 16	Thiepval Memorial.
29238	Jones. W.H.	Liverpool.	Pte.	17th	K.I.A	10 07 16	Thiepval Memorial.
29154	McEachern. J.J.	South Boston, Mass. USA.	A/Cpl.	20th	K.I.A	10 07 16	Thiepval Memorial.
33216	Miller. H.	Liverpool.	L/Cpl.	17th	K.I.A	10 07 16	Thiepval Memorial.
24975	Roberts. J.	Liverpool.	Pte.	17th	K.I.A	10 07 16	Bernafay Wood, Brit Cem.
23829	Sherry. J.A.	Liverpool.	Sgt	20th	D.O.W	10 07 16	Abbeville C.C.
15741	Stephens. J.Mc.	Birkenhead, Cheshire.	Pte.	17th	K.I.A	10 07 16	Thiepval Memorial.
21939	Taylor. A.	Prescot, Lancs.	Pte.	17th	K.I.A	10 07 16	Thiepval Memorial.
31137	Tyndall. V.J.	Liverpool.	Pte.	17th	K.I.A	10 07 16	Thiepval Memorial.
29720	Alleyne. A.S.	Liverpool.	Pte.	19th	K.I.A	11 07 16	Thiepval Memorial.
23916	Bradshaw. A.	Runcorn, Cheshire.	Pte.	20th	D.O.W	11 07 16	Etaples Mil Cem
21689	Carew. F.	Liverpool.	L/Cpl.	19th	K.I.A	11 07 16	Thiepval Memorial.
16512	Clarke. D.H.	Wavertree, Liverpool.	Pte.	18th	D.O.W	11 07 16	Abbeville C.C.
17802	Colyer. T.	Liverpool.	Pte.	19th	K.I.A	11 07 16	Thiepval Memorial.
22675	Dobson. J.J.	Bootle, Liverpool.	Pte.	20th	K.I.A	11 07 16	Thiepval Memorial.
25569	Duckett. W.M.	Liverpool.	Pte.	19th	K.I.A	11 07 16	Thiepval Memorial.
25587	Edwards. G.E.	Garston, Liverpool.	Pte.	19th	K.I.A	11 07 16	Thiepval Memorial.
17820	Elliott. F.	Fairfield, Liverpool.	Pte.	19th	K.I.A	11 07 16	Thiepval Memorial.
17818	Ellis. K.H.	Liverpool.	Sgt.	19th	K.I.A	11 07 16	Thiepval Memorial.
21407	Fraser. C.B	Liverpool.	Pte.	19th	K.I.A	11 07 16	Thiepval Memorial.
15846	Gleave. H.G.	Wallasey, Cheshire.	Pte.	17th	K.I.A	11 07 16	Bernafay Wood, Brit Cem.
22094	Gordon. J.J.	Widnes, Lancs.	Pte.	20th	D.O.W	11 07 16	Etaples Mil Cem
17845	Harper. H.	Bootle, Liverpool.	L/Cpl.	19th	K.I.A	11 07 16	Thiepval Memorial.
23027	Harrison. J.P.	Liverpool.	Pte.	20th	K.I.A	11 07 16	Flatiron Copse Cem, Mametz.
17842	Hunt. H.	Liverpool.	Pte.	19th	K.I.A	11 07 16	Thiepval Memorial.
17872	Irvine. D.W.	Liverpool.	Sgt.	19th	K.I.A	11 07 16	Thiepval Memorial.
21412	Kelly. H.F.	Liverpool.	Pte.	19th	K.I.A	11 07 16	Thiepval Memorial.
21413	Kingston. H.	Southport, Lancs.	Pte.	19th	K.I.A	11 07 16	Bernafay Wood, Brit Cem.
17906	McRobie. H.D.	Liverpool.	L/Cpl.	19th	K.I.A	11 07 16	Thiepval Memorial.
15961	Miles. W.J.	Liverpool.	Sgt.	17th	K.I.A	11 07 16	Thiepval Memorial.
15833	Pearson. K.	Birkenhead, Cheshire.	Pte.	17th	D.O.W	11 07 16	Peronne Rd Cem, Maricourt.
17916	Peart. H.	Thornage, Norfolk.	Pte.	19th	K.I.A	11 07 16	Thiepval Memorial.
21688	Slater. W.T.W.	Liverpool.	Pte.	19th	K.I.A	11 07 16	Thiepval Memorial.
	Small. H.A. M.C.	Belfast.	2nd Lieut.	20th	K.I.A	11 07 16	Thiepval Memorial.
27315	Smyth. D.	Liverpool.	Pte.	17th	K.I.A	11 07 16	Thiepval Memorial.
	Sproat. J.Mc. M.C.	Rock Ferry, Cheshire.	2nd Lieut.	17th	K.I.A	11 07 16	Thiepval Memorial.
17962	Thomas. J.	Liverpool.	Pte.	19th	K.I.A	11 07 16	Thiepval Memorial.
17959	Turner. A.	Stockport, Cheshire.	Pte.	19th	K.I.A	11 07 16	Thiepval Memorial.
17984	Warren. J.H.	Wallasey. Ches.	Pte.	19th	K.I.A	11 07 16	Thiepval Memorial.
17979	Williams. N.L	Hoylake, Cheshire.	Pte.	19th	K.I.A	11 07 16	Thiepval Memorial.
15874	Askew. W.H.C.	Rock Ferry, Cheshire.	Cpl.	17th	K.I.A	12 07 16	Peronne Rd Cem, Maricourt.
32807	Carmichael. J.A.	Liverpool.	Pte.	17th	K.I.A	12 07 16	Peronne Rd Cem, Maricourt.
16087	Cowman. R.J.	Liverpool.	Cpl.	17th	K.I.A	12 07 16	Thiepval Memorial.
26663	Crowe. J.W.	Anfield, Liverpool.	Pte.	17th	K.I.A	12 07 16	Bernafay Wood, Brit Cem.
15424	Croxson. C.G.	Bromborough, Cheshire.	L/Cpl.	17th	K.I.A	12 07 16	Bernafay Wood, Brit Cem.
15241	Damsell. A.S.	Liverpool.	Pte.	17th	K.I.A	12 07 16	Dive Copse B.C. Sailly-le-Sec.
31521	Ecroyd. F.H.	Accrington, Lancs.	Pte.	17th	K.I.A 10/12 07 16		Thiepval Memorial.
15436	Hancock. H.	Liverpool.	Pte.	17th	K.I.A 10/12 07 16		Thiepval Memorial.
31164	Hankin. E.	Liverpool.	Pte.	17th	K.I.A 10/12 07 16		Thiepval Memorial.
24834	Harvey. H.W.	Wallasey, Cheshire.	L/Cpl.	17th	K.I.A 10/12 07 16		Thiepval Memorial.
26113	Harvey. NT.	Liverpool.	Pte.	17th	K.I.A	12 07 16	Bernafay Wood, Brit Cem.
15002	Henshaw. A.J.	West Kirby, Cheshire.	L/Cpl.	17th	K.I.A	12 07 16	Delville Wood Cem, Longueval.
22404	Hewitson. H.	Liverpool.	Pte.	20th	K.I.A	12 07 16	Guillemont Rd Cem.
	Hick. H.C.		2nd Lieut	19th	K.I.A	12 07 16	Peronne Rd Cem, Maricourt.
22415	Holt. J.D.	Runcorn, Cheshire.	L/Cpl.	20th	K.I.A	12 07 16	Thiepval Memorial.
24484	Houseman. W.	Liverpool.	Pte.	20th	D.O.W	12 07 16	Abbeville C.C.

15065	Hunt. T.H.	Crewe, Cheshire.	L/Cpl.	17th	K.I.A 10/12 07 16	Thiepval Memorial.
27364	Kerr. J.G.	Liverpool.	Pte.	17th	K.I.A 10/12 07 16	Thiepval Memorial.
26115	Kniveton. J.K.	West Kirby, Cheshire.	Pte.	17th	K.I.A 12 07 16	Bernafay Wood, Brit Cem.
15880	Lenthall. W.H.	Liverpool.	Pte.	17th	K.I.A 10/12 07 16	Thiepval Memorial.
26619	Lumb. A.S.	Liverpool.	Pte.	17th	K.I.A 12 07 16	Thiepval Memorial.
15237	Mather. E.	Cheltenham, Glos.	Pte.	17th	K.I.A 10/12 07 16	Thiepval Memorial.
24932	McDonald. J.	Southport, Lancs.	Pte.	17th	K.I.A 12 07 16	Bernafay Wood, Brit Cem.
15149	McDowell. G.	Liverpool.	Pte.	17th	K.I.A 10/12 07 16	Thiepval Memorial.
29782	Mercer. E.H.	Seven Kings, Essex.	Pte.	17th	K.I.A 10/12 07 16	Thiepval Memorial.
23614	Morley. G.	Leyton, Essex.	C.S.M.	17th	K.I.A 10/12 07 16	Thiepval Memorial.
15006	Moses. F.	New Brighton, Cheshire.	Sgt.	17th	K.I.A 10/12 07 16	Thiepval Memorial.
24979	Penkman. E.	Liverpool.	Pte.	17th	K.I.A 12 07 16	Thiepval Memorial.
15789	Ryder. C.	Liverpool.	Pte.	17th	K.I.A 10/12 07 16	Thiepval Memorial.
15878	Sampson. J.F.	New Brighton, Ches.	Pte.	17th	K.I.A 12 07 16	Peronne Rd Cem, Maricourt.
15904	Spence. R.L	Liverpool.	Cpl.	17th	K.I.A 12 07 16	Serre Rd No.2 Cem, B-Hamel.
25720	Waterworth. A.	Seacombe, Cheshire.	L/Cpl.	17th	K.I.A 10/12 07 16	Thiepval Memorial.
15748	Williams. F.G.	Liverpool.	Pte.	17th	K.I.A 12 07 16	Thiepval Memorial.
15320	Williams. G.R.	Birkenhead, Cheshire.	Pte.	17th	K.I.A 10/12 07 16	Thiepval Memorial.
15500	Woodney. H.	Liverpool.	Pte.	17th	K.I.A 12 07 16	Bernafay Wood, Brit Cem.
22730	Holland. J.	Liverpool.	Pte.	20th	K.I.A 13 07 16	Guillemont Rd Cem.
26068	Johnson. E.	Liverpool.	Pte.	17th	D.O.W 13 07 16	Peronne Rd Cem, Maricourt.
23636	Skirrow. J.	Wallasey, Cheshire.	Pte.	17th	D.O.W 13 07 16	Wallasey, Rake Lane Cem.
16848	Truby. W.J.	Liverpool.	Pte.	18th	D.O.W 13 07 16	Heilly Station Cem.Mericourt
23989	Usherwood. W.G.	Liverpool.	Pte.	20th	D.O.W 13 07 16	Corbie CC Ext.
34808	Furber. J.H.	Liverpool.	Pte.	17th	D.O.W 15 07 16	La Neuville Brit Cem, Corbie
	Griffin. D.M.	Hampstead, London.	2nd Lieut.	18th	D.O.W 16 07 16	Abbeville C.C.
16890	Brotherston. W.J.	Liscard, Cheshire.	Pte.	18th	D.O.W 17 07 16	Guildford, (Old Stoke) Surrey.
29186	Greener. E.J.	Liverpool.	Pte.	19th	D.O.W 17 07 16	St Sever Cem, Rouen.
27346	Mawdsley. J.	Liverpool.	Pte.	18th	D.O.W 18 07 16	St Sever Cem, Rouen.
	Tomlinson. R.H.	Liverpool.	2nd Lieut.	18th	D.O.W 19 07 16	St Sever Cem, Rouen.
15437	Jones. W.F.	Liverpool.	Sgt.	17th	D.O.W 20 07 16	Daours CC Ext.
16376	Kirkpatrick. G.H.A	Bootle, Liverpool.	Pte.	18th	D.O.W 20 07 16	Bootle Cem, Liverpool.
24882	Mercer. J.A.	Runcorn, Cheshire	Pte.	18th	D.O.W 20 07 16	Ashton-by-Sutton,(St.Peters CY)
29673	O'Neill. J.	Dublin.	Pte.	19th	D.O.W 20 07 16	Abbeville C.C.
16744	Redhead. J.	Liverpool.	Pte.	18th	D.O.W 20 07 16	Boulogne East Cem.
23897	Wright. J.H.	Liverpool.	Pte.	20th	K.I.A 20 07 16	Thiepval Memorial.
22921	Barclay. R.	Chippenham, Wilts.	Pte.	20th	D.O.W 21 07 16	Bronfay Farm Mil Cem.
16855	Massam. J.	Seacombe, Cheshire	Pte.	18th	D.O.W 22 07 16	Liverpool, Yew Tree CY.
32076	Durkin. J.	Haslingden, Lancs	Pte.	18th	K.I.A 23 07 16	Thiepval Memorial.
16069	Dickinson. W.	Liverpool.	Pte.	17th	D.O.W 24 07 16	Abbeville C.C.
17087	Prescott. F.	Maghull, Liverpool.	Pte.	18th	D.O.W 25 07 16	Etaples Mil Cem
26521	Winter. T.J.	Liverpool.	Pte.	19th	Died 25 07 16	Kensal Green Cem, London.
24202	Livesay. J.	Chorley, Lancs.	Pte.	19th	K.I.A 29 07 16	Thiepval Memorial.
16428	Nixon. H.	Stoke-on-Trent, Staffs.	Pte.	18th	D.O.W 29 07 16	Hanley Cem, Stoke-on-Trent.
17681	Plaskett. W.C.	Wheelock, Cheshire.	Pte.	19th	K.I.A 29 07 16	Thiepval Memorial.
15199	Abell. C.	Anfield, Liverpool.	Pte.	17th	K.I.A 30 07 16	Thiepval Memorial.
22598	Ainsworth. C.H.	Liverpool.	Pte.	20th	K.I.A 30 07 16	Thiepval Memorial.
17776	Allan. G.W.	Prenton, Cheshire.	L/Cpl.	19th	K.I.A 30 07 16	Thiepval Memorial.
22306	Allen. R.E.	Brockton, Salop.	Pte.	20th	K.I.A 30 07 16	Guillemont Rd Cem.
21673	Alsop. C.L.	Liverpool.	Pte.	19th	K.I.A 30 07 16	Thiepval Memorial.
21825	Andrews. A.	Edge Hill, Liverpool.	Pte.	19th	K.I.A 30 07 16	Thiepval Memorial.
22311	Atherton. S.	Birkenhead, Cheshire.	L/Cpl.	20th	K.I.A 30 07 16	Guillemont Rd Cem.
22607	Atherton. W.	Birkenhead, Cheshire.	Sgt.	20th	K.I.A 30 07 16	Thiepval Memorial.
22312	Austin. H.S.	Liverpool.	Pte.	20th	K.I.A 30 07 16	Guillemont Rd Cem.
17998	Bagnall. H.G.	Liverpool.	L/Cpl.	19th	K.I.A 30 07 16	Thiepval Memorial.
32732	Baker. C.T.P.	Stanley, Liverpool.	Pte.	19th	K.I.A 30 07 16	Thiepval Memorial.
25347	Baker. W.H.P.	Kirkdale, Liverpool.	L/Cpl.	19th	K.I.A 30 07 16	Thiepval Memorial.
30255	Ball. L.Y.	Runcorn, Cheshire.	Pte.	19th	K.I.A 30 07 16	Thiepval Memorial.

35762	Bantoft. T.	Droylsden, Manchester.	Pte.	20th	K.I.A	30 07 16	Delville Wood Cem, Longueval.
17785	Barlow. G.H.	Liverpool.	Sgt.	19th	K.I.A	30 07 16	Guillemont Rd Cem.
21443	Bartlett. E.	Liverpool.	Pte.	19th	K.I.A	30 07 16	Thiepval Memorial.
22320	Bason. T.	Liverpool.	Pte.	20th	K.I.A	30 07 16	Thiepval Memorial.
15389	Beard. W.L.	New Brighton, Cheshire.	Sgt.	17th	K.I.A	30 07 16	Thiepval Memorial.
21446	Bebbington. G.A.	Knutsford, Cheshire.	L/Cpl.	19th	K.I.A	30 07 16	Guillemont Rd Cem.
22925	Bell. W.E.	Liverpool.	Pte.	20th	K.I.A	30 07 16	Thiepval Memorial.
29256	Bellion. G.	Wavertree, Liverpool.	Pte.	17th	K.I.A	30 07 16	Guillemont Rd Cem.
21684	Bennion. O.E.	Llanfyllin, N.Wales	Pte.	19th	K.I.A	30 07 16	Thiepval Memorial.
21737	Benson. W.J.	Liverpool.	Pte.	19th	K.I.A	30 07 16	Thiepval Memorial.
13889	Bestwick. T.	West Hampton, Staffs.	Sgt.	20th	K.I.A	30 07 16	Thiepval Memorial.
27521	Bickerton.J.	Crewe, Cheshire.	Pte.	17th	K.I.A	30 07 16	Guillemont Rd Cem.
22014	Biglands. W.J.	Liverpool.	Pte.	20th	K.I.A	30 07 16	Thiepval Memorial.
27305	Billingsley. S.	Liverpool.	Pte.	19th	K.I.A	30 07 16	Thiepval Memorial.
21744	Black. L.	Liverpool.	Pte.	19th	K.I.A	30 07 16	Thiepval Memorial.
17550	Bolton. S.R.	Liverpool.	Pte.	19th	K.I.A	30 07 16	Thiepval Memorial.
15335	Bond. H.	Tranmere, Cheshire.	Pte.	17th	K.I.A	30 07 16	Thiepval Memorial.
24296	Bond. S.	Liverpool.	Pte.	19th	K.I.A	30 07 16	Thiepval Memorial.
33214	Boote. W.	Liverpool.	Pte.	17th	K.I.A	30 07 16	Thiepval Memorial.
	Boundy. F.E.	Liverpool.	Lieut.	17th	D.O.W	30 07 16	Guillemont Rd Cem.
30223	Bradley. J.	Walton, Liverpool.	Pte.	19th	K.I.A	30 07 16	Thiepval Memorial.
15246	Bradley. P.	Liverpool.	Cpl.	17th	D.O.W	30 07 16	Thiepval Memorial.
24481	Brady. H.	Liverpool.	Pte.	19th	K.I.A	30 07 16	Thiepval Memorial.
25930	Bricknell. F.C.	Liverpool.	Pte.	19th	K.I.A	30 07 16	Thiepval Memorial.
24909	Buckland. R.	Birkenhead, Cheshire.	Pte.	17th	K.I.A	30 07 16	Thiepval Memorial.
22029	Buckley. B.	Birkenhead, Cheshire.	Pte.	20th	K.I.A	30 07 16	Guillemont Rd Cem.
22032	Bunn. J.	Warrington, Cheshire.	Pte.	20th	K.I.A	30 07 16	Thiepval Memorial.
17798	Bunting. C.V.	Egremont, Cheshire.	Pte.	19th	D.O.W	30 07 16	Guillemont Rd Cem.
25572	Burnage. W.J.	Biggleswade, Beds.	L/Cpl.	19th	K.I.A	30 07 16	Guillemont Rd Cem.
24828	Burrell. J.	Liverpool.	Pte.	17th	K.I.A	30 07 16	Thiepval Memorial.
24585	Burrows. R.	Liverpool.	Pte.	18th	K.I.A	30 07 16	Citadel New Mil Cem, Fricourt.
30188	Carbery. C.E.	Liverpool.	Pte.	19th	K.I.A	30 07 16	Thiepval Memorial.
31606	Carter. F.	Liverpool.	Pte.	19th	K.I.A	30 07 16	Thiepval Memorial.
31607	Carter. W.G.	Dingle, Liverpool.	Pte.	19th	K.I.A	30 07 16	Thiepval Memorial.
17801	Cartmell. J.D.	Colwyn Bay, N.Wales.	Pte.	19th	K.I.A	30 07 16	Thiepval Memorial.
	Carver. H.Q.	N.S.W. Australia.	Lieut.	19th	K.I.A	30 07 16	Guillemont Rd Cem.
21944	Chamock. E.	Rainhill, Lancs.	L/Cpl.	20th	K.I.A	30 07 16	Thiepval Memorial.
22647	Clark. F.L.	Liverpool.	Pte.	20th	K.I.A	30 07 16	Thiepval Memorial.
22954	Clark. J.	Liverpool.	Pte.	20th	K.I.A	30 07 16	Thiepval Memorial.
15652	Clarke. R.H.	Liverpool.	Cpl.	17th	K.I.A	30 07 16	Guillemont Rd Cem.
21746	Clarke. T.	Liverpool.	Pte.	19th	K.I.A	30 07 16	Bernafay Wood, Brit Cem.
17306	Clayton. A.J.	Liverpool.	L/Cpl.	19th	K.I.A	30 07 16	Guillemont Rd Cem.
34591	Clough. A.	Liverpool.	Pte.	17th	K.I.A	30 07 16	Thiepval Memorial.
	Cockey. J.E.P.	London.	2nd Lieut.	20th	K.I.A	30 07 16	Thiepval Memorial.
15057	Coe. J.D.	New Brighton, Cheshire.	Pte.	17th	K.I.A	30 07 16	Thiepval Memorial.
29615	Coghlan. J.	Liverpool.	Cpl.	20th	K.I.A	30 07 16	Thiepval Memorial.
21714	Cole. S.	Liverpool.	Pte.	19th	K.I.A	30 07 16	Thiepval Memorial.
24201	Cole. W.S.	Liverpool.	Pte.	19th	K.I.A	30 07 16	Guillemont Rd Cem.
15902	Coles. F.A.	Liverpool.	Pte.	17th	K.I.A	30 07 16	Thiepval Memorial.
22658	Collett. A.	Liverpool.	Pte.	20th	K.I.A	30 07 16	Thiepval Memorial.
16019	Collett. H.W.H.	Hoylake, Cheshire.	Pte.	17th	K.I.A	30 07 16	Thiepval Memorial.
17304	Colligan. J.	Liverpool.	Pte.	19th	K.I.A	30 07 16	Thiepval Memorial.
15665	Collins. P.W.	Liverpool.	Pte.	17th	K.I.A	30 07 16	Guillemont Rd Cem.
31158	Collins. S.F.	Liverpool.	Cpl.	17th	K.I.A	30 07 16	Guillemont Rd Cem.
15501	Collinson. A.W.	Liverpool.	L/Cpl.	17th	K.I.A	30 07 16	Thiepval Memorial.
21808	Colton. R.	Liverpool.	Pte.	19th	K.I.A	30 07 16	Thiepval Memorial.
29614	Connor. A.	Liverpool.	Pte.	20th	K.I.A	30 07 16	Thiepval Memorial.
15257	Cook. P.H. M.M.	Liverpool.	Pte.	17th	K.I.A	30 07 16	Thiepval Memorial.

17289	Cook. S.I.	Cramer, Norfolk.	Pte.	19th	K.I.A	30 07 16	Thiepval Memorial.
26177	Cooke. S.	Bootle, Liverpool.	Pte.	17th	K.I.A	30 07 16	Thiepval Memorial.
23925	Costello. J.	Liverpool.	Pte.	20th	K.I.A	30 07 16	Thiepval Memorial.
15209	Cotton. E.B.	Birkenhead, Cheshire.	Pte.	17th	K.I.A	30 07 16	Thiepval Memorial.
21470	Coulthard. E.	Seacombe, Cheshire.	Pte.	19th	K.I.A	30 07 16	Thiepval Memorial.
24448	Crawford. A.	Liverpool.	Pte.	20th	K.I.A	30 07 16	Thiepval Memorial.
21462	Crawford. D.W.	Waterloo, Liverpool.	Cpl.	19th	K.I.A	30 07 16	Thiepval Memorial.
15523	Crellin. J.S.	Liverpool.	L/Cpl.	19th	K.I.A	30 07 16	Thiepval Memorial.
24499	Cross. F.	Liverpool.	L/Cpl.	20th	K.I.A	30 07 16	Thiepval Memorial.
24298	Crossley. F.	Willaston, Ches.	Pte.	19th	K.I.A	30 07 16	Guillemont Rd Cem.
17561	Cunningham. J.	Liverpool.	Pte.	19th	K.I.A	30 07 16	Guillemont Rd Cem.
29197	Cunningham. P.	Liverpool.	Pte.	19th	K.I.A	30 07 16	Guillemont Rd Cem.
22968	Curley. S.	Liverpool.	Pte.	20th	K.I.A	30 07 16	Thiepval Memorial.
17296	Curwen. T.(MM)	Walton, Liverpool.	Cpl.	19th	K.I.A	30 07 16	Combles CC Ext.
34569	Cutts. A.R.	(Ascroft.C) Liverpool.	Pte.	19th	K.I.A	30 07 16	Thiepval Memorial.
24453	Darcy. H.	Upton, Cheshire.	Pte.	20th	K.I.A	30 07 16	Guillemont Rd Cem.
17816	Davidson. T.R.	Liverpool.	L/Cpl.	19th	K.I.A	30 07 16	Thiepval Memorial.
32652	Davies. G.H.	Liverpool.	Pte.	17th	K.I.A	30 07 16	Guillemont Rd Cem.
17319	Davies. G.L.	Egremont, Cheshire.	Pte.	19th	K.I.A	30 07 16	Thiepval Memorial.
35985	Davis. T.	Lees, Manchester.	L/Cpl.	20th	K.I.A	30 07 16	Thiepval Memorial.
29606	Dawe. S.	Liverpool.	Pte.	19th	K.I.A	30 07 16	Thiepval Memorial.
24479	De-La-Cruz. N.	Liverpool.	Pte.	20th	K.I.A	30 07 16	Thiepval Memorial.
17574	Delaney. T.	Liverpool.	Pte.	19th	K.I.A	30 07 16	Thiepval Memorial.
22362	Delbanco. J.	Southport, Lancs.	Pte.	20th	K.I.A	30 07 16	Thiepval Memorial.
22671	Dempsey.G.	Liverpool.	Sgt.	20th	K.I.A	30 07 16	Guillemont Rd Cem.
15915	Dening. G.	Liverpool.	Sgt.	17th	K.I.A	30 07 16	Thiepval Memorial.
15753	Devine. T.M.	Liverpool.	Pte.	17th	K.I.A	30 07 16	Thiepval Memorial.
27329	Dick. W.H.	Bootle, Liverpool.	Pte.	19th	K.I.A	30 07 16	Thiepval Memorial.
34816	Dickinson. J.E.	Everton, Liverpool.	Pte.	20th	K.I.A	30 07 16	Thiepval Memorial.
15239	Dix. W.	Liverpool.	Pte.	17th	K.I.A	30 07 16	Guillemont Rd Cem.
21489	Dixon. R.M.	Preston, Lancs.	Pte.	19th	K.I.A	30 07 16	Thiepval Memorial.
32257	Doherty. A.H.	Liverpool.	Pte.	20th	K.I.A	30 07 16	Thiepval Memorial.
15859	Drury. P.C.	Poulton, Cheshire.	Pte.	17th	K.I.A	30 07 16	Thiepval Memorial.
21751	Duggan. W.J.	Liverpool.	Pte.	20th	K.I.A	30 07 16	Guillemont Rd Cem.
22679	Easton. C.H.	Birkenhead, Cheshire.	Sgt.	20th	K.I.A	30 07 16	Thiepval Memorial.
	Eddison. T.D.	Eastbourne, Sussex.	Lieut.	19th	K.I.A	30 07 16	Thiepval Memorial.
25335	Edgerton. W.	Liverpool.	Pte.	20th	K.I.A	30 07 16	Thiepval Memorial.
32619	Edwards. E.	Walton, Liverpool.	Pte.	20th	K.I.A	30 07 16	Thiepval Memorial.
17325	Edwards. F	West Derby, Liverpool.	Pte.	19th	K.I.A	30 07 16	Guillemont Rd Cem.
24959	Eggleston. F.	Liverpool.	Pte.	17th	K.I.A	30 07 16	Serre Rd No.2 Cem, B-Hamel.
16057	Elliott. A.	Liverpool.	Pte.	17th	K.I.A	30 07 16	Thiepval Memorial.
17575	Ellison. F.H.	Liverpool.	Pte.	19th	K.I.A	30 07 16	Guillemont Rd Cem.
31673	Evans. A.	Liverpool.	Pte.	20th	K.I.A	30 07 16	Thiepval Memorial.
21492	Evans. G.L.	Liverpool.	Pte.	19th	K.I.A	30 07 16	Guillemont Rd Cem.
29243	Evans. R.	Poulton, Cheshire.	Pte.	17th	K.I.A	30 07 16	Thiepval Memorial.
21404	Evans. R.R.	Rhyl, N.Wales.	Pte.	19th	K.I.A	30 07 16	Guillemont Rd Cem.
15934	Eyden. H.	St. Helens, Lancs.	Sgt.	17th	K.I.A	30 07 16	Peronne Rd Cem, Maricourt.
15942	Fairclough. H.J.	Liverpool.	Pte.	17th	K.I.A	30 07 16	Peronne Rd Cem, Maricourt.
	Paris. S.J.		2nd Lieut.	20th	K.I.A	30 07 16	Guillemont Rd Cem.
21497	Fawcett. A.	Liverpool.	L/Cpl.	19th	K.I.A	30 07 16	Thiepval Memorial.
17828	Finn. A. M.M	Liverpool.	L/Cpl.	19th	K.I.A	30 07 16	Thiepval Memorial.
22076	Fishwick. C.D.	Liverpool.	Pte.	20th	K.I.A	30 07 16	Thiepval Memorial.
17827	Fitzsimmons. F.A.	Liverpool.	Pte.	19th	K.I.A	30 07 16	Thiepval Memorial.
22081	Fleming. B.J.C.	Liverpool.	Sgt.	20th	K.I.A	30 07 16	Thiepval Memorial.
15799	Fletcher. A.E.	Liverpool.	L/Cpl.	17th	K.I.A	30 07 16	Thiepval Memorial.
32335	Ford. A.	Liverpool.	Cpl.	20th	K.I.A	30 07 16	Thiepval Memorial.
17580	Forrest. E.	Liverpool.	L/Cpl.	19th	K.I.A	30 07 16	Thiepval Memorial.
22377	Fox. J.	Liverpool.	Pte.	20th	K.I.A	30 07 16	Thiepval Memorial.

22688	France. C.	Runcorn, Cheshire.	L/Cpl.	20th	K.I.A	30 07 16	Thiepval Memorial.
21778	Francis. A.	Birkenhead, Cheshire.	L/Cpl.	19th	K.I.A	30 07 16	Thiepval Memorial.
	Fraser. W.	Bootle, Liverpool.	Capt.	19th	K.I.A	30 07 16	Guillemont Rd Cem.
22378	Fraser. W.G.	Liverpool.	Pte.	20th	K.I.A	30 07 16	Thiepval Memorial.
24245	Frear. F.	Liverpool.	Pte.	19th	K.I.A	30 07 16	Thiepval Memorial.
22381	Fryer. G.W.	Runcorn, Cheshire.	Sgt.	20th	K.I.A	30 07 16	Thiepval Memorial.
15433	Fullerton. W.J.	Liverpool.	Pte.	17th	K.I.A	30 07 16	Thiepval Memorial.
	Furlong. P.J.	Dublin, Ireland.	2nd Lieut	19th	K.I.A	30 07 16	Thiepval Memorial.
23936	Gaffney. J.	Garston, Liverpool.	Pte.	20th	K.I.A	30 07 16	Thiepval Memorial.
31084	Gallagher. C.H.	Woolton, Liverpool.	Pte.	19th	K.I.A	30 07 16	Guillemont Rd Cem.
35957	Gamett. A.E.	Northwich, Ches.	Pte.	20th	K.I.A	30 07 16	Thiepval Memorial.
31794	Garrett. W.	Laxey, I.O.M.	Pte.	17th	K.I.A	30 07 16	Guillemont Rd Cem.
21507	Garside. F.	Crewe.	Pte.	19th	K.I.A	30 07 16	Serre Rd No.2 Cem, B-Hamel.
25325	George. E.	Seaforth, Liverpool.	Pte.	19th	K.I.A	30 07 16	Thiepval Memorial.
32668	Georgeson. R.	Kirkdale, Liverpool.	Pte.	19th	K.I.A	30 07 16	Thiepval Memorial.
17353	Gerrits. J.E.	Liverpool.	Pte.	19th	K.I.A	30 07 16	Thiepval Memorial.
19516	Glassey. T.J.	Stockport, Cheshire.	Pte.	19th	K.I.A	30 07 16	Thiepval Memorial.
17611	Golding. G.S.	Liverpool.	Pte.	19th	K.I.A	30 07 16	Thiepval Memorial.
22695	Good. J.	Liverpool.	Pte.	20th	K.I.A	30 07 16	Thiepval Memorial.
15736	Goodfellow. J.G.	Bristol.	L/Cpl.	17th	K.I.A	30 07 16	Guillemont Rd Cem.
15711	Goodwin. F.	Liverpool.	Pte.	17th	K.I.A	30 07 16	Thiepval Memorial.
21785	Gorman. J.	Liverpool.	Pte.	19th	K.I.A	30 07 16	Thiepval Memorial.
22697	Goudie. R.D.	Liverpool.	Pte.	20th	K.I.A	30 07 16	Thiepval Memorial.
15955	Goulboum. R.H.	Litherland, Liverpool.	Pte.	17th	K.I.A	30 07 16	Thiepval Memorial.
23941	Grantham. G.E.	Birkenhead, Cheshire.	Cpl.	20th	K.I.A	30 07 16	Guillemont Rd Cem.
23017	Green. C.	Widnes, Lancs.	Pte.	20th	K.I.A	30 07 16	Guillemont Rd Cem.
29176	Green. C.	Liverpool.	Pte.	19th	K.I.A	30 07 16	Guillemont Rd Cem.
14636	Green. S.	St.Helens, Lancs.	Pte.	20th	K.I.A	30 07 16	Thiepval Memorial.
21951	Greenwood. A.	St.Helens, Lancs.	Cpl.	20th	K.I.A	30 07 16	Thiepval Memorial.
30236	Greer. G.	Orrell, Wigan, Lancs.	Pte.	20th	K.I.A	30 07 16	Thiepval Memorial.
19639	Gregory. J.	Swinton, Manchester.	Pte.	19th	K.I.A	30 07 16	Thiepval Memorial.
21501	Griffiths. R.	Anfield, Liverpool.	Sgt.	19th	K.I.A	30 07 16	Thiepval Memorial.
15442	Hamilton. D.	Liverpool.	Pte.	17th	K.I.A	30 07 16	Thiepval Memorial.
15829	Hancock. P.	Birkenhead.	Sgt.	17th	K.I.A	30 07 16	Thiepval Memorial.
15774	Handley. J.A.	Runcorn, Cheshire.	Cpl.	17th	K.I.A	30 07 16	Thiepval Memorial.
15896	Harper. F.F.	Birkenhead, Cheshire.	Pte.	17th	K.I.A	30 07 16	Thiepval Memorial.
21530	Harris. W.	Edge Hill, Liverpool.	Pte.	19th	K.I.A	30 07 16	Thiepval Memorial.
23028	Harrison. C.N.	Liverpool.	Pte.	20th	K.I.A	30 07 16	Thiepval Memorial.
22099	Harrison. J.A.	Liverpool.	Pte.	20th	K.I.A	30 07 16	Thiepval Memorial.
16033	Harrison. W.R.	Liverpool.	Pte.	17th	K.I.A	30 07 16	Thiepval Memorial.
17379	Hartley. G.	Liverpool.	Pte.	19th	K.I.A	30 07 16	Thiepval Memorial.
23947	Haselden. P.	Liverpool.	Pte.	20th	K.I.A	30 07 16	Thiepval Memorial.
27344	Hawley. E.A.	Liverpool.	Pte.	19th	K.I.A	30 07 16	Thiepval Memorial.
22400	Hayden. H.	Liverpool.	Pte.	20th	K.I.A	30 07 16	Guillemont Rd Cem.
23029	Haynes. T.	Widnes, Lancs.	L/Cpl.	20th	K.I.A	30 07 16	Thiepval Memorial.
34295	Hayworth. H.	Liverpool.	Pte.	19th	K.I.A	30 07 16	Thiepval Memorial.
21518	Heath. C.	Chester.	Pte.	19th	K.I.A	30 07 16	Thiepval Memorial.
23949	Hedley. R.	Liverpool.	Pte.	20th	K.I.A	30 07 16	Thiepval Memorial.
21814	Hellon. J.	Walton, Liverpool.	Cpl.	17th	K.I.A	30 07 16	Guillemont Rd Cem.
26137	Henton. H.C.	Liverpool.	Pte.	17th	K.I.A	30 07 16	Thiepval Memorial.
17368	Henwood. W.	Liverpool.	Pte.	19th	K.I.A	30 07 16	Thiepval Memorial.
21520	Hosier. J.	Seacombe, Cheshire.	Pte.	19th	K.I.A	30 07 16	Thiepval Memorial.
17592	Hiatt. F.W.	Liverpool.	Pte.	19th	D.O.W	30 07 16	Thiepval Memorial.
22111	Hibbert. B.	Seacombe, Cheshire.	L/Cpl.	20th	K.I.A	30 07 16	Thiepval Memorial.
24257	Higgins. E.	Bootle, Liverpool.	Pte.	19th	K.I.A	30 07 16	Thiepval Memorial.
21512	Hilton. J.C.	Hoylake.	Pte.	19th	K.I.A	30 07 16	Thiepval Memorial.
23034	Hindle. W.	Burnley, Lancs.	Pte.	20th	K.I.A	30 07 16	Thiepval Memorial.
30827	Hitchmough. J.E.		Pte.	20th	K.I.A	30 07 16	Thiepval Memorial.

21529	Hobbs. J.	Gt.Crosby, Liverpool.	Pte.	19th	K.I.A	30 07 16	Thiepval Memorial.
22411	Hodgers. J.	Liverpool.	Pte.	20th	K.I.A	30 07 16	Thiepval Memorial.
22412	Hodgson. H.W.	Liverpool.	Pte.	20th	K.I.A	30 07 16	Guillemont Rd Cem.
35366	Hodson. A.	New Bedford, USA.	Pte.	17th	K.I.A	30 07 16	Thiepval Memorial.
21733	Holden. G.	Hoylake, Cheshire.	Pte.	19th	K.I.A	30 07 16	Thiepval Memorial.
24962	Holden. J.M.	Upholland, Lancs.	Pte.	17th	K.I.A	30 07 16	Guillemont Rd Cem.
17361	Holden. W.F.	Liverpool.	Pte.	19th	K.I.A	30 07 16	Guillemont Rd Cem.
38097	Holdship. H.	Salford, Manchester.	Pte.	17th	K.I.A	30 07 16	Thiepval Memorial.
29950	Holt. J.	Clayton-le-Moors, Lancs.	Pte.	17th	K.I.A	30 07 16	Guillemont Rd Cem.
30263	Hoole. F.A.	Liverpool.	Pte.	17th	K.I.A	30 07 16	Thiepval Memorial.
35761	Hopwood. J.	Denton, Manchester.	Pte.	20th	K.I.A	30 07 16	Thiepval Memorial.
23036	Horn. A.	Garstang, Lancs.	Pte.	20th	K.I.A	30 07 16	Thiepval Memorial.
22119	Hough. A.	Liverpool.	Pte.	20th	K.I.A	30 07 16	Guillemont Rd Cem.
15818	Houldsworth. G.E.	Liverpool.	Pte.	17th	K.I.A	30 07 16	Guillemont Rd Cem.
22724	Howard. A.J.	Newton-le-Willows, Lancs.	Pte.	20th	K.I.A	30 07 16	Thiepval Memorial.
29663	Hughes. A.	Liverpool.	Pte.	20th	K.I.A	30 07 16	Flatiron Copse Cem, Mametz.
17601	Hughes. A.E.	Eastham, Ches.	Pte.	19th	K.I.A	30 07 16	Thiepval Memorial.
17847	Hughes. W.P.	Colwyn Bay, N.Wales.	Pte.	19th	K.I.A	30 07 16	Thiepval Memorial.
17354	Humphreys. R.A.	Liverpool.	Pte.	19th	K.I.A	30 07 16	Thiepval Memorial.
29123	Hunt. J.	Liverpool.	Pte.	19th	K.I.A	30 07 16	Thiepval Memorial.
36667	Hunts.	Chortey, Lancs.	Pte.	20th	K.I.A	30 07 16	Thiepval Memorial.
23041	Hurrell. A.J.	Seacombe, Cheshire.	Pte.	20th	K.I.A	30 07 16	Thiepval Memorial.
22424	Ingham. E.V.	Hoylake, Cheshire.	L/Cpl.	20th	K.I.A	30 07 16	Thiepval Memorial.
21832	Irwin. S.D.	Liverpool.	Pte.	19th	K.I.A	30 07 16	Thiepval Memorial.
22427	Jaques. R.H.	Liverpool.	Pte.	20th	K.I.A	30 07 16	Thiepval Memorial.
13162	Jaundrill.W.	Prescot, Lancs.	Pte.	17th	K.I.A	30 07 16	Thiepval Memorial.
30144	Jenkins. W.J.	Chester.	Pte.	20th	K.I.A	30 07 16	Thiepval Memorial.
21844	Johnson. G.	Liverpool.	L/Cpl.	19th	K.I.A	30 07 16	Thiepval Memorial.
31075	Johnson. G.S.	Liverpool.	Sgt.	19th	K.I.A	30 07 16	Thiepval Memorial.
26621	Johnson J.E.	Liverpool.	Pte.	19th	K.I.A	30 07 16	Thiepval Memorial.
17630	Johnston. H.	Liverpool.	L/Cpl.	19th	K.I.A	30 07 16	Thiepval Memorial.
22138	Johnston. S.	Anfield, Liverpool.	Pte.	20th	K.I.A	30 07 16	Thiepval Memorial.
	Johnston. W.H.	Belfast, Ireland.	2nd Lieut.	17th	K.I.A	30 07 16	Guillemont Rd Cem.
16144	Jones. C.P.	Rhuddlan, Flints.	Pte.	17th	K.I.A	30 07 16	Thiepval Memorial.
34302	Jones. E.	Liverpool.	Pte.	17th	K.I.A	30 07 16	Thiepval Memorial.
16008	Jones. H.	Liverpool.	Pte.	17th	K.I.A	30 07 16	Thiepval Memorial.
	Jones. H.M.	Youghal, Cork, Ireland.	Lieut.	19th	K.I.A	30 07 16	Thiepval Memorial.
22743	Jones. J.B.	Liverpool.	Sgt.	20th	K.I.A	30 07 16	Thiepval Memorial.
22741	Jones. J.P.	Liverpool.	Pte.	20th	K.I.A	30 07 16	Thiepval Memorial.
15661	Jones. L.M.	Underhill, Denbigh.	Pte.	17th	K.I.A	30 07 16	Guillemont Rd Cem.
15960	Jones. R.T.	Liverpool.	Sgt.	17th	K.I.A	30 07 16	Thiepval Memorial.
22130	Jones. T.F.	Egremont, Cheshire.	Pte.	20th	K.I.A	30 07 16	Thiepval Memorial.
17393	Jones. W.D.	Liverpool.	Pte.	19th	K.I.A	30 07 16	Thiepval Memorial.
22437	Jones. W.G.	Liverpool.	Pte.	20th	K.I.A	30 07 16	Guillemont Rd Cem.
17864	Jowitt. J.H.	Liverpool.	L/Sgt.	19th	K.I.A	30 07 16	Thiepval Memorial.
23957	Joynson. A.	Liverpool.	Pte.	20th	K.I.A	30 07 16	Thiepval Memorial.
22438	Judd. W.C.	Crewe, Cheshire.	Pte.	20th	K.I.A	30 07 16	Thiepval Memorial.
24336	Judge W	Liverpool.	Pte.	19th	K.I.A	30 07 16	Thiepval Memorial.
21544	Kandler. C.Mc.	Gloucester.	Pte.	20th	K.I.A	30 07 16	Thiepval Memorial.
35872	Kelly. A.	Marown, I.O.M.	Pte.	17th	K.I.A	30 07 16	Guillemont Rd Cem.
29255	Kelly. W.J.	Liscard, Cheshire.	L/Cpl.	17th	K.I.A	30 07 16	Thiepval Memorial.
15925	Kenworthy. T.K.	Liverpool.	Pte.	17th	K.I.A	30 07 16	Peronne Rd Cem, Maricourt.
17878	Kinder. J.E.	Liverpool.	Pte.	19th	K.I.A	30 07 16	Thiepval Memorial.
27593	Knight. E.	Liverpool.	Pte.	17th	K.I.A	30 07 16	Thiepval Memorial.
17632	Knowles. R.P.	Warrington, Cheshire.	L/Cpl.	19th	K.I.A	30 07 16	Thiepval Memorial.
21756	Lambert JW	Liverpool.	Pte.	19th	K.I.A	30 07 16	Thiepval Memorial.
23078	Lawrenson. H.	Liverpool.	Sgt.	20th	K.I.A	30 07 16	Thiepval Memorial.
15217	Ledger. D.	Seacombe, Cheshire.	Pte.	17th	K.I.A	30 07 16	Guillemont Rd Cem.

22453	Leonard. B.	Walton, Liverpool.	C.S.M.	20th	K.I.A	30 07 16	Guillemont Rd Cem.
23081	Lewis. F.	Birkenhead, Cheshire.	L/Cpl.	20th	K.I.A	30 07 16	Thiepval Memorial.
25541	Lines. H.	Liverpool.	Pte.	19th	K.I.A	30 07 16	Thiepval Memorial.
21961	Litchfield J.	Liverpool.	Pte.	20th	K.I.A	30 07 16	Guillemont Rd Cem.
22760	Little. C.	Douglas, I.O.M.	Pte.	20th	K.I.A	30 07 16	Thiepval Memorial.
17416	Little. T.W.	Birkenhead, Cheshire.	Pte.	19th	K.I.A	30 07 16	Thiepval Memorial.
25349	Lloyd. A.K.	Liverpool.	Pte.	19th	K.I.A	30 07 16	Guillemont Rd Cem.
17411	Lockhart. H.	Liverpool.	Pte.	19th	K.I.A	30 07 16	Thiepval Memorial.
36187	Lohrenz. A.	Liverpool.	Pte.	17th	K.I.A	30 07 16	Serre Rd No.2 Cem, B-Hamel.
17879	Lover. H.	Birkenhead, Cheshire.	Pte.	19th	K.I.A	30 07 16	Thiepval Memorial.
21551	Lowe. F.C.	Prescot, Lancs.	Pte.	19th	K.I.A	30 07 16	Thiepval Memorial.
29717	Lunt. E.	Liverpool.	Pte.	19th	K.I.A	30 07 16	Thiepval Memorial.
14596	Lynch. T.J.	Liverpool.	Pte.	20th	K.I.A	30 07 16	Thiepval Memorial.
23097	Macklin. W.	Birkenhead, Cheshire.	L/Cpl.	20th	K.I.A	30 07 16	Thiepval Memorial.
17664	Maiden. A.	Willaston, Cheshire.	Pte.	19th	K.I.A	30 07 16	Thiepval Memorial.
30139	Malone. W.	Liverpool.	Pte.	19th	K.I.A	30 07 16	Thiepval Memorial.
31611	Mansley. F.L	Liverpool.	L/Cpl.	19th	K.I.A	30 07 16	Thiepval Memorial.
15728	Manson. M.	Liverpool.	Pte.	17th	K.I.A	30 07 16	Thiepval Memorial.
22767	Martin. J.H.	Liverpool.	Pte.	20th	K.I.A	30 07 16	Thiepval Memorial.
27361	Martin. W.G.	Liverpool.	Pte.	19th	K.I.A	30 07 16	Thiepval Memorial.
23095	Mason. E.	Liverpool.	Pte.	20th	K.I.A	30 07 16	Thiepval Memotijal.
30235	Matthews. A.	Liverpool.	Pte.	20th	K.I.A	30 07 16	Guillemont Rd Cem.
23628	Maybrick. L.	Southport, Lancs.	L/Cpl.	17th	K.I.A	30 07 16	Thiepval Memorial.
15909	Maybury. T.W.	Liverpool.	Pte.	17th	K.I.A	30 07 16	Thiepval Memorial.
32636	McDonald. J.J.	Poulton, Cheshire.	Pte.	19th	K.I.A	30 07 16	Thiepval Memorial.
15308	McDonnell. J.H.	Liscard, Cheshire.	Pte.	17th	K.I.A	30 07 16	Thiepval Memorial.
17436	McGough. G.A.	Egremont, Cheshire.	Pte.	19th	K.I.A	30 07 16	Thiepval Memorial.
24627	McNally. T.	Liverpool.	Pte.	17th	K.I.A	30 07 16	Guillemont Rd Cem.
27536	McNichol. F.	Liverpool.	Pte.	19th	K.I.A	30 07 16	Thiepval Memorial.
23111	McNiven. C.	Liverpool.	Pte.	20th	K.I.A	30 07 16	Thiepval Memorial.
29125	McRae. S.	Bootle, Liverpool.	Pte.	19th	K.I.A	30 07 16	Guillemont Rd Cem.
17426	McWhirr. A.	Liverpool.	Pte.	19th	K.I.A	30 07 16	Thiepval Memorial.
21762	Meadows. W.	Liverpool.	Pte.	19th	K.I.A	30 07 16	Thiepval Memorial.
17431	Meeson. W.	Exeter, Devon.	Pte.	19th	K.I.A	30 07 16	Guillemont Rd Cem.
23112	Meinig. T.	Liverpool.	L/Cpl.	20th	K.I.A	30 07 16	Thiepval Memorial.
	Melly. R.E.	Meridan, Warwicks.	Lieut.	20th	K.I.A	30 07 16	Thiepval Memorial.
21562	Meredith. F.P.	Chester.	Pte.	19th	K.I.A	30 07 16	Guillemont Rd Cem.
16082	Miller. R.	Liverpool.	L/Cpl.	17th	K.I.A	30 07 16	Thiepval Memorial.
26698	Milsom. G.	Liverpool.	Pte.	17th	K.I.A	30 07 16	Guillemont Rd Cem.
23115	Mitchell. N.L	Blundellsands, Liverpool	A/Sgt.	20th	K.I.A	30 07 16	Thiepval Memorial.
24222	Molyneux. L	Liverpool.	Pte.	19th	K.I.A	30 07 16	Thiepval Memorial.
29716	Moore. A.	Liverpool.	Pte.	20th	K.I.A	30 07 16	Thiepval Memorial.
23966	Moran. E.	Liverpool.	L/Cpl.	20th	K.I.A	30 07 16	Thiepval Memorial.
22784	Morley. J.	Great Crosby, Liverpool.	Pte.	20th	K.I.A	30 07 16	Thiepval Memorial.
22779	Morris A.	Ruabon, N.Waies.	Pte.	20th	K.I.A	30 07 16	Thiepval Memorial.
21561	Morris. A.H.	Liverpool.	Pte.	19th	D.O.W	30 07 16	Dive Copse B.C, Sailly-le-Sec.
22175	Morris. D.	Llandegfan, Anglesey.	Pte.	20th	K.I.A	30 07 16	Thiepval Memorial.
17424	Morris. F.	Rock Ferry, Cheshire.	Pte.	19th	K.I.A	30 07 16	Thiepval Memorial.
17902	Morris. F.	Westminster, London.	Pte.	19th	K.I.A	30 07 16	Thiepval Memorial.
27941	Moss C.	Liverpool.	Pte.	17th	K.I.A	30 07 16	Thiepval Memorial.
24362	Moss. G.	Liverpool.	Pte.	19th	K.I.A	30 07 16	Thiepval Memorial.
22186	Munday. J.S.	Liverpool.	Pte.	20th	K.I.A	30 07 16	Thiepval Memorial.
17435	Murphy. G.C.	Widnes, Lancs.	Pte.	19th	K.I.A	30 07 16	Thiepval Memorial.
30124	Murphy. P.	Liverpool.	Pte.	19th	K.I.A	30 07 16	Thiepval Memorial.
22787	Murphy. T.	Liverpool.	A/Sgt.	20th	K.I.A	30 07 16	Thiepval Memorial.
15576	Murray. W.A.	Heswell, Cheshire.	Pte.	17th	K.I.A	30 07 16	Guillemont Rd Cem.
	Musker. J.W.	Bootle, Liverpool.	2nd Lieut.	20th	K.I.A	30 07 16	Thiepval Memorial.
21578	Negus. J.W.	Crewe, Cheshire.	L/Cpl.	19th	K.I.A	30 07 16	Thiepval Memorial.

17911	Neill. M.	Wallasey, Cheshire.	Pte.	19th	K.I.A	30 07 16	Thiepval Memorial.
27303	Nelson. S.	Douglas, I.O.M.	Pte.	19th	K.I.A	30 07 16	Thiepval Memorial.
10626	Nesbitt. J.	Liverpool.	Pte.	17th	K.I.A	30 07 16	Guillemont Rd Gem.
22189	Newall. A.S.	Northwich, Cheshire.	L/Cpl.	20th	K.I.A	30 07 16	Thiepval Memorial.
22188	Newbold. C.	Liverpool.	Pte.	20th	K.I.A	30 07 16	Thiepval Memorial.
21579	Newbould. T.	Bootle, Liverpool.	Pte.	19th	K.I.A	30 07 16	Thiepval Memorial.
17913	Newell. H.J.	Liverpool.	Pte.	19th	K.I.A	30 07 16	Thiepval Memorial.
21576	Newsam. E.J.	Liverpool.	Pte.	19th	K.I.A	30 07 16	Thiepval Memorial.
22792	Nichol. S.F.	Wallasey, Cheshire.	A/L/Sgt	20th	K.I.A	30 07 16	Thiepval Memorial.
22191	Nickle. J.J.	Liscard, Cheshire.	L/Cpl.	20th	K.I.A	30 07 16	Thiepval Memorial.
	Nickson. W.	Birkenhead, Cheshire.	Capt.	19th	K.I.A	30 07 16	Serre Rd No.2 Cem, B-Hamel *
21577	Nooney. J.	Wallasey, Cheshire.	L/Cpl.	19th	K.I.A	30 07 16	Thiepval Memorial.
15282	Norcliffe. W.W.	Wallasey, Cheshire.	Pte.	17th	K.I.A	30 07 16	Thiepval Memorial.
34264	Norris. W.	Knowsley, Lancs.	Pte.	20th	K.I.A	30 07 16	Thiepval Memorial.
25513	O'Connor. O.	Liverpool.	Pte.	19th	K.I.A	30 07 16	Guillemont Rd Cem.
26592	O'Neill. J.	Thatto Heath, St.Helens.Lancs	Pte.	17th	K.I.A	30 07 16	Thiepval Memorial.
	Orford. E.G.	Liverpool.	Capt.	20th	K.I.A	30 07 16	Thiepval Memorial.
23132	Osborne. J.	Nottingham.	Sgt.	20th	K.I.A	30 07 16	Thiepval Memorial.
17677	Osborne. W.	Liverpool.	Pte.	19th	K.I.A	30 07 16	Thiepval Memorial.
24385	Owen. H.K.	Liverpool.	Pte.	19th	K.I.A	30 07 16	Thiepval Memorial.
17932	Owen. I.	Aintree, Liverpool.	L/Cpl.	19th	K.I.A	30 07 16	Guillemont Rd Cem.
25576	Owens. W.B.	Seaforth, Liverpool.	Pte.	19th	D.O.W	30 07 16	Dive Copse B.C, Sailly-le-Sec.
22507	Padfield. H.J.	Liverpool.	A/Sgt.	20th	K.I.A	30 07 16	Thiepval Memorial.
21706	Parsons. F.J.	Liverpool.	Pte.	19th	K.I.A	30 07 16	Thiepval Memorial.
17697	Partridge. B.	Liverpool.	Sgt.	19th	K.I.A	30 07 16	Thiepval Memorial.
22212	Pearson. C.F.	Runcorn, Cheshire.	Pte.	20th	K.I.A	30 07 16	Thiepval Memorial.
21424	Pearson. H.O.	Liverpool.	L/Cpl.	19th	K.I.A	30 07 16	Thiepval Memorial.
24353	Polling. G.	Liverpool.	Pte.	19th	K.I.A	30 07 16	Thiepval Memorial.
22214	Penketh. W.	Liverpool.	Pte.	20th	K.I.A	30 07 16	Thiepval Memorial.
23135	Peters. W.T.	Birkenhead, Cheshire.	Pte.	20th	K.I.A	30 07 16	Thiepval Memorial.
31005	Phillips. W.J.	Liverpool.	Pte.	20th	K.I.A	30 07 16	Thiepval Memorial.
17926	Pierce. E.	Liverpool.	Pte.	19th	K.I.A	30 07 16	Thiepval Memorial.
16016	Pierce F.A	Chester.	L/Cpl.	17th	K.I.A	30 07 16	Thiepval Memorial.
21589	Pinches. N.G.	Chester.	L/Cpl.	19th	K.I.A	30 07 16	Guillemont Rd Cem.
29627	Pink. G.	Seacombe, Cheshire.	Pte.	20th	K.I.A	30 07 16	Thiepval Memorial.
23144	Poole. A.	Northwich, Cheshire.	Pte.	20th	K.I.A	30 07 16	Thiepval Memorial.
	Porritt. E.R.	Chester.	2nd Lieut.	17th	K.I.A	30 07 16	Serre Rd No.2 Cem, B-Hamel.
17443	Power. V.M.	Bristol, Glos.	Pte.	19th	K.I.A	30 07 16	Guillemont Rd Cem.
15889	Pratt. J.F.	Liverpool.	Pte.	17th	K.I.A	30 07 16	Thiepval Memorial.
17691	Prophett. L.	Liverpool.	Pte.	19th	K.I.A	30 07 16	Thiepval Memorial.
17450	Pugh. T.	Liverpool.	Sgt.	19th	K.I.A	30 07 16	Thiepval Memorial.
33196	Pulford. A.	Seacombe, Cheshire.	Pte.	17th	K.I.A	30 07 16	Guillemont Rd Cem.
22511	Quarters. E.	Bootle, Liverpool.	Pte.	20th	K.I.A	30 07 16	Thiepval Memorial.
29715	Quayle. R.J.	Liverpool.	Pte.	19th	K.I.A	30 07 16	Guillemont Rd Cem.
26540	Quinn. J.	Liverpool.	L/Cpl.	20th	K.I.A	30 07 16	Thiepval Memorial.
35165	Ramsbottom. J.	Southport, Lancs.	Cpl.	20th	K.I.A	30 07 16	Serre Rd No.2 Cem, B-Hamel.*
22227	Ramsden. R.	Hartford, Cheshire.	Pte.	20th	K.I.A	30 07 16	Thiepval Memorial.
21607	Rees. J.M.	Queensferry, Ches.	Pte.	19th	K.I.A	30 07 16	Guillemont Rd Cem.
22512	Reid. A.M.	Glasgow.	Pte.	20th	K.I.A	30 07 16	Thiepval Memorial.
31127	Ritson. A.E.	Liverpool.	Pte.	17th	K.I.A	30 07 16	Thiepval Memorial.
24379	Roberts. J.H.	Holywell, Flints.	Pte.	19th	K.I.A	30 07 16	Thiepval Memorial.
21617	Roberts. J.R.	Liverpool.	L/Cpl.	19th	K.I.A	30 07 16	Guillemont Rd Cem.*
15691	Roberts. M.	Abersoch, Caernarvon.	Pte.	17th	K.I.A	30 07 16	Guillemont Rd Cem.
23815	Roberts. R.	Grocolon, Caernarvon.	Pte.	20th	K.I.A	30 07 16	Thiepval Memorial.
22236	Roberts. W.S.	Liverpool.	Pte.	20th	K.I.A	30 07 16	Guillemont Rd Cem.
24924	Rodgers. R.D.	Liverpool.	A/Sgt.	17th	K.I.A	30 07 16	Thiepval Memorial.
24997	Rogers. E.	Tarporley, Cheshire.	Pte.	17th	K.I.A	30 07 16	Thiepval Memorial.
17703	Rogers. G.G.	Wrexham.	Sgt.	19th	K.I.A	30 07 16	Serre Rd No.2 Cem, B-Hanqel.*

16054	Ronson. G.R.	Walton, Liverpool.	Pte.	17th	K.I.A	30 07 16	Heath Cem, Harbonnieres.
10668	Roots. H.H.	Snodland, Kent.	Pte.	19th	K.I.A	30 07 16	Thiepval Memorial.
24246	Roper. H.J.	Liverpool.	Pte.	19th	K.I.A	30 07 16	Thiepval Memorial.
15578	Rowatt. E.	Liverpool.	Pte.	17th	K.I.A	30 07 16	Thiepval Memorial.
35531	Russell. P.J.	Onchan, I.O.M.	Pte.	17th	K.I.A	30 07 16	Guillemont Rd Cem.
27579	Sampson. H.	Duncannon Fort, Wexford.	Pte.	17th	K.I.A	30 07 16	Serre Rd No.2 Cem, B-Hamel.
29280	Saywell. E.	London.	Pte.	20th	K.I.A	30 07 16	Thiepval Memorial.
24849	Scaife. F.	Liverpool.	Pte.	17th	K.I.A	30 07 16	Thiepval Memorial.
27534	Scarth L.	Liverpool.	Pte.	19th	K.I.A	30 07 16	Thiepval Memorial.
23979	Schofield. C.C.	Liverpool.	L/Cpl.	20th	K.I.A	30 07 16	Thiepval Memorial.
17948	Scott. J.V.	Crosby. Liverpool.	Pte.	19th	K.I.A	30 07 16	Thiepval Memorial.
15441	Scott. N.V.	Liscard, Cheshire.	Pte.	17th	K.I.A	30 07 16	Thiepval Memorial.
26185	Scott S.	Liverpool.	L/Cpl.	17th	K.I.A	30 07 16	Guillemont Rd Cem.
	Sergiades. J.N.	Liverpool.	Lieut.	19th	K.I.A	30 07 16	Thiepval Memorial.
24822	Sheppard H.	Liverpool.	L/Cpl.	17th	K.I.A	30 07 16	Thiepval Memorial.
22835	Shercliff. H.	Burton-on-Trent, Staffs.	A/Sgt.	20th	K.I.A	30 07 16	Thiepval Memorial.
17481	Shipman. H.A.	Nottingham.	L/Sgt.	19th	K.I.A	30 07 16	Thiepval Memorial.
23634	Simcock.W.	Tarporley, Cheshire.	Pte.	17th	K.I.A	30 07 16	Thiepval Memorial.
21727	Slater. J.A.H.	Liverpool.	Pte.	19th	K.I.A	30 07 16	Thiepval Memorial.
23831	Slater. P.A.	Liverpool.	Pte.	20th	K.I.A	30 07 16	Thiepval Memorial.
24899	Smith A	Liverpool.	Pte.	17th	K.I.A	30 07 16	Guillemont Rd Cem.
32972	Smith. C.G.	St-Helens, Lancs.,	Pte.	17th	K.I.A	30 07 16	Thiepval Memorial.
36380	Smith. J.E.	Knowsley, Lancs.	Pte.	20th	K.I.A	30 07 16	Thiepval Memorial.
	Smith. R.H.	Birkenhead, Cheshire.	2nd Lieut.	17th	K.I.A	30 07 16	Guillemont Rd Cem.
22257	Smith. T. D.C.M.	Liverpool.	Pte.	20th	K.I.A	30 07 16	Thiepval Memorial.
22260	Soden H.R.	Liverpool.	Pte.	20th	K.I.A	30 07 16	Guillemont Rd Cem.
17462	Sowerby. A.	Anfield, Liverpool.	L/Cpl.	19th	K.I.A	30 07 16	Thiepval Memorial.
17479	Sowerby.J.A.	Liverpool.	Pte.	19th	K.I.A	30 07 16	Thiepval Memorial.
17716	Starkey. T.	Widnes, Lancs.	Pte.	19th	K.I.A	30 07 16	Thiepval Memorial.
26533	Stephenson. S.	St. Helens, Lancs.	Pte.	19th	K.I.A	30 07 16	Thiepval Memorial.
15569	Stocker. E.J.	Liverpool.	Pte.	17th	K.I.A	30 07 16	Thiepval Memorial.
22849	Stringer. J.H.	Runcorn, Cheshire.	Pte.	20th	K.I.A	30 07 16	Thiepval Memorial.
17472	Sullivan. J.H.	Bootle, Liverpool.	Pte.	19th	K.I.A	30 07 16	Thiepval Memorial.
22545	Sullivan. J.J.	Liverpool.	L/Cpl.	20th	K.I.A	30 07 16	Thiepval Memorial.
16338	Sykes. J.C.	New Brighton, Cheshire.	Pte.	17th	K.I.A	30 07 16	Thiepval Memorial.
23840	Tait. W.	Liverpool.	A/L/Sgt.	20th	K.I.A	30 07 16	Thiepval Memorial.
29659	Taylor. A.	Liverpool.	Pte.	19th	K.I.A	30 07 16	Serre Rd No.2 Cem, B-Hamel.*
22268	Taylor. E.	Heywood, Lancs.	Pte.	20th	K.I.A	30 07 16	Thiepval Memorial.
33229	Taylor. E.H.	Liverpool.	L/Cpl.	19th	K.I.A	30 07 16	Thiepval Memorial.
15861	Thomas. F.	Liverpool.	A/Cpl.	17th	K.I.A	30 07 16	Thiepval Memorial.
27479	Thomas. J.	Southport, Lancs.	Pte.	17th	K.I.A	30 07 16	Thiepval Memorial.
32485	Thomas. J.A.	Liverpool.	Pte.	17th	K.I.A	30 07 16	Thiepval Memorial.
17513	Thomas. J.B.	Liverpool.	L/Cpl.	17th	K.I.A	30 07 16	Thiepval Memorial.
23986	Thomas. R.O.	Liverpool.	Pte.	20th	K.I.A	30 07 16	Thiepval Memorial.
15938	Thomas. S.H.	Frodsham, Cheshire.	Pte.	17th	K.I.A	30 07 16	Thiepval Memorial.
24280	Thompson. G.F.	Liverpool.	Pte.	19th	K.I.A	30 07 16	Thiepval Memorial.
22860	Thompson. H.	Liverpool.	A/Cpl.	20th	K.I.A	30 07 16	Guillemont Rd Cem.
21634	Thorp. F.	Southport, Lancs.	Pte.	19th	K.I.A	30 07 16	Thiepval Memorial.
24764	Titley. W.	Liverpool.	Pte.	17th	K.I.A	30 07 16	Thiepval Memorial.
15746	Tomkinson. C.W.	Liverpool.	L/Cpl.	17th	K.I.A	30 07 16	Thiepval Memorial.
29120	Tunna. R.	Liverpool.	Pte.	19th	K.I.A	30 07 16	Thiepval Memorial.
16098	Tunnington. N.J.	Liverpool.	A/Cpl.	17th	K.I.A	30 07 16	Thiepval Memorial.
15977	Udall. W.E.	Liverpool.	A/Sgt.	17th	K.I.A	30 07 16	Thiepval Memorial.
21930	Unsworth. J.W.N.	Abergele, N.Wales.	Pte.	17th	K.I.A	30 07 16	Guillemont Rd Cem.
17968	Usher. G.E.	Liverpool.	C.S.M.	19th	K.I.A	30 07 16	Thiepval Memorial.
	Vaughan. J.	Blundellsands, Liverpool.	2nd Lieut.	20th	K.I.A	30 07 16	Thiepval Memorial.
	Vaughan-Roberts. R.W.	Blundellsands, Liverpool.	Lieut.	19th	K.I.A	30 07 16	Bernafay Wood, Brit Cem.
21641	Voce. H.I.	Liverpool.	Pte.	19th	K.I.A	30 07 16	Thiepval Memorial.

23853	Wagstaff. R.	Hyde, Cheshire.	Pte.	20th	K.I.A	30 07 16	Thiepval Memorial.
15714	Walker. F.	Liverpool.	Pte.	17th	K.I.A	30 07 16	Thiepval Memorial.
26554	Wallace. R.	Liverpool.	Pte.	19th	K.I.A	30 07 16	Thiepval Memorial.
17759	Walshe. H.C.L.	Liverpool.	Pte.	19th	K.I.A	30 07 16	Thiepval Memorial.
22566	Warburton. H.	Liverpool.	Pte.	20th	K.I.A	30 07 16	Thiepval Memorial.
22282	Waring. C.		Pte.	20th	K.I.A	30 07 16	Thiepval Memorial.
22879	Waring. J.	Liverpool.	Pte.	20th	K.I.A	30 07 16	Thiepval Memorial.
15277	Wealthy. J.H.	Liverpool.	Pte.	17th	K.I.A	30 07 16	Thiepval Memorial.
35348	Webster. J.A.	Upholland, Wigan.	Pte.	20th	K.I.A	30 07 16	Thiepval Memorial.
17500	Weston. H.A.	Liverpool.	Pte.	19th	K.I.A	30 07 16	Thiepval Memorial.
21973	Weston. S.G.	Runcorn, Cheshire.	Pte.	17th	K.I.A	30 07 16	Guillemont Rd Cem.
21654	Whipp. W.	Gt. Saughall, Cheshire.	L/Cpl.	19th	K.I.A	30 07 16	Guillemont Rd Cem.
32664	Whitby. W.	Birkenhead, Cheshire.	Pte.	19th	K.I.A	30 07 16	Guillemont Rd Cem.
17509	White J C.	Liverpool.	C.S.M.	19th	K.I.A	30 07 16	Thiepval Memorial.
22577	Whitehead. F.	Liverpool.	Pte.	20th	K.I.A	30 07 16	Thiepval Memorial.
29298	Whitehouse. F.	Liverpool.	Pte.	20th	K.I.A	30 07 16	Thiepval Memorial.
	Whiting. T.	Liverpool.	Capt.	20th	K.I.A	30 07 16	Thiepval Memorial.
	Wilkinson. G.E.	Liverpool.	2nd Lieut.	20th	K.I.A	30 07 16	Thiepval Memorial.
17503	Wilkinson. H.	Liscard, Cheshire.	Pte.	19th	K.I.A	30 07 16	Thiepval Memorial.
17498	Wilkinson. L.H.	Liverpool.	Pte.	19th	K.I.A	30 07 16	Thiepval Memorial.
15488	Williams. A.R.	Liverpool.	Sgt.	17th	K.I.A	30 07 16	Guillemont Rd Cem.
21649	Williams. F.	Chester.	Pte.	19th	K.I.A	30 07 16	Thiepval Memorial.
21435	Williams. G.	Walton, Liverpool.	Pte.	19th	K.I.A	30 07 16	Thiepval Memorial.
17971	Williams. J.	Birkenhead, Ches.	Pte.	19th	K.I.A	30 07 16	Thiepval Memorial.
23998	Williams. R.	Liverpool.	Pte.	20th	K.I.A	30 07 16	Thiepval Memorial.
29602	Willis. G.	Leighton Buzzard.	Pte.	20th	K.I.A	30 07 16	Thiepval Memorial.
	Willmer. W.	Birkenhead, Cheshire.	Capt.	19th	K.I.A	30 07 16	Thiepval Memorial.
34357	Wilson. C.D.	Liverpool.	Pte.	20th	K.I.A	30 07 16	Thiepval Memorial.
17762	Wilson. F.	Liverpool.	Pte.	19th	K.I.A	30 07 16	Thiepval Memorial.
35949	Wilson. H.	Colne, Lancs.	Pte.	20th	K.I.A	30 07 16	Thiepval Memorial.
15597	Wilson. J.W.	Birkenhead. Cheshire.	Pte.	17th	K.I.A	30 07 16	Thiepval Memorial.
21796	Winkle. E.E.	Liverpool.	L/Cpl.	19th	K.I.A	30 07 16	Thiepval Memorial.
36428	Wolfenden. J.	Liverpool.	Pte.	17th	K.I.A	30 07 16	Thiepval Memorial.
17747	Wood. E.H.	Liverpool.	Pte.	19th	K.I.A	30 07 16	Thiepval Memorial.
30179	Wood. T.	Runcorn, Cheshire.	L/Cpl.	19th	K.I.A	30 07 16	Thiepval Memorial.
	Woodin. W.G.	Rock Ferry, Cheshire.	2nd Lieut.	20th	K.I.A	30 07 16	Thiepval Memorial.
34614	Woodward. H.H.	Liverpool.	Pte.	17th	K.I.A	30 07 16	Flatiron Copse Cem, Mametz.
17746	Woolfenden. B.	Liverpool.	Sgt.	19th	K.I.A	30 07 16	Delville Wood Cem, Longueval.
21937	Wraight. G.F.	Faversham, Kent.	Pte.	17th	K.I.A	30 07 16	Thiepval Memorial.
23871	Wright. H.H.J.	Liverpool.	Pte.	20th	K.I.A	30 07 16	Thiepval Memorial.
24260	Wright. H.L.	Liverpool.	Pte.	19th	K.I.A	30 07 16	Thiepval Memorial.
21651	Wrigley. W.J.	Liverpool.	Pte.	19th	K.I.A	30 07 16	Guillemont Rd Cem.
29699	Blake. C.L	Liverpool.	Pte.	19th	D.O.W	31 07 16	Dive Copse B.C. Sailly-le-Sec.
24961	Johnson. R.	Upholland, Lancs.	Pte.	19th	D.O.W	31 07 16	Carnoy Military Cem.
22740	Jones. F.	Liverpool.	C.S.M.	20th	D.O.W	31 07 16	Corbie CC Ext.
23955	Jones. H.	Liverpool.	Pte.	20th	D.O.W	31 07 16	La Neuville Brit Cem Corbie
17954	Smith. H.T.	Oxton, Birkenhead.	Pte.	19th	K.I.A	31 07 16	Guillemont Rd Cem.
17327	Flockhart. J.H.	Liverpool.	Sgt.	19th	D.O.W	01 08 16	Corbie CC Ext.
22813	Porteous. J.	Carlisle.	Pte.	20th	D.O.W	01 08 16	Corbie CC Ext.
30278	Woods J	Liverpool.	Pte.	19th	D.O.W	01 08 16	La Neuville Brit Cem, Corbie
32703	Gibson. G.H.	Crewe, Cheshire.	Pte.	17th	D.O.W	02 08 16	Corbie CC Ext.
17750	Wadman. T.	Liverpool.	Pte.	19th	D.O.W	02 08 16	Corbie CC Ext.
21735	Hunt. A.	Fazackerley, Liverpool.	Pte.	19th	D.O.W	03 08 16	Corbie CC Ext.
25540	Blackhouse. J.	Liverpool.	Pte.	20th	K.I.A	05 08 16	Thiepval Memorial.
	Butcher. R.N.	Luton, Beds.	2nd Lieut.	20th	D.O.W	05 08 16	La Neuville CC Corbie
26558	Circuit. F.J.	Everton, Liverpool.	Cpl.	17th	D.O.W	05 08 16	Hollybrook Mem. Southampton.
16042	Holliday. A.	Liverpool.	Pte.	17th	D.O.W	05 08 16	Anfield Cem, Liverpool.
22845	Smith. H.J.	Liverpool.	Pte.	20th	D.O.W	05 08 16	Abbeville C.C.

34960	Fumess. J.	Higher Walton, Preston.	Pte.	20th	K.I.A	06 08 16	Peronne CC Ext.
36640	Jones. J.	Liverpool.	Pte.	20th	K.I.A	06 08 16	Loos Memorial.
24664	Lloyd. H.	Liverpool.	Pte.	18th	K.I.A	06 08 16	Robecq CC
22521	Scaife. R.H.	Liverpool.	Pte.	20th	D.O.W	06 08 16	Abbeville C.C.
36765	Buckley. W.H.	Liverpool.	Pte.	17th	K.I.A	07 08 16	Ratiron Copse Cem, Mametz.
36654	Hickling. A.E.	Anfield, Liverpool.	L/Cpl.	20th	K.I.A	07 08 16	Corbie CC Ext.
24216	Martlew. R.	Prescot, Lancs.	Pte.	19th	D.O.W	07 08 16	Wimereux CC
15718	Imison. C.K.	Chester.	Sgt.	17th	D.O.W	09 08 16	Abbeville C.C.
36764	Jervis. T.E.	Liverpool.	Pte.	17th	D.O.W	09 08 16	Danzig Alley Cem, Mametz.
22247	Saul. T.P.	Preston, Lancs.	Pte.	20th	D.O.W	09 08 16	St Sever Cem, Rouen.
32943	Leach. E.	Liverpool.	Pte.	17th	D.O.W	11 08 16	Abbeville C.C.
31020	Daltow. J.W.	Eastham, Cheshire.	Pte.	18th	K.I.A	12 08 16	Gorre Brit Cem, Beuvry
27369	Jones. H.	Bootle, Liverpool.	Pte.	18th	D.O.W	12 08 16	Exeter High Cem, Devon
30800	Fleming. G.V.	Liverpool.	Pte.	17th	D.O.W	13 08 16	Anfield Cem, Liverpool.
17248	Sheppey. A.	Liverpool.	Pte.	18th	D.O.W	13 08 16	Toxteth Park Cem, Liverpool.
17479	Watling. W.	Liverpool.	Pte.	19th	D.O.W	13 08 16	Kirkdale Cem, Liverpool.
26141	Higham. A.	Wavertree, Liverpool.	Pte.	17th	D.O.W	15 08 16	Allerton Cem, Liverpool.
32718	Street. G.R.L.	Liverpool.	Pte.	18th	K.I.A	15 08 16	Gorre Brit Cem, Beuvry
15784	Bibby. C.L.	Bidston, Cheshire.	Pte.	17th	D.O.W	17 08 16	Abbeville C.C.
16646	Campbell. A.J.	Liverpool.	Cpl.	18th	D.O.W	20 08 16	Chocques Mil Cem
16799	Currie. J.	Aintree, Liverpool.	L/Cpl.	18th	K.I.A	20 08 16	Gorre Brit Cem, Beuvry
23385	Hyland. G.	Kirkdale, Liverpool.	Pte.	18th	K.I.A	20 08 16	Gorre Brit Cem, Beuvry
22484	Moody. F.	Liverpool.	Pte.	20th	D.O.W	20 08 16	Anfield Cem, Liverpool.
26101	Quinn. J.	Liverpool.	Pte.	18th	K.I.A	20 08 16	Loos Memorial.
11621	Rockcliffe. A.	Southport, Lancs.	Pte.	18th	K.I.A	20 08 16	Loos Memorial.
32618	Roe. S.M.	Liverpool.	Pte.	18th	K.I.A	20 08 16	Loos Memorial.
	Slaughter. A.C.	Norfolk.	2nd Lieut.	18th	D.O.W	23 08 16	Gorre Brit Cem, Beuvry.
26088	Dodson. J.	Liverpool.	Pte.	18th	K.I.A	26 08 16	Lonsdale Cem, Aveluy.
17835	Hallwood. A.	Warrington, Cheshire.	Pte.	19th	K.I.A	26 08 16	Lonsdale Cem, Aveluy.
24977	Charlton. J.W.	Everton, Liverpool.	Pte.	17th	K.I.A	28 08 16	Cerisy-Gailly Mil Cem.
300256	Gwilt. R.	Liverpool.	Pte.	18th	K.I.A	28 08 16	Berlin South West Cem.
38028	Cannon. D.	Lezayre, I.O.M.	Pte.	19th	K.I.A	30 08 16	Gorre Brit Cem, Beuvry
4489	Hughes. W.H.	Liverpool.	Pte.	19th	K.I.A	30 08 16	Gorre Brit Cem, Beuvry
16414	Gray. A.E.	Rock Ferry, Cheshire	Sgt.	18th	D.O.W	31 08 16	St.Marie Cem, Le Havre.
34611	Bowers. E.	Warrington, Cheshire.	Pte.	17th	K.I.A	04 09 16	Thiepval Memorial.
15198	Schollick. R.	Liverpool.	Pte.	17th	K.I.A	04 09 16	Gorre Brit Cem, Beuvry
21568	Mahar. W.	Liverpool.	Pte.	19th	K.I.A	05 09 16	Loos Memorial.
22179	Moore. D.G.	Birmingham.	Pte.	20th	Died.	06 09 16	Point 110 O M Cem.Fricourt
	Morris. G.M.	London.	2nd Lieut.	17th	D.O.W	07 09 16	Brompton Cem, London.
25551	Boyde. R.N.	Dingle, Liverpool.	Pte.	20th	D.O.W	10 09 16	Corbie CC Ext.
35940	Watson. G.A. (J.Graham)	Liverpool.	Pte.	18th	K.I.A	11 09 16	Gorre Brit Gem, Beuvry
35420	Stephens. J.J.	Liverpool.	Pte.	18th	D.O.W	12 09 16	La Neuville Brit Cem, Corbie
17800	Concannon. E.	Liverpool.	C.S.M.	19th	K.I.A	13 09 16	Loos Memorial.
52076	Jordan. T.	Castlereagh, Ireland.	Pte.	19th	K.I.A	13 09 16	Loos Memorial.
	Lloyd. R.G. M.C.	Southport, Lancs.	Lieut.	19th	K.I.A	13 09 16	Loos Memorial.
5440	Worsley. A.R.	Seaforth, Liverpool.	Pte.	19th	K.I.A	13 09 16	Gorre Brit Cem, Beuvry
26366	Whittle. A.	Leigh, Lancs.	Pte.	18th	K.I.A	15 09 16	Bernafay Wood. Brit Cem
22393	Hanley. J.	Liverpool.	Pte.	20th	K.I.A	16 09 16	Loos Memorial.
	Taylor. N.L	Roscommon, Ireland.	2nd Lieut	19th	D.O.W	18 09 16	Bethune Town Cem.
23842	Taggart. H.J.	Liverpool.	Pte.	20th	D.O.W	21 09 16	St. Helens Cem, Lancs.
5955	Cregeen. C.	Southport, Lancs.	Pte.	17th	D.O.W	28 09 16	Kirk Christ Rushen(H.T.CY)IOM
17377	Hubbard. F.J.	Liverpool.	Pte.	19th	D.O.W	02 10 16	Longuenesse Cem, St.Omer.
27032	Nester. M.	Liverpool.	Pte.	19th	K.I.A	07 10 16	Thiepval Memorial.
57045	Hickling. T.	Liverpool.	Pte.	18th	K.I.A	10 10 16	Vis-en-Artois Memorial.
52804	Abrahams. H.	Broughton, Manchester	Pte.	17th	K.I.A	12 10 16	Thiepval Memorial.
52480	Adams. F.E.	Wood Green, Middlesex.	Pte.	18th	K.I.A	12 10 16	Thiepval Memorial.
51684	Allen. S.H	Liverpool.	Pte.	19th	K.I.A	12 10 16	Thiepval Memorial.
15695	Almond. J.	Liverpool.	Pte.	17th	K.I.A	12 10 16	Thiepval Memorial.

52802	Asbury. J.	Bootle, Liverpool.	Pte.	17th	K.I.A	12 10 16	Warlencourt Brit Cem.
17233	Barber. T.	Liverpool.	Pte.	17th	K.I.A	12 10 16	Warlencourt Brit Cem.
21975	Barker. H.	Southport, Lancs.	Pte.	20th	K.I.A	12 10 16	Thiepval Memorial.
51961	Barnes. W.	Liverpool.	Pte.	17th	K.I.A	12 .10 16	Warlencourt Brit Cem.
26195	Barrett. W.	Liverpool.	Pte.	17th	K.I.A	12 10 16	Warlencourt Brit Cem.
52809	Barton. R.	Birkdale, Southport, Lancs.	Pte.	17th	K.I.A	12 10 16	St Sever Cem Ext, Rouen.
51656	Bellis. J.G.	Liscard, Cheshire.	Pte.	17th	K.I.A	12 10 16	Warlencourt Brit Cem.
26147	Billyard. H.	Walton, Liverpool.	Pte.	17th	K.I.A	12 10 16	Thiepval Memorial.
51648	Blanchard. J.	Bootle, Liverpool.	Pte.	17th	D.O.W	12 10 16	Heilly Station Cem.Mericourt
52827	Bolger. T.	Liverpool.	Pte.	17th	K.I.A	12 10 16	Warlencourt Brit Cem.
52019	Broughton. B.	Blackpool, Lancs.	Pte.	17th	K.I.A	12 10 16	Thiepval Memorial.
31091	Brown. J.	Liverpool.	Pte.	17th	K.I.A	12 10 16	Thiepval Memorial.
51593	Brown. T.	Liverpool.	Pte.	17th	K.I.A	12 10 16	Warlencourt Brit Cem.
29141	Burns. W.	Liverpool.	Pte.	20th	K.I.A	12 10 16	Thiepval Memorial.
51965	Campbell. C.	Litherland, Liverpool.	Pte.	17th	K.I.A	12 10 16	Thiepval Memorial.
51773	Chamberlain. T.	Liverpool.	Pte.	19th	K.I.A	12 10 16	Thiepval Memorial.
38132	Chamock. J.	Preston. Lancs.	Pte.	17th	K.I.A	12 10 16	Thiepval Memorial. «
36889	Christian. R.C.	Castletown, I.O.M.	Pte.	20th	K.I.A	12 10 16	Thiepval Memorial.
51997	Clark. J.D.	Birkenhead, Cheshire.	Pte.	17th	K.I.A	12 10 16	Warlencourt Brit Cem.
30177	Clarke. E.	Liverpool.	Pte.	17th	K.I.A	12 10 16	Thiepval Memorial.
51649	Clay. H.	Waterloo, Liverpool.	Pte.	17th	K.I.A	12 10 16	Thiepval Memorial.
	Collin. K.G.	Birkenhead, Cheshire.	2nd Lieut.	17th	K.I.A	12 10 16	Warlencourt Brit Cem.
36327	Corlett. A.	Andrews, IOM	Pte.	18th	K.I.A	12 10 16	Thiepval Memorial.
15352	D'Arcy. R.	Liverpool.	Pte.	17th	K.I.A	12 10 16	Warlencourt Brit Cem.
36489	Davies. P.	Bolton, Lancs.	Pte.	18th	K.I.A	12 10 16	Thiepval Memorial.
51952	Deeley. F.A.	Warrington, Cheshire.	Pte.	17th	K.I.A	12 10 16	Thiepval Memorial.
14489	Delaney. M.	Liverpool.	Pte.	18th	K.I.A	12 10 16	Thiepval Memorial.
31657	Dewar. A.A.	Liverpool.	Pte.	20th	K.I.A	12 10 16	Warlencourt Brit Cem.
	Dixon. J.G.	Southport, Lancs.	2nd Lieut.	17th	K.I.A	12 10 16	Thiepval Memorial.
51053	Ducksbury. J.W.	Bootle, Liverpool.	Pte.	17th	K.I.A	12 10 16	Thiepval Memorial.
24950	Duffy. E.	Liverpool.	Pte.	17th	K.I.A	12 10 16	London Rd Cem Ext, Longueval
24638	Eccleston. G.E.	Dingle, Liverpool.	L/Cpl.	17th	K.I.A	12 10 16	Warlencourt Brit Cem.
52797	Emery. R. Everton,	Liverpool.	Pte.	17th	K.I.A	12 10 16	Warlencourt Brit Cem.
52859	Evans. A.E.	Waterloo, Liverpool.	Pte.	17th	K.I.A	12 10 16	Thiepval Memorial.
15985	Eyes. J.O.	Chester.	L/Cpl.	17th	K.I.A	12 10 16	Thiepval Memorial.
26051	Field. E.L.	Bedford.	Pte.	17th	K.I.A	12 10 16	Thiepval Memorial.
52689	Flynn. J.	Bangor, N.Wales.	Pte.	17th	K.I.A	12 10 16	Thiepval Memorial.
24993	Foden. F.R.F.	Seacombe, Cheshire.	Pte.	17th	K.I.A	12 10 16	Warlencourt Brit Cem.
51990	Fox. E.	Macclesfield, Cheshire.	Pte.	17th	K.I.A	12 10 16	Warlencourt Brit Cem.
26197	Fursland. J.	Liverpool.	Pte.	17th	K.I.A	12 10 16	Warlencourt Brit Cem.
52829	Gerrard. R. Bootle.	Liverpool.	Pte.	17th	K.I.A	12 10 16	Thiepval Memorial.
51650	Gibbins. A.A.	Liverpool.	Pte.	17th	K.I.A	12 10 16	Warlencourt Brit Cem.
34309	Gittins. B.	Liverpool.	Pte.	18th	K.I.A	12 10 16	Thiepval Memorial.
51561	Goldstone. A.	Liverpool.	Pte.	17th	K.I.A	12 10 16	Warlencourt Brit Cem.
51905	Goodwin. J.	Seaforth, Liverpool.	Pte.	17th	K.I.A	12 10 16	Thiepval Memorial.
	Grennan. G.L.	Southport, Lancs.	2nd Lieut.	20th	K.I.A	12 10 16	Thiepval Memorial.
51932	Griffiths. W.H.	Ormskirk, Lancs.	Pte.	17th	K.I.A	12 10 16	Thiepval Memorial.
52953	Hammond. T.B.	Romford, Essex.	Pte.	17th	K.I.A	12 10 16	Warlencourt Brit Cem.
23024	Handley. J.W.	Liverpool.	L/Cpl.	20th	K.I.A	12 10 16	Warlencourt Brit Cem.
51967	Hargreaves. S.	Darwen, Lancs.	Pte.	17th	K.I.A	12 10 16	Warlencourt Brit Cem.
52806	Harris. W.D.	Southport, Lancs.	Pte.	17th	K.I.A	12 10 16	Warlencourt Brit Cem.
15870	Haworth. R.S.	Liverpool.	L/Cpl.	17th	K.I.A	12 10 16	Warlencourt Brit Cem.
52954	Hayden. S.P.	Newport, Essex.	L/Cpl.	17th	K.I.A	12 10 16	Thiepval Memorial.
27183	Hayes. T.	Manchester.	Pte.	20th	K.I.A	12 10 16	Thiepval Memorial.
52037	Henderson. F.	Everton, Liverpool.	Pte.	19th	K.I.A	12 10 16	Warlencourt Brit Cem.
24592	Henshaw. J.	Liverpool.	Pte.	17th	K.I.A	12 10 16	Thiepval Memorial.
51963	Hoos. W.J.	Kirkdale, Liverpool.	Pte.	17th	K.I.A	12 10 16	Warlencourt Brit Cem.
	Hornby. W.	Garstang, Lancs.	2nd Lieut.	17th	K.I.A	12 10 16	Warlencourt Brit Cem.

	Horser. S.C.S.		Capt.	17th	K.I.A	12 10 16	Thiepval Memorial.
16245	Hughes. H.	New Brighton, Cheshire	Pte.	18th	D.O.W	12 10 16	Thiepval Memorial.
15996	Huntley. A.S.	Liverpool.	Pte.	17th	K.I.A	12 10 16	Thiepval Memorial.
52919	Ince. A.	Chorley, Lancs.	Cpl.	17th	K.I.A	12 10 16	Warlencourt Brit Cem.
51603	Ingoldsby. J.	Liverpool.	Pte.	17th	K.I.A	12 10 16	Warlencourt Brit Cem.
	Ireland. S.J.	Cornwall.	2nd Lieut.	17th	K.I.A	12 10 16	Thiepval Memorial.
51591	Jeffreys. A.R.	Wembley, London.	Pte.	17th	K.I.A	12 10 16	Thiepval Memorial.
15700	Johnson. T.R.	Liverpool.	Pte.	17th	K.I.A	12 10 16	Warlencourt Brit Cem.
15557	Johnson. W.	Liverpool.	Pte.	17th	K.I.A	12 10 16	Thiepval Memorial.
51569	Jones. A.	Liverpool.	Pte.	17th	K.I.A	12 10 16	Warlencourt Brit Cem.
51964	Jones. A.E.	Liverpool.	Pte.	17th	K.I.A	12 10 16	Thiepval Memorial.
32659	Jones. G.	Liverpool.	Pte.	18th	K.I.A	12 10 16	Thiepval Memorial.
51968	Jones. G.F.	Abergele, N.Wales.	Pte.	17th	K.I.A	12 10 16	Thiepval Memorial.
32775	Jones. H.	Crewe, Cheshire.	Pte.	18th	K.I.A	12 10 16	Thiepval Memorial.
53133	Jones. J.	Ironbridge, Salop.	Pte.	19th	K.I.A	12 10 16	Caterpillar Valley, Longueval.
16002	Jones. J.A.	Liverpool.	Pte.	17th	K.I.A	12 10 16	Thiepval Memorial.
51624	Jones. J.E.	Liverpool.	Pte.	17th	K.I.A	12 10 16	Thiepval Memorial.
31658	Jones. J.R.C.	Liverpool.	Pte.	20th	K.I.A	12 10 16	Thiepval Memorial.
30822	Jones. J.W.	Crewe, Cheshire	Pte.	18th	D.O.W	12 10 16	Dartmoor Cem.Becordel-Becourt.
51958	Kane. J.P.	Barrow-in-Furness.	Pte.	17th	K.I.A	12 10 16	Thiepval Memorial.
52847	Kelly. J.	Liverpool.	Pte.	17th	K.I.A	12 10 16	Thiepval Memorial.
21738	Kelly. W.J.	Laxey, I.O.M.	Pte.	20th	K.I.A	12 10 16	Thiepval Memorial.
15793	Kennedy. H.	Ormskirk, Lancs.	Pte.	17th	K.I.A	12 10 16	Thiepval Memorial.
51670	Lawrence. E.O.	Liverpool.	Pte.	17th	K.I.A	12 10 16	Thiepval Memorial.
37395	Lomas. J.	Ashton-under-Lyne, Lancs.	Pte.	20th	K.I.A	12 10 16	Warlencourt Brit Cem.
51951	Longshaw. F.	Warrington, Cheshire.	Pte.	17th	K.I.A	12 10 16	Warlencourt Brit Cem.
52803	Lovelady H	Great Crosby, Liverpool.	L/Cpl.	17th	K.I.A	12 10 16	Warlencourt Brit Cem.
51915	Madden. J.H.	Liverpool.	Pte.	17th	K.I.A	12 10 16	Thiepval Memorial.
22464	Maine. J.G.	Liverpool.	Cpl.	20th	K.I.A	12 10 16	Thiepval Memorial.
51651	Mann. P.	Liverpool.	Pte.	17th	K.I.A	12 10 16	Warlencourt Brit Cem.
22765	Mawdsley. H.	Crosby, Liverpool.	Pte.	20th	K.I.A	12 10 16	Warlencourt Brit Cem.
52828	Mawdsley. J.	Southport, Lancs.	Pte.	17th	K.I.A	12 10 16	Thiepval Memorial.
52871	Maxwell. J.	Southport, Lancs.	Pte.	17th	K.I.A	12 10 16	Thiepval Memorial.
24910	McCarthy. D.	Fazackerley, Liverpool.	Pte.	17th	K.I.A	12 10 16	Thiepval Memorial.
52157	McPhee. J.	Manchester.	Pte.	19th	K.I.A	12 10 16	Warlencourt Brit Cem.
15608	Moore. R.	Liverpool.	Pte.	17th	K.I.A	12 10 16	Thiepval Memorial.
32700	Morgan. F.J.	Liverpool.	Pte.	20th	K.I.A	12 10 16	Thiepval Memorial.
51647	Morris. A.	Liverpool.	Pte.	17th	K.I.A	12 10 16	Warlencourt Brit Cem.
22488	Mortimore R	Seacombe, Cheshire.	Pte.	20th	K.I.A	12 10 16	Thiepval Memorial.
36948	Mosses. R.H.	Liverpool.	Pte.	20th	K.I.A	12 10 16	Warlencourt Brit Cem.
51665	Murdock. W.	Liverpool.	Pte.	17th	K.I.A	12 10 16	Thiepval Memorial.
27548	Myers. G.	Liverpool.	L/Cpl.	17th	K.I.A	12 10 16	Warlencourt Brit Cem.
51641	Nickson. A.J.	Cambridge.	Pte.	17th	K.I.A	12 10 16	Warlencourt Brit Cem.
30160	O'Born. G.	Liverpool.	Pte.	17th	K.I.A	12 10 16	Thiepval Memorial.
17680	Parry. T.H.	Liscard, Cheshire.	Pte.	19th	K.I.A	12 10 16	Thiepval Memorial.
	Paterson. R.D.	Rock Ferry, Cheshire.	Lieut.	20th	K.I.A	12 10 16	Caterpillar Valley, Longueval.
24240	Pennock. H.	Liverpool.	Pte.	20th	K.I.A	12 10 16	Thiepval Memorial.
52014	Perry. J.	Keswick, Cumberland.	Pte.	18th	K.I.A	12 10 16	Warlencourt Brit Cem.
24835	Phillips. E.L.	Ormskirk, Lancs.	Pte.	17th	K.I.A	12 10 16	Thiepval Memorial.
29657	Piercy. A.	Liverpool.	Pte.	20th	K.I.A	12 10 16	Thiepval Memorial.
12052	Plaice. F.	Downham, Norfolk.	Pte.	17th	K.I.A	12 10 16	Thiepval Memorial.
51587	Preston. H.	Liverpool.	Pte.	17th	K.I.A	12 10 16	Thiepval Memorial.
21998	Pulford. F.I.	Fairfield, Liverpool.	Pte.	17th	K.I.A	12 10 16	Thiepval Memorial.
51922	Rainford. J.	Skelmersdale, Lancs.	Pte.	17th	K.I.A	12 10 16	Warlencourt Brit Cem.
15634	Rawlands. A.B.	Liscard, Cheshire.	L/Cpl.	17th	K.I.A	12 10 16	Thiepval Memorial.
52007	Reynolds. A.E.	Gresford, Denbigh, N.Wales.	Pte.	17th	K.I.A	12 10 16	Thiepval Memorial.
51600	Richards. R.A.	Southport, Lancs.	Pte.	17th	K.I.A	12 10 16	Warlencourt Brit Cem.
51579	Roberts. C.	Wavertree, Liverpool.	Pte.	17th	K.I.A	12 10 16	Warlencourt Brit Cem.

31174	Robinson. J.	Liverpool.	Pte.	17th	K.I.A	12 10 16	Warlencourt Brit Cem.
31628	Roche. B.	Liverpool.	Pte.	20th	K.I.A	12 10 16	Thiepval Memorial.
53140	Sands. S.J.	Thornage, Norfolk.	Pte.	20th	K.I.A	12 10 16	Thiepval Memorial.
52951	Saunders. W.	Broomfield, Essex.	Pte.	17th	K.I.A	12 10 16	Thiepval Memorial.
52937	Schofield. F.	Rochdale, Lancs.	Pte.	17th	K.I.A	12 10 16	Warlencourt Brit Cem.
29622	Sidebotham. R.F.	Llandudno, N.Wales.	Pte.	20th	K.I.A	12 10 16	Thiepval Memorial.
51949	Silvester. C.	Aintree, Liverpool.	Pte.	17th	K.I.A	12 10 16	Warlencourt Brit Cem.
51607	Singleton. J.	Liscard, Cheshire.	Pte.	17th	K.I.A	12 10 16	Warlencourt Brit Cem.
26026	Slee. G.W.	New Brighton, Ches.	Pte.	17th	K.I.A	12 10 16	Warlencourt Brit Cem.
38140	Smallshaw. J.	Liverpool.	Pte.	20th	K.I.A	12 10 16	Thiepval Memorial.
51594	Smith. L.H.	Philadelphia, USA.	Pte.	17th	K.I.A	12 10 16	Thiepval Memorial.
22266	Stringer. J.	Widnes, Lancs.	Pte.	20th	K.I.A	12 10 16	Thiepval Memorial.
22547	Swift H.	New York, USA.	Pte.	20th	K.I.A	12 10 16	Thiepval Memorial.
36733	Swift. R.	Preston, Lancs.	Pte.	20th	K.I.A	12 10 16	Thiepval Memorial.
23839	Tallis. G.F.	Liverpool.	Sgt.	20th	K.I.A	12 10 16	Caterpillar Valley, Longueval.
29253	Thompson. F.	Liverpool.	Pte.	20th	K.I.A	12 10 16	Thiepval Memorial.
37221	Toft. C.	Manchester.	Pte.	20th	K.I.A	12 10 16	Warlencourt Brit Cem.
52836	Tynan. R.	Walton, Liverpool.	Pte.	17th	K.I.A	12 10 16	Warlencourt Brit Cem.
51986	Venables. H.	Wrexham, N.Wales.	Pte.	17th	K.I.A	12 10 16	Thiepval Memorial.
15314	Waddington. L	Sefton Park, Liverpool.	Pte.	17th	K.I.A	12 10 16	Warlencourt Brit Cem.
22878	Waine. H.	Liverpool.	Pte.	20th	K.I.A	12 10 16	Thiepval Memorial.
52092	Walker. W.J.	Liverpool.	Pte.	19th	K.I.A	12 10 16	Serre Rd No.2 Cem, B-Hamel.*
15627	Ward. G. M.M.	Pickering, Yorks.	Sgt.	17th	K.I.A	12 10 16	Thiepval Memorial.
32346	Warren. J.W.	Liverpool.	Pte.	20th	K.I.A	12 10 16	Warlencourt Brit Cem.
8852	Welsh. J.M.	Liverpool.	Pte.	20th	Died.	12 10 16	Ford Cem, Liverpool.
51582	Williams. E.A.	New Brighton, Cheshire.	Pte.	17th	K.I.A	12 10 16	Thiepval Memorial.
51616	Williams. R.	Liverpool.	Pte.	17th	K.I.A	12 10 16	Thiepval Memorial.
52266	Wood. A.	Manchester.	Pte.	20th	K.I.A	12 10 16	Thiepval Memorial.
15286	Wright. G.F.	Liverpool.	L/Cpl.	17th	K.I.A	12 10 16	Thiepval Memorial.
15022	Young. H.G.	New Brighton, Cheshire.	Sgt.	17th	K.I.A	12 10 16	Thiepval Memorial.
36026	Young. J.	Southport, Lancs.	Pte.	18th	K.I.A	12 10 16	Thiepval Memorial.
51716	Young. J.F.	Liverpool.	Pte.	20th	K.I.A	12 10 16	Thiepval Memorial.
52038	Mayoh. R.	Bolton, Lancs.	Pte.	19th	D.O.W	13 10 16	Caterpillar Valley, Longueval.
17227	Smith. E.	Birkenhead, Cheshire.	Cpl.	18th	K.I.A	13 10 16	Thiepval Memorial.
	Watson. A.P.	Waterloo, Liverpool.	2nd Lieut.	17th	D.O.W	13 10 16	Heilly Station Cem.Mericourt
15781	Alderson. R.H.	Liverpool.	Sgt.	17th	D.O.W	14 10 16	Heilly Station Cem.Mericourt
17285	Bellwood. E.G.	Liverpool.	Pte.	19th	D.O.W	14 10 16	Heilly Station Cem.Mericourt
32667	Fallding. J.	Liverpool.	Pte.	18th	K.I.A	14 10 16	Thiepval Memorial.
25103	Mayoh. W.	Bolton, Lancs.	Pte.	17th	D.O.W	14 10 16	Flatiron Copse Cem, Mametz.
30280	Doyle. J.P.	Liverpool.	Pte.	20th	D.O.W	15 10 16	St Sever Cem, Rouen.
34608	Foster. T.	Ormskirk, Lancs.	Pte.	18th	K.I.A	15 10 16	Thiepval Memorial.
53107	Green. J.	Leigh-on-Sea, Essex	Pte.	18th	K.I.A	15 10 16	Bulls Rd Cem, Flers.
11282	Hynes. S.	Liverpool.	Pte.	17th	D.O.W	15 10 16	Heilly Station Cem.Mericourt
32717	Lee. E.	Leigh, Lancs.	Pte.	20th	D.O.W	15 10 16	Heilly Station Cem.Mericourt
34271	Vickers. J.	Crewe, Cheshire.	Pte.	18th	K.I.A	15 10 16	Bulls Rd Cem, Flers.
22929	Bird. E.	Liverpool.	Pte.	20th	D.O.W	16 10 16	Heilly Station Cem.Mericourt
16795	Collision W.F.	Douglas IOM	Pte.	18th	K.I.A	16 10 16	Thiepval Memorial.
21665	Morton. W.B.	Banbridge, Co.Down.	Pte.	19th	K.I.A	17 10 16	Thiepval Memorial.
16984	Orrett. J.S.	Prescot, Lancs.	Sgt.	18th	K.I.A	17 10 16	Thiepval Memorial.
	Ravenscroft. G.	Birkenhead, Cheshire.	Capt.	18th	K.I.A	17 10 16	Warlencourt Brit Cem.
32713	Blayney. W.J.	Oswestry, N.Wales.	Pte.	18th	K.I.A	18 10 16	Thiepval Memorial.
52464	Bradford. W.	Brentwood, Essex	Pte.	18th	K.I.A	18 10 16	Thiepval Memorial.
17046	Broady. W.T.	Egremont, Cheshire.	Pte.	18th	K.I.A	18 10 16	Thiepval Memorial.
24646	Bulfield. T.	Liverpool.	Pte.	18th	K.I.A	18 10 16	Thiepval Memorial.
17272	Burke. M.J.	Liverpool.	Pte.	19th	K.I.A	18 10 16	Thiepval Memorial.
29492	Cadwell. J.J.	Churchtown, Southport	Pte.	18th	K.I.A	18 10 16	Warlencourt Brit Cem.
52443	Caine. J.	St.Helens, Lancs.	L/Cpl.	18th	K.I.A	18 10 16	Thiepval Memorial.
25828	Cooper. T.	Ellesmere Port, Cheshire	Pte.	18th	K.I.A	18 10 16	Thiepval Memorial.

17180	Cork. A.W.S.	Hanley, Staffs	Cpl.	18th	D.O.W	18 10 16	Dartmoor Cem.Becordel-Becourt.
16706	Cromwell. T.L	Hunts Cross, Liverpool.	Cpl.	18th	K.I.A	18 10 16	Thiepval Memorial.
32741	Fitzroy. F.	Liverpool.	Pte.	18th	K.I.A	18 10 16	Thiepval Memorial.
32690	Golothan. W.P.	Chester.	Sgt.	18th	K.I.A	18 10 16	Thiepval Memorial.
32795	Griffiths. J.W.	Kirkdale, Liverpool.	Pte.	18th	K.I.A	18 10 16	Warlencourt Brit Cem.
	Handyside. J.	Leith, Edinburgh.	2nd Lieut.	18th		18 10 16	Thistle Dump Cem, Longueval.
17057	Hilditch. A.J.	Liverpool.	Pte.	18th	K.I.A	18 10 16	Warlencourt Brit Cem.
52460	Jackson. F.	Pendleton, Manchester	Pte.	18th	K.I.A	18 10 16	Warlencourt Brit Cem.
17197	Jones. T.L	Liverpool.	C.S.M.	18th	K.I.A	18 10 16	Warlencourt Brit Cem.
53115	Langford. H.G.	Willesden Green, London	Pte.	18th	D.O.W	18 10 16	Warlencourt Brit Cem.
39054	Leadsom. S.	Liverpool.	Pte.	18th	K.I.A	18 10 16	Warlencourt Brit Cem.
26640	Linton. W.E.	Liverpool.	Pte.	18th	K.I.A	18 10 16	Thiepval Memorial.
52136	Morrell. A.	Failsworth, Manchester.	Pte.	19th	K.I.A	18 10 16	Thiepval Memorial.
16981	Morris. J.P.	Hoylake, Cheshire.	Cpl.	18th	K.I.A	18 10 16	Warlencourt Brit Cem.
17080	Narracott. F.V.	Liverpool.	Pte.	18th	K.I.A	18 10 16	Thiepval Memorial.
17933	O'Keeffe. G.	Liverpool.	Pte.	19th	K.I.A	18 10 16	Guillemont Rd Cem.
52163	Owen. H.	Hulme, Manchester.	Pte.	19th	K.I.A	18 10 16	Thiepval Memorial.
33265	Poole. F.	Shrewsbury, Salop.	Pte.	18th	K.I.A	18 10 16	Thiepval Memorial.
36914	Powderly. T.	Bootle, Liverpool.	Pte.	20th	D.O.W	18 10 16	Etaples Mil Cem
52102	Reilly. T.	Manchester.	Pte.	19th	K.I.A	18 10 16	Thiepval Memorial.
16992	Rench. T.F.	Liverpool.	Pte.	18th	K.I.A	18 10 16	Thiepval Memorial.
33698	Reynolds. G.H.	Craven Arms, Salop.	Pte.	18th	K.I.A	18 10 16	Thiepval Memorial.
22821	Roberts. E.	Seacombe, Cheshire.	Pte.	20th	K.I.A	18 10 16	Thiepval Memorial.
52186	Sawer. T.A.	Salford, Manchester.	Pte.	19th	K.I.A	18 10 16	Caterpillar Valley, Longueval.
	Sinclair. W.	Hull, Yorks.	2nd Lieut.	18th	K.I.A	18 10 16	Thiepval Memorial.
26587	Speed. R.	Liverpool.	Pte.	18th	K.I.A	18 10 16	Thiepval Memorial.
21786	Stead. M.H.	Liverpool.	L/Cpl.	19th	K.I.A	18 10 16	Thiepval Memorial.
33175	Taylor. A.	Liverpool.	Pte.	18th	K.I.A	18 10 16	Warlencourt Brit Cem.
	Twemlow-Allen. W.A.		2nd Lieut.	18th	K.I.A	18 10 16	Warlencourt Brit Cem.
	Wane. H.	Lytham St. Annes, Lancs.	2nd Lieut.	18th	K.I.A	18 10 16	Thiepval Memorial.
32676	Williams. F.	Birkenhead, Cheshire.	L/Cpl.	18th	K.I.A	18 10 16	Warlencourt Brit Cem.
16790	Winstanley. E.	Liscard, Cheshire.	Pte.	18th	K.I.A	18 10 16	Thiepval Memorial.
35808	Winterbottom. S.	Ashton-under-Lyne, Lancs.	Pte.	18th	K.I.A	18 10 16	Thiepval Memorial.
34935	Yates. A.	Bolton, Lancs.	Pte.	18th	K.I.A	18 10 16	Thiepval Memorial.
15156	Beckett. F.P.	Liverpool.	C.S.M.	17th	D.O.W	19 10 16	Dernancourt CC Ext.
53077	Bullimore. W.	Melton Mowbray, Leics.	Pte.	20th	K.I.A	19 10 16	Thiepval Memorial.
33720	Clarke. W.	Ashton-on-Clun, Salop	Pte.	18th	K.I.A	19 10 16	Thiepval Memorial.
52010	Cross. T.	Birkenhead, Cheshire.	Pte.	17th	D.O.W	19 10 16	St Sever Cem, Rouen.
16381	Leyland. W.	Liverpool.	L/Cpl.	18th	K.I.A	19 10 16	Thiepval Memorial.
53196	Mannion. S.	Hednesford, Staffs.	Pte.	20th	K.I.A	19 10 16	A.I.F Burial Ground, Flers.
52028	Vernon. G.	Liverpool.	Pte.	20th	K.I.A	19 10 16	Thiepval Memorial.
22676	Dunstan. A.	Huddersfield, Yorks.	L/Cpl.	20th	K.I.A	20 10 16	Thiepval Memorial.
22677	Dutton. A.	Runcorn, Cheshire.	Pte.	20th	K.I.A	20 10 16	Thiepval Memorial.
21574	McDowall. H.	Liverpool.	Pte.	19th	D.O.W	20 10 16	Heilly Station Cem.Mericourt
22782	Molyneux. E.	Kirkdale, Liverpool.	Sgt.	20th	K.I.A	20 10 16	Thiepval Memorial.
22181	Moulton. R.W.	Liverpool.	A/CSM.	20th	K.I.A	20 10 16	Thiepval Memorial.
35948	Newton. A.	Oldham, Lancs.	Pte.	20th	K.I.A	20 10 16	A.I.F Burial Ground, Flers.
23127	Nickson. C.H.	Runcorn, Cheshire.	Pte.	20th	K.I.A	20 10 16	Thiepval Memorial.
52217	Rowe. N.S.	Chorlton-o-M, Manchester.	Pte.	19th	K.I.A	20 10 16	Thiepval Memorial.
23841	Taylor. A.	Liverpool.	Pte.	20th	K.I.A	20 10 16	Thiepval Memorial.
23856	Webber. S.J.	Bootle, Liverpool.	Pte.	18th	K.I.A	20 10 16	Thiepval Memorial.
51609	Whitehead. G.	Bromborough, Ches.	Pte.	17th	D.O.W	20 10 16	St Sever Cem, Rouen.
27905	Williams. W.	Garston, Liverpool.	Pte.	20th	K.I.A	20 10 16	Thiepval Memorial.
16767	Beatie. J.R.	Liverpool.	Pte.	18th	D.O.W	21 10 16	Heilly Station Cem.Mericourt
32259	Woods. J.	Liverpool.	Pte.	20th	K.I.A	21 10 16	Thiepval Memorial.
37895	Bibby. J.	Liverpool.	Pte.	20th	K.I.A	22 10 16	Thiepval Memorial.
36883	Cowin. W.J.	Douglas, I.O.M.	Pte.	20th	K.I.A	22 10 16	Thiepval Memorial.
38487	Crebbin. E.	Liverpool.	Pte.	20th	K.I.A	22 10 16	Thiepval Memorial.

22989	Elcock. A.W.	Droylsden, Manchester.	Pte.	20th	K.I.A	22 10 16	Thiepval Memorial.
16238	Hamilton. S. Bootle,	Liverpool.	Pte.	18th	D.O.W	22 10 16	Etaples Mil Cem
22820	Richardson. S.	Liverpool.	Pte.	20th	K.I.A	22 10 16	Thiepval Memorial.
26885	Shimmin. J.A.C.	Douglas, I.O.M.	Pte.	20th	K.I.A	22 10 16	Thiepval Memorial.
52857	Cookson. A.	Southport, Lancs.	Pte.	17th	K.I.A	23 10 16	Thiepval Memorial.
37251	Sanderson. J.	Southport, Lancs.	Pte.	20th	D.O.W	23 10 16	Thistle Dump Cem, Longueval.
33220	Bracken. D.	Liverpool.	Pte.	18th	D.O.W	24 10 16	Porte-de-Paris Cem, Cambrai
29248	Williams. J.S.	Liverpool.	Pte.	17th	D.O.W	25 10 16	Heilly Station Cem.Mericourt
38230	McEvoy. B.	Dundalk, Ireland.	Pte.	20th	D.O.W	26 10 16	Heilly Station Cem.Mericourt
53064	Graham. C.J.	Leicester.	Pte.	20th	D.O.W	27 10 16	Etaples Mil Cem
33019	Hotson. A.M.	Liverpool.	Cpl.	18th	D.O.W	29 10 16	Etaples Mil Cem
22457	Lines. J.C. M.M.	Liverpool.	L/Cpl.	20th	D.O.W	29 10 16	Etaples Mil Cem
16539	Lipson. S.W.	Liverpool.	Pte.	18th	K.I.A	01 11 16	De Cusine Ravine B.C.Basseux
31166	Evans. E.H.	Seacombe, Cheshire.	Pte.	18th	D.O.W	02 11 16	Etaples Mil Cem
52069	Gwinnell. J.	Bootle, Liverpool.	Pte.	19th	Died	05 11 16	Etaples Mil Cem
34370	Bebbington. F.	Nantwich, Cheshire.	Pte.	17th	K.I.A	06 11 16	Bienvillers Mil Cem.
17836	Heyes. R.C.	Anfield, Liverpool.	Pte.	19th	K.I.A	08 11 16	Pommier CC
52027	Smith. W.J.	Liverpool.	L/Cpl.	20th	K.I.A	09 11 16	Berles Position Cem.
15034	Slater. W.	Birkenhead, Cheshire.	Pte.	17th	K.I.A	12 11 16	Bienvillers Mil Cem.
51667	Schola. J.	Liverpool.	Pte.	17th	D.O.W	13 11 16	Warlincourt Halte Brit Cem.
52805	Kirkham. A.	Preston. Lancs.	Pte.	17th	D.O.W	15 11 16	Warlincourt Halte Brit Cem.
52229	Jackson. E.	Northwich, Cheshire.	Pte.	19th	D.O.W	16 11 16	Humbercamps CC Ext
52177	Meakin. F.	Manchester.	L/Cpl.	19th	K.I.A	16 11 16	Bienvillers Mil Cem.
53141	Rogers. E.A.	Great Yarmouth, Norfolk.	L/Cpl.	20th	K.I.A	20 11 16	Berles Position Cem.
21618	Riley. C.A.	Liverpool.	Pte.	19th	K.I.A	21 11 16	Warlincourt Halte Brit Cem.
34257	Lake. G.	Bootle, Liverpool.	Pte.	20th	D.O.W	22 11 16	Warlincourt Halte Brit Cem.
21736	Allday. J	Liverpool.	Pte.	19th	D.O.W	24 11 16	Warlincourt Halte Brit Cem.
52252	Hickson. F.O.	Manchester.	Pte.	19th	K.I.A	24 11 16	Berles Position Cem.
51681	Jeffries. A. E.	New Brighton, Cheshire.	Pte.	19th	D.O.W	24 11 16	Warlincourt Halte Brit Cem.
21798	Wainwright. H.	Liverpool.	Pte.	19th	K.I.A	24 11 16	Berles Position Cem.
13200	Lewis. P.	Glasgow.	Pte.	18th	K.I.A	26 11 16	De Cusine Ravine B.C.Basseux
49244	Phillipson. W.S.	Lancaster, Lancs.	Pte.	18th	Died	26 11 16	De Cusine Ravine B.C.Basseux
24526	Turton. J.E.	Liverpool.	Pte.	18th	Died	26 11 16	De Cusine Ravine B.C.Basseux
51893	Arnold. J.	Burscough, Lancs.	Pte.	17th	K.I.A	27 11 16	Berles-au-Bois C.Y Ext.
49090	Anderson. A.	Chadderton, Lancs.	Pte.	20th	D.O.W	30 11 16	Warlincourt Halte Brit Cem.
26178	Benion. S.	Warrington, Cheshire.	Pte.	17th	K.I.A	30 11 16	Thiepval Memorial.
21809	Ingram. E.R.	Egremont, Liverpool.	Pte.	17th	Died	04 12 16	Kirkdale Cem, Liverpool.
22315	Bolton. J.	Liverpool.	Pte.	20th	D.O.W	07 12 16	Everton Cem, Liverpool
21637	Turner. R.	Seaforth, Liverpool.	Pte.	19th	D.O.W	07 12 16	West Derby Cem, Liverpool.
51586	Bond. T.	Kirkdale, Liverpool.	L/Cpl.	17th	Died	09 12 16	Warlincourt Halte Brit Cem.
35374	Curtis. F.J.	Bootle. Liverpool.	Pte.	17th	K.I.A	10 12 16	Berles-au-Bois C.Y Ext.
37379	Cropper. J.	Hesketh Bank, Lancs.	Pte.	20th	K.I.A	13 12 16	Arras Memorial.
49059	Entwistle. T.	Liverpool.	Pte.	20th	K.I.A	13 12 16	Arras Memorial.
10645	Goldman. H.	London.	Pte.	20th	K.I.A	13 12 16	Arras Memorial.
42326	Harrison. W.	Liverpool.	Pte.	17th	Died	13 12 16	Mont Huon Cem, Le Treport.
49076	Hayes. S.	Liverpool.	Pte.	20th	K.I.A	13 12 16	Berles Position Cem.
53075	Houlson. T.	Loughborough. Leics.	Pte.	20th	K.I.A	13 12 16	Berles Position Cem.
23994	Wear. W.J.	Liverpool.	Pte.	20th	K.I.A	13 12 16	Berles Position Cem.
42776	Slater. G.	Liverpool.	Pte.	17th	D.O.W	19 12 16	Etaples Mil Cem
49022	Barlow. G.	Salford, Manchester.	Pte.	17th	K.I.A	21 12 16	Arras Memorial.
48998	Batty. J.	Blackburn, Lancs.	Pte.	17th	K.I.A	21 12 16	Arras Memorial.
52167	Almond. A.	Altrincham, Ches.	L/Cpl.	19th	D.O.W	22 12 16	Etaples Mil Cem
26199	Collins. W.J.	Liverpool.	Pte.	20th	D.O.W	24 12 16	Warlincourt Halte Brit Cem.
49115	Pawson. E.	Birkenhead, Cheshire.	Pte.	17th	D.O.W	26 12 16	Etaples Mil Cem
31674	Chisnell. J.	Liverpool.	Pte.	20th	Died	28 12 16	Berles-au-Bois C.Y Ext.
23025	Hancock. H.I.	Liverpool.	Pte.	20th	K.I.A	29 12 16	Berles Position Cem.
22296	Wilson. E.	Liverpool.	Pte.	20th	D.O.W	29 12 16	Warlincourt Halte Brit Cem.
23849	Turton. G.A.	Widnes, Lancs.	Pte.	20th	Died.	30 12 16	Widnes Cem, Lancs.

57867	Scott. J.A.	New Hartley, Northumberland	Pte.	18th	K.I.A	02 01 17	De Cusine Ravine B.C.Basseux
23864	Williams. T.C.	Liverpool.	Sgt.	20th	D.O.W	02 01 17	Chipilly CC.
27378	Jones. A.H.	Liverpool.	Pte.	19th	K.I.A	03 01 17	Berles-au-Bois C.Y Ext.
57443	McRitchie.J.	Glasgow.	Pte.	20th	D.O.W	05 01 17	Warlincourt Halte Brit Cem.
24616	Reeves. G.A.	Liverpool.	Pte.	18th	Died	05 01 17	West Derby Cem, Liverpool.
41813	Minton. A.	Liverpool.	Pte.	19th	D.O.W	07 01 17	West Derby Cem, Liverpool.
42405	Lloyd. A.	Liverpool.	Pte.	17th	Died	18 01 17	Doullens CC Ext 1
15752	Olley. A.C.	Liverpool.	Pte.	17th	D.O.W	25 01 17	Etaples Mil Cem,
49007	Murray. H.C.	Birmingham.	Pte.	17th	K.I.A	28 01 17	Cherisy Rd East Cem, Heninel.
22102	Hancock. E.	Little Neston, Cheshire.	Pte.	20th	Died.	29 01 17	St.Marie Cem, Le Havre.
49205	Losh. J.	Great Harwood, Lancs.	Pte.	20th	Died.	03 02 17	Mont Huon Cem, Le Treport.
46761	Power. J.	Liverpool.	Pte.	17th	Died.	04 02 17	St.Pol CC Ext.
29513	Nutbrown. A.E.	Liverpool.	Pte.	17th	K.I.A	17 02 17	Arras Memorial.
48088	Kirby. T.	Liverpool.	Pte.	19th	K.I.A	18 02 17	Agny New Military Cem.
49590	Berry. J.	Manchester.	Pte.	20th	K.I.A	19 02 17	Arras Memorial.
52462	Ainsworth. R.	Bury, Lancs.	Pte.	18th	Died	21 02 17	Warlincourt Halte Brit Cem.
	Bolton. C.R.	Birkenhead, Cheshire.	Capt.	19th	K.I.A	22 02 17	Agny New Military Cem.
17208	Proctor. H.	Liverpool.	Pte.	18th	Died	22 02 17	Anfield Cem, Liverpool.
15568	Smith. G.	Liverpool.	Pte.	17th	K.I.A	22 02 17	Agny New Military Cem.
57707	Flinders. A.	Dunkirk, Notts.	Pte.	18th	D.O.W	27 02 17	Avesnes Comte C.C.Ext.
19270	Moores. J.	Manchester.	L/Cpl.	19th	Died	01 03 17	Warlincourt Halte Brit Cem.
26043	Murray. J.	Liverpool.	Pte.	18th	Died	02 03 17	Agny New Military Cem.
57907	Brooks. D.H.	Harlesden, London.	Pte.	17th	D.O.W	05 03 17	Avesnes Comte C.C.Ext.
15409	Grey-Smith. W. (Smith)	Aughton, Lancs.	L/Cpl.	17th	K.I.A	06 03 17	Agny New Military Cem.
58698	Hardwick. J.	Walsall, Staffs.	Pte.	20th	K.I.A	06 03 17	Agny New Military Cem.
52811	Rigby. E.S.	Southport, Lancs.	Pte.	17th	K.I.A	06 03 17	Agny New Military Cem.
22798	Parr. F.G.	Liverpool.	Pte.	20th	Died.	07 03 17	Warlincourt Halte Brit Cem.
57569	Rodger. R.N.	Dundee.	Pte.	18th	D.O.W	08 03 17	St.Hilaire CC, Frevant
51578	Ramson. F.H.	Liverpool.	Pte.	17th	Died.	09 03 17	Vis-en-Artois Brit Cem. Haucourt
32750	Wall. LJ.	Ellesmere Port, Cheshire.	Pte.	18th	K.I.A	11 03 17	Agny New Military Cem.
66708	Jackson. A.	Bollington, Cheshire.	Pte.	17th	Died	14 03 17	Bollington. Ches.(St.J the B)
29287	Makinson. H.J.	Liverpool.	A/L/Cpl	20th	K.I.A	14 03 17	Agny New Military Cem.
36759	Webster. J.W.	Litherland. Liverpool.	Pte.	20th	K.I.A	14 03 17	Agny New Military Cem.
16566	Eardley. J.	Liverpool.	Cpl.	18th	K.I.A	15 03 17	Agny New Military Cem.
69634	Richardson. W.	Warrington, Cheshire.	Pte.	17th	Died.	16 03 17	St Sever Cem Ext, Rouen.
22906	Adams. R.H.	Toxteth, Liverpool.	L/Cpl.	20th	K.I.A	18 03 17	Agny New Military Cem.
22059	Downey. F.C.	Liverpool.	Pte.	20th	K.I.A	18 03 17	Agny New Military Cem.
48290	Durr. J.	Liverpool.	Pte.	20th	D.O.W	18 03 17	Avesnes Comte C.C.Ext.
51727	Sanders. A.E.H.	Wallasey, Cheshire.	Pte.	20th	K.I.A	19 03 17	London Cem, Neuville-Vitasse
41810	Mahoney. J.M.	Liverpool.	Pte.	20th	D.O.W	20 03 17	Avesnes Comte C.C.Ext.
58676	Rimmer.J.		Pte.	20th	K.I.A	20 03 17	Arras Memorial.
	Green. R.E.		2nd Lieut.	20th	K.I.A	21 03 17	Arras Memorial.
22470	McArdle. R.	Liverpool.	Cpl.	20th	K.I.A	21 03 17	London Cem. Neuville-Vitasse
24471	McEvoy. E.	Liverpool.	L/Cpl.	20th	K.I.A	21 03 17	London Cem. Neuville-Vitasse
17427	Mollison. C.S. M.M.	Goole, Yorks.	L/Cpl.	19th	K.I.A	21 03 17	Bucquoy Rd Cem, Ficheux.
39279	Parkes. D.	Liverpool.	Pte.	20th	D.O.W	21 03 17	Avesnes Comte C.C.Ext.
24244	Bobbins. W.J.K.	Bristol, Glos.	Cpl.	19th	K.I.A	21 03 17	Bucquoy Rd Cem, Ficheux.
49551	Wilkinson. J.J.	Liverpool.	Pte.	20th	K.I.A	21 03 17	Arras Memorial.
17855	Ireland. J.M.	Seaforth, Lancs.	Pte.	19th	D.O.W	22 03 17	Warlincourt Halte Brit Cem.
21683	Thomas. C.R.	Bootle. Liverpool.	L/Cpl.	19th	D.O.W	26 03 17	Perth Cem, Zillebeke.
51906	Huyton. J.R.	Scarisbrick, Lancs.	Pte.	17th	D.O.W	27 03 17	Mont Huon Cem, Le Treport.
57600	Gallagher. C.	Dewsbury, Yorks.	Pte.	18th	D.O.W	31 03 17	Etaples Mil Cem
52470	Lawless. T.	Heywood, Lancs.	Pte.	18th	K.I.A	01 04 17	London Cem, Neuville-Vitasse
16345	Woollam. J.A.	Chester.	L/Sgt.	18th	K.I.A	01 04 17	London Cem, Neuville-Vitasse
13597	Brown. J.	Everton, Liverpool.	Pte.	18th	K.I.A	02 04 17	Arras Memorial.
23664	Hemming. W.	Aughton, Lancs.	L/Cpl.	18th	K.I.A	02 04 17	Arras Memorial.
26012	Whittaker.'B.	Walton, Liverpool.	Pte.	18th	D.O.W	02 04 17	Henin CC Ext.
73794	Morgan. T.	Builth Wells, N.Wales	Pte.	18th	Died	04 04 17	Varennes Mil Cem.

51722	Barrington. G.	Liverpool.	Pte.	20th	Died	05 04 17	Kirkdale Cem, Liverpool.
52104	Hanks. F.	Chorlton-o-M, Manchester.	Pte.	19th	D.O.W	05 04 17	Etaples Mil Cem.
49044	Corcoran. W.	Bury, Lancs.	Pte.	17th	K.I.A	08 04 17	Menin Gate Memorial.
57384	Stewart. M.	Glasgow.	Pte.	18th	D.O.W	08 04 17	Warlincourt Halte Brit Cem.
57389	Adams. J.	Greenock, Renfrew.	Pte.	20th	K.I.A	09 04 17	Henin Crucifix Cem.
29753	Adams. W	Liverpool.	L/Cpl.	19th	K.I.A	09 04 17	Arras Memorial.
57515	Allan. A.C.	Dundee.	Pte.	18th	K.I.A	09 04 17	Bucquoy Rd Cem, Ficheux.
13380	Allman. S.L	Liverpool.	Pte.	19th	K.I.A	09 04 17	St. Martin Calvaire Brit Cem.
15213	Alsop. L	Liverpool.	Pte.	17th	K.I.A	09 04 17	St. Martin Calvaire Brit Cem.
57691	Aram. J.	Long Eaton, Derbyshire	Pte.	18th	K.I.A	09 04 17	Arras Memorial.
	Ashcroft. F.	Birkenhead, Cheshire.	2nd Lieut.	18th	K.I.A	09 04 17	Neuville-Vitasse Rd Cem.
52036	Atherton. J.	Liverpool.	Pte.	19th	K.I.A	09 04 17	St. Martin Calvaire Brit Cem.
51576	Baggett. C.T.	Egremont, Cheshire.	L/Cpl.	17th	K.I.A	09 04 17	St. Martin Calvaire Brit Cem.
57974	Barlow. A.H.	Fulham, London	Pte.	18th	D.O.W	09 04 17	Bucquoy Rd Cem, Ficheux.
57625	Bell. J.A.C.	Dewsbury, Yorks	Pte.	18th	K.I.A	09 04 17	Wancourt Brit Cem.
57912	Bennett. W.H.	Holloway, London	Pte.	18th	K.I.A	09 04 17	Arras Memorial.
21865	Bennett. W.J. M.M.	Liverpool.	Sgt.	18th	K.I.A	09 04 17	London Cem, Neuville-Vitasse
14932	Sevan. G.	Liverpool.	Sgt.	18th	K.I.A	09 04 17	Neuville-Vitasse Rd Cem.
52040	Bly. G.H.	Liverpool.	Pte.	19th	K.I.A	09 04 17	St.Martin Calvaire Brit Cem.
57380	Boyd. J.	Glasgow.	L/Cpl.	20th	K.I.A	09 04 17	Henin Crucifix Cem.
58001	Brede. F.H.	Ponders End, London.	Pte.	17th	K.I.A	09 04 17	St.Martin Calvaire Brit Cem.
58772	Briggs. E.	Leeds.	Pte.	18th	K.I.A	09 04 17	Arras Memorial.
16354	Briggs. G.	Waterloo, Liverpool.	Cpl.	18th	K.I.A	09 04 17	Wancourt Brit Cem.
17172	Brown. G.	Everton, Liverpool.	Pte.	18th	K.I.A	09 04 17	London Cem, Neuville-Vitasse
57696	Buckingham. H.	Beeston, Notts	Pte.	18th	K.I.A	09 04 17	Neuville-Vitasse Rd Cem.
42710	Cadwell. J.C.	Southport, Lancs.	Pte.	17th	K.I.A	09 04 17	Henin Crucifix Cem.
57446	Callan. F.J.	Druncondra, Dublin.	Pte.	20th	K.I.A	09 04 17	Henin Crucifix Cem.
	Carr. A.R.	Sale, Cheshire.	2nd Lieut.	17th	K.I.A	09 04 17	Henin Crucifix Cem.
49499	Chamock. R.	Waterloo, Liverpool.	Pte.	20th	K.I.A	09 04 17	Henin Crucifix Cem.
57399	Chattell.A.	Glasgow	Pte.	18th	K.I.A	09 04 17	Wancourt Brit Cem.
53091	Clapinson. F.	London	Pte.	18th	K.I.A	09 04 17	Neuville-Vitasse Rd Cem.
48032	Clarke. J.	Liverpool.	Pte.	19th	K.I.A	09 04 17	St-Martin Calvaire Brit Cem.
31299	Cochrane. W.	Manchester.	Pte.	19th	K.I.A	09 04 17	St.Martin Calvaire Brit Cem.
35761	Collins. R.	Liverpool.	Pte.	18th	K.I.A	09 04 17	Arras Memorial.
57504	Cooper. H.Mc.	Dundee	Pte.	18th	K.I.A	09 04 17	Neuville-Vitasse Rd Cem.
58713	Crowther. J.	Rochdale, Lancs.	Pte.	20th	K.I.A	09 04 17	Henin Crucifix Cem.
51663	Cuthbert. R.N.	Seaforth, Liverpool.	Pte.	17th	K.I.A	09 04 17	St. Martin Calvaire Brit Cem.
21889	Davies. AT.	Liverpool.	Pte.	18th	K.I.A	09 04 17	Arras Memorial.
52239	Davies. R.V.	Manchester.	Pte.	19th	K.I.A	09 04 17	St. Martin Calvaire Brit Cem.
	Davis. B.S.	Duffield, Yorks.	2nd Lieut.	17th	K.I.A	09 04 17	Bucquoy Rd Cem, Ficheux.
57528	Dignam. J.V.	Dundee	Pte.	18th	K.I.A	09 04 17	Neuville-Vitasse Rd Cem.
47129	Downs. C.	Leigh, Lancs.	Pte.	19th	K.I.A	09 04 17	Arras Memorial.
57740	Duff. D.	Whitley Bay.	Pte.	19th	D.O.W	09 04 17	Bucquoy Rd Cem, Ficheux.
32263	Dufresne. V.	Ottawa, Canada.	Pte.	19th	K.I.A	09 04 17	Arras Memorial.
48288	Dunbar. P.J.	Liverpool.	Pte.	19th	K.I.A	09 04 17	St.Martin Calvaire Brit Cem.
52449	Edwins. E.	Ammanford, Camarthen	Pte.	18th	K.I.A	09 04 17	Wancourt Brit Cem.
58659	Elliott. J.	Bolton, Lancs.	Pte.	20th	K.I.A	09 04 17	Henin Crucifix Cem.
31086	Ellis. J.A.	Liverpool.	Pte.	17th	K.I.A	09 04 17	Arras Memorial.
	Ewing. H.G.	Chester.	2nd Lieut.	18th	K.I.A	09 04 17	Arras Memorial.
17052	Fazackerley. A.E. (MM)	Birkenhead, Cheshire.	Pte.	18th	K.I.A	09 04 17	Wancourt Brit Cem.
57647	Fieldhouse. J.A.	Otley, Yorks	Pte.	18th	K.I.A	09 04 17	Wancourt Brit Cem.
57407	Finlay. A.	Glasgow.	Pte.	19th	K.I.A	09 04 17	Arras Memorial.
29690	Floyd. G.J.	Orrell, Wigan, Lancs.	Pte.	18th	K.I.A	09 04 17	Arras Memorial.
36933	Forster. W.	Liverpool.	Pte.	20th	K.I.A	09 04 17	Arras Memorial.
57756	Foster. J.G.	Sunderland	Pte.	18th	K.I.A	09 04 17	Arras Memorial.
57927	Francis. A.	Southsea, Hants	Pte.	18th	K.I.A	09 04 17	Arras Memorial.
57925	Freeman. W.	Cambridge.	Pte.	18th	D.O.W	09 04 17	Bucquoy Rd Cem, Ficheux.
21771	Fullerton. T.D.	Liverpool.	Sgt.	19th	K.I.A	09 04 17	Henin Crucifix Cem.

57922	Fullilove. T.E.	Walworth, London	Pte.	18th	K.I.A	09 04 17	London Cem, Neuville-Vitasse
32319	Grice. H.	Crewe, Cheshire.	Pte.	20th	K.I.A	09 04 17	Henin Crucifix Cem.
36686	Gwinnell. C.	Bootle, Liverpool.	Pte.	20th	K.I.A	09 04 17	Henin Crucifix Cem.
38815	Hampton. H.	Liverpool.	Pte.	20th	K.I.A	09 04 17	Henin Crucifix Cem.
21828	Hankin, J.	Liverpool.	Pte.	18th	K.I.A	09 04 17	Neuville-Vitasse Rd Cem.
30252	Hanson. A.	Liverpool.	L/Cpl.	18th	K.I.A	09 04 17	Arras Memorial.
58061	Harrison. G.	Hyde, Cheshire.	Pte.	18th	K.I.A	09 04 17	Arras Memorial.
49077	Hayden. C.	Liverpool.	Pte.	20th	K.I.A	09 04 17	Henin Crucifix Cem.
52241	Hinchcliffe. N.	Lintwistle, Derbys.	Pte.	19th	K.I.A	09 04 17	St.Martin Calvaire Brit Cem.
38051	Hobson. L.	Liverpool.	Pte.	20th	K.I.A	09 04 17	Henin Crucifix Cem.
52168	Horton. F.	Bramhall, Ches.	Pte.	19th	K.I.A	09 04 17	Arras Memorial.
49243	Hurst. J.	Liverpool.	Pte.	18th	K.I.A	09 04 17	Wancourt Brit Cem.
57793	Irving. R.	Seaham Harbour, Durham.	Pte.	19th	K.I.A	09 04 17	Henin Crucifix Cem.
57935	Johnson. W.J.	London.	Pte.	18th	K.I.A	09 04 17	Arras Memorial.
51746	Johnston. W.D.	London.	Pte.	20th	K.I.A	09 04 17	Henin Crucifix Cem.
47456	Kearns. J.	Liverpool.	Pte.	19th	K.I.A	09 04 17	St.Martin Calvaire Brit Cem.
31130	Keary. M.	Liverpool.	Pte.	17th	K.I.A	09 04 17	Arras Memorial.
57723	Lane. J.W.	Nottingham	Pte.	18th	K.I.A	09 04 17	Wancourt Brit Cem.
16973	Laycock. E.M	Liverpool.	Pte.	18th	K.I.A	09 04 17	Bucquoy Rd Cem, Ficheux.
15525	Lessells. R.V.	Egremont, Cheshire.	Pte.	17th	K.I.A	09 04 17	Arras Memorial.
16251	Lewis. H.J.	Port Sunlight, Cheshire	L/Cpl.	18th	K.I.A	09 04 17	Wancourt Brit Cem.
30956	Lloyd. J.	Altcar, Lancs.	Pte.	17th	K.I.A	09 04 17	Henin Crucifix Cem.
57431	Lunn. W.G.	London	Cpl.	18th	K.I.A	09 04 17	Neuville-Vitasse Rd Cem.
57654	Lupton. E.H.	Otley, Yorks	Pte.	18th	K.I.A	09 04 17	Neuville-Vitasse Rd Cem.
57551	MacDonald. R.F.	Dundee	Pte.	18th	K.I.A	09 04 17	Wancourt Brit Cem.
57552	Mackay. A.W.	Dundee	Pte.	18th	K.I.A	09 04 17	Neuville-Vitasse Rd Cem.
49565	Mackie. D.	Liverpool.	Pte.	20th	K.I.A	09 04 17	Henin Crucifix Cem.
49252	Martin. J.A.	Padiham, Lancs.	Pte.	18th	K.I.A	09 04 17	Neuville-Vitasse Rd Cem.
	Mason. G.W.		Lieut.	19th	K.I.A	09 04 17	St. Martin Calvaire Brit Cem.
16786	Mason. R.	Liverpool.	Pte.	18th	K.I.A	09 04 17	Neuville-Vitasse Rd Cem.
48405	McAdam. T.	Liverpool.	Pte.	19th	K.I.A	09 04 17	St.Martin Calvaire Brit Cem.
57513	McIntosh. W.	Arbroath	Pte.	18th	K.I.A	09 04 17	Bucquoy Rd Cem, Ficheux.
52427	Mercer. A.	Church, Lancs.	Pte.	18th	K.I.A	09 04 17	Wancourt Brit Cem.
57450	Mitchell. R.	Gavelock, Dumbarton.	Pte.	20th	K.I.A	09 04 17	Henin Crucifix Cem.
57559	Mitchell. W.M.	Dundee	Pte.	18th	K.I.A	09 04 17	Wancourt Brit Cem.
31716	Moore. J.E.	Douglas, I.O.M.	Pte.	17th	K.I.A	09 04 17	Henin Crucifix Cem.
52054	Morson. S.J.	Liverpool.	Pte.	17th	D.O.W	09 04 17	St.Martin Calvaire Brit Cem.
25745	Mortimer. A.	Boston. Mass. USA	Pte.	18th	K.I.A	09 04 17	Wancourt Brit Cem.
34583	Munsey. J.	Liverpool.	Pte.	18th	K.I.A	09 04 17	Arras Memorial.
49087	Murphy. G.	Liverpool.	Pte.	20th	K.I.A	09 04 17	Henin Crucifix Cem.
42878	Olver. F.	Liverpool.	Pte.	19th	K.I.A	09 04 17	St.Martin Calvaire Brit Cem.
31695	Parker. J.	Tuebrook, Liverpool.	Pte.	20th	K.I.A	09 04 17	Henin Crucifix Cem.
57728	Parkinson. H.	Seedley, Manchester	Pte.	18th	K.I.A	09 04 28	Bucquoy Rd Cem, Ficheux.
29265	Parry. A.W.	Liverpool.	L/Cpl.	17th	K.I.A	09 04 17	Arras Memorial.
53122	Patten. P.	Southend.	Pte.	18th	K.I.A	09 04 17	Neuville-Vitasse Rd Cem.
57565	Pattie. C.	Dundee.	Pte.	18th	K.I.A	09 04 17	Neuville-Vitasse Rd Cem.
51584	Peers. W.	Bromborough, Cheshire.	Pte.	17th	K.I.A	09 04 17	St.Martin Calvaire Brit Cem.
57069	Petthrick. C.	Sheffield., Yorks.	A/Cpl.	18th	K.I.A	09 04 17	Wancourt Brit Cem.
78295	Pritchard. T.	West Kirby, Cheshire.	Pte.	20th	Died.	09 04 17	Rochin CC
51771	Proctor. J.T.	Blackpool, Lancs.	Pte.	19th	K.I.A	09 04 17	Bucquoy Rd Cem, Ficheux.
57658	Prust. H.	Rothwell Haigh, Leeds.	Pte.	18th	K.I.A	09 04 17	Wancourt Brit Cem.
58043	Quibell. H.T.	Barnsbury, London.	Pte.	18th	K.I.A	09 04 17	Wancourt Brit Cem.
57825	Redshaw. W.P.	Seaham Harbour, Durham.	Pte.	19th	K.I.A	09 04 17	St. Martin Calvaire Brit Cem.
47477	Reed. P.	Liverpool.	Pte.	18th	K.I.A	09 04 17	Arras Memorial.
57828	Richardson. W.S.	Hetton-le-Hole, Durham	Pte.	18th	K.I.A	09 04 17	Neuville-Vitasse Rd Cem.
32632	Roberts. D.	Liverpool.	Pte.	17th	D.O.W	09 04 17	Bucquoy Rd Cem, Ficheux.
17142	Roberts. E.J.	Liverpool.	Pte.	18th	K.I.A	09 04 17	Wancourt Brit Cem.
23814	Roberts. G.	Llangollen, N.Wales.	Pte.	20th	K.I.A	09 04 17	Henin Crucifix Cem.

48457	Roberts. J.	Birkdale, Southport, Lancs.	Pte.	19th	K.I.A	09 04 17	Arras Memorial.
57500	Ross.. W.Mc.	Garmouth, Morayshire.	Cpl.	19th	K.I.A	09 04 17	St. Martin Calvaire Brit Cem.
34301	Rowlands. D.C.	Colwyn Bay, N.Wales.	Pte.	20th	K.I.A	09 04 17	Henin Crucifix Cem.
32963	Sefton. J.	Liverpool.	Pte.	18th	K.I.A	09 04 17	Neuville-Vitasse Rd Cem.
57834	Sharman. H.		Pte.	19th	K.I.A	09 04 17	Arras Memorial.
52055	Shaw. W.J.	Liverpool.	Sgt.	19th	K.I.A	09 04 17	Arras Memorial.
52086	Shipton. F.W.	Bootle, Liverpool.	Pte.	19th	K.I.A	09 04 17	Arras Memorial.
36949	Sloan. D.	Liverpool.	Pte.	20th	K.I.A	09 04 17	Arras Memorial.
57836	Smart. J.		Pte.	19th	K.I.A	09 04 17	St. Martin Calvaire Brit Cem.
57572	Smith. J.	Dundee.	Pte.	18th	K.I.A	09 04 17	Neuville-Vitasse Rd Cem.
58046	Smith. J.G.	London.	Pte.	17th	K.I.A	09 04 17	Arras Memorial.
27913	Spencer. W.	Liverpool.	L/Cpl.	20th	K.I.A	09 04 17	Henin Crucifix Cem.
57893	Stanley. A.	Islington, London.	Pte.	19th	K.I.A	09 04 17	St. Martin Calvaire Brit Cem.
57843	Suggett. W.		Pte.	19th	K.I.A	09 04 17	Arras Memorial.
57879	Sutherland. J.	Denistoun, Glasgow.	A/Cpl.	18th	K.I.A	09 04 17	Arras Memorial.
31847	Taylor. K.	Darwen, Lancs.	Pte.	19th	K.I.A	09 04 17	St.Martin Calvaire Brit Cem.
21670	Telford. T.W.	Penrith, Cumb.	Pte.	19th	K.I.A	09 04 17	Henin Crucifix Cem.
15403	Thomlinson. W.I.	Liverpool.	Pte.	17th	K.I.A	09 04 17	Henin Crucifix Cem.
31188	Tillinghurst. G.E.	Liverpool.	Pte.	17th	K.I.A	09 04 17	Henin Crucifix Cem.
36213	Travis. J.	Nelson, Colne.	Pte.	19th	K.I.A	09 04 17	Arras Memorial.
49557	Traynor. J.	Liverpool.	Pte.	20th	K.I.A	09 04 17	Henin Crucifix Cem.
31620	Vaughan. F.E.	Birkenhead, Cheshire.	Pte.	18th	K.I.A	09 04 17	Arras Memorial.
52481	Walker. A.	Peterborough, Northants.	Pte.	18th	K.I.A	09 04 17	Arras Memorial. *
35337	Walker. G.	Bolton, Lancs.	Pte.	20th	K.I.A	09 04 17	Henin Crucifix Cem.
17221	Wells. O.W.	Ruthin, N.Wales.	Pte.	18th	K.I.A	09 04 17	Arras Memorial.
57851	Wigham. R.	Sunderland.	Pte.	19th	K.I.A	09 04 17	St. Martin Calvaire Brit Cem.
42772	Wolfe. H.H.	Liverpool.	Pte.	17th	K.I.A	09 04 17	St. Martin Calvaire Brit Cem.
26659	Wright. H.	St. Helens, Lancs.	Pte.	17th	K.I.A	09 04 17	Arras Memorial.
49488	Anderson. W.	Bootle, Liverpool.	Pte.	20th	K.I.A	10 04 17	Henin Crucifix Cem.
19597	Buss. W.	Manchester	Pte.	18th	D.O.W	10 04 17	Gouy-en-Artois CC Ext.
57524	Clow. D.	Dundee	Pte.	18th	D.O.W	10 04 17	Gouy-en-Artois CC Ext.
49570	Lockerbye. G.	Castle Douglas, Kirkcudbright	Pte.	20th	K.I.A	10 04 17	Henin Crucifix Cem.
49086	McCool. J.J.	Liverpool.	Pte.	20th	K.I.A	10 04 17	Arras Memorial.
46144	Riozzie. A.	Liverpool.	Pte.	19th	D.O.W	10 04 17	Warlincourt Halte Brit Cem.
57677	Willars. A.E.	Annesley, Notts.	A/Cpl.	18th	D.O.W	10 04 17	Warlincourt Halte Brit Cem.
51601	Booth. G.O.	Liverpool.	Pte.	17th	D.O.W	11 04 17	Warlincourt Halte Brit Cem.
43486	Broomnead. R.	Southport, Lancs.	Pte.	17th	D.O.W	11 04 17	St Sever Cem, Rouen.
48343	Buck. J.	Southport, Lancs.	Pte.	19th	D.O.W	11 04 17	Warlincourt Halte Brit Cem.
58003	Cameron. F.G.	Plumstead, London.	Pte.	17th	D.O.W	11 04 17	Warlincourt Halte Brit Cem.
58648	Clarkson. A.	Liverpool.	Pte.	20th	D.O.W	11 04 17	Warlincourt Hate Brit Cem.
57599	Cockram. F.	Normanton, Yorks.	Pte.	18th	D.O.W	11 04 17	Warlincourt Halte Brit Cem.
17123	Gunson.W.D.	Llangollen. N.Wales	Sgt.	18th	D.O.W	11 04 17	Warlincourt Hatte Brit Cem.
21343	Jackson. E.	Manchester.	Pte.	20th	K.I.A	11 04 17	Henin Crucifix Cem.
34502	Jenkins. A.	Burnley, Lancs.	Pte.	19th	D.O.W	11 04 17	Warlincourt Halte Brit Cem.
51728	Merrills. A.	Liverpool.	Pte.	20th	D.O.W	11 04 17	Gouy-en-Artois CC Ext.
52053	Proctor. T.A.	Dingle, Liverpool.	L/Sgt.	19th	D.O.W	11 04 17	Abbeville C.C. Ext.
24648	Roberts. J.	Ellesmere Port, Cheshire.	Pte.	18th	D.O.W	11 04 17	Warlincourt Halte Brit Cem.
22310	Armstrong. H.	Crewe, Cheshire.	Pte.	20th	D.O.W	12 04 17	Warlincourt Halte Brit Cem.
57700	Colley. L.V.	Bilston, Staffs.	Pte.	19th	D.O.W	12 04 17	Warlincourt Halte Brit Cem.
52467	Gabb. L.	Birmingham	Pte.	18th	D.O.W	12 04 17	Warlincourt Halte Brit Cem.
52099	Neal. D.	Manchester.	Pte.	19th	D.O.W	12 04 17	Warlincourt Halte Brit Cem.
37410	Sucksmith. A.	Shaw, Lancs.	Pte.	20th	K.I.A	12 04 17	Arras Memorial.
57693	Brabben. M.	Wisbech, Cambs.	Pte.	18th	D.O.W	13 04 17	Warlincourt Halte Brit Cem.
22715	Heaney. J.	Seaforth, Liverpool.	Pte.	20th	D.O.W	13 04 17	St Sever Cem Ext, Rouen.
27749	Miller. H.	Chorley, Lancs.	Pte.	19th	D.O.W	13 04 17	Warlincourt Halte Brit Cem.
17569	Davies. J.	Liverpool.	L/Cpl.	19th	D.O.W	14 04 17	St Sever Cem Ext, Rouen.
48367	Gallagher. F.	Liverpool.	Pte.	19th	K.I.A	14 04 17	Warlincourt Halte Brit Cem.
16250	Lally. W.R.	Manchester	L/Sgt.	18th	D.O.W	14 04 17	Warlincourt Halte Brit Cem.

258

	Mahon. O.S.W.	Wallasey, Cheshire.	2nd Lieut.	17th	D.O.W 14 04 17	Mont Huon Cem, Le Treport.
18210	Welbourne. S.	Garth, Nr. Meath.S.Wales.	Pte.	19th	D.O.W 14 04 17	Etaples Mil Cem
47317	Lumb. C.F.	Birkdale, Southport, Lancs.	Pte.	19th	D.O.W 15 04 17	Etaples Mil Cem
57597	Picken.J.	Glasgow.	A/Cpl.	20th	D.O.W 15 04 17	Etaples Mil Cem
16994	Roberts. E.	Liverpool.	L/Cpl.	18th	D.O.W 15 04 17	Warlincourt Halte Brit Cem.
22358	Davies. G.H.	Birkenhead, Cheshire.	L/Cpl.	20th	D.O.W 17 04 17	St Sever Cem, Rouen.
52026	Peterson. P.	Liverpool.	Pte.	20th	D.O.W 18 04 17	Abbeville C.C. Ext.
	Statton. P.G.	Devonport, Devon.	2nd Lieut.	18th	D.O.W 18 04 17	Abbeville C.C. Ext.
57857	Carter. J.	South Hylton, Durham	Pte.	18th	D.O.W 19 04 17	Etaples Mil Cem
57792	Howarth. A.	Sunderland	Pte.	18th	D.O.W 19 04 17	Abbeville C.C. Ext.
48091	King.W.	Liverpool.	Pte.	20th	D.O.W 19 04 17	Abbeville C.C. Ext.
57400	Chisholm. R.	Glasgow.	Pte.	20th	D.O.W 20 04 17	St Sever Cem Ext, Rouen.
73671	Marsden. E.	Rusholme, Manchester	Pte.	18th	Died 20 04 17	Puchevillers Brit Cem
48040	Cook. H.	Liverpool.	Pte.	18th	K.I.A 22 04 17	Cherisy Rd East Cem, Heninel.
52865	Gadansky. H.B.	Manchester.	Pte.	17th	Died 22 04 17	Etaples Mil Cem
17242	Adams. W.E.	Liverpool.	Pte.	18th	K.I.A 23 04 17	Arras Memorial.
16763	Allmark. W.T.	Anfield, Liverpool.	Cpl.	18th	K.I.A 23 04 17	Arras Memorial.
53085	Bell. P.P.	Kingston, Surrey.	Cpl.	18th	D.O.W 23 04 17	St Sever Cem Ext, Rouen
	Calcott. C.D.	Shrewsbury, Salop.	Lieut.	18th	K.I.A 23 04 17	Rookery Brit Cem, Heninel.
57916	Cooper. C.W.	Paris, France	Pte.	18th	K.I.A 23 04 17	Arras Memorial.
53100	Evans. E.E.	Golders Green, London	Pte.	18th	K.I.A 23 04 17	Arras Memorial.
56724	Hignett. W.A.	Liverpool.	Pte.	18th	K.I.A 23 04 17	Arras Memorial.
57713	Hodgkinson. J	Beeston, Notts	Pte.	18th	K.I.A 23 04 17	Arras Memorial.
17602	Hollis. W.A.	Eastham, Cheshire	Pte.	18th	K.I.A 23 04 17	Arras Memorial.
25114	Hughes. J.R.	Openshaw, Manchester	Sgt.	18th	K.I.A 23 04 17	Arras Memorial.
17060	Jones. J.D.	Liverpool.	C.S.M	18th	K.I.A 23 04 17	Rookery Brit Cem, Heninel.
54140	McGarry. H.E.	Peel, I.O.M.	Pte.	18th	K.I.A 23 04 17	Wancourt Brit Cem.
17422	Moxley. F.	Birkenhead, Cheshire.	Pte.	18th	K.I.A 23 04 17	Arras Memorial.
34579	Oliver. J.E.	Seacombe, Cheshire	Pte.	18th	K.I.A 23 04 17	Cherisy Rd East Cem, Heninel.
53120	Pearce. J.	Brentford, Middlesex.	Pte.	18th	K.I.A 23 04 17	Arras Memorial.
27508	Rea. W.	Liverpool.	Pte.	18th	K.I.A 23 04 17	Arras Memorial.
57686	Sims. W.E.	Sherwood, Notts.	Pte.	18th	K.I.A 23 04 17	Arras Memorial.
57675	Stray. L.	Nottingham.	A/Cpl.	18th	K.I.A 23 04 17	Arras Memorial.
16342	Thomas. W.O.	Liverpool.	Pte.	18th	K.I.A 23 04 17	Heninel CC Ext.
16404	Trevitt.T.H.	Liverpool.	Pte.	18th	K.I.A 23 04 17	Heninel CC Ext.
16627	Wilson. F.	Liverpool.	Pte.	18th	K.I.A 23 04 17	Cherisy Rd East Cem, Heninel.
57542	Jamieson. D.F.	Arbroath	Pte.	18th	K.I.A 24 04 17	Cherisy Rd East Cem, Heninel.
52830	Peacock. J.	Birkdale, Southport, Lancs.	Pte.	17th	K.I.A 24 04 17	Arras Memorial.
22928	Billington. H.	Northwich, Cheshire.	Cpl.	20th	D.O.W 25 04 17	London Cem, Neuville-Vitasse
48282	Blanchard. J.	Liverpool.	Pte.	20th	K.I.A 25 04 17	Arras Memorial.
58658	Denton. W.	Aintree, Liverpool.	Pte.	20th	K.I.A 25 04 17	Cherisy Rd East Cem, Heninel.
22714	Healy. I.E.	Anfield, Liverpool.	Sgt.	20th	K.I.A 25 04 17	Arras Memorial.
57815	Milne. J.T.	Sunderland.	Pte.	19th	K.I.A 25 04 17	Mont Huon Cem, Le Treport.
58673	Owen. E.G.	Liverpool.	Pte.	20th	K.I.A 25 04 17	Arras Memorial.
16853	Young. G.N.H.	Liverpool.	Pte.	18th	K.I.A 25 04 17	Cherisy Rd East Cem, Heninel.
17166	Barber. LJ.	Liverpool.	Pte.	18th	K.I.A 26 04 17	Arras Memorial.
57751	Boustead. E.P.	Hetton-le-Hole, Durham	Pte.	18th	K.I.A 26 04 17	Arras Memorial.
38448	Calland. J.	Liverpool.	Pte.	18th	K.I.A 26 04 17	Arras Memorial.
56659	Davies. W.S.	Huyton Quarry, Lancs.	Pte.	18th	K.I.A 26 04 17	Cherisy Rd East Cem, Heninel.
57651	Kemp. B.	Leeds, Yorks	Pte.	18th	K.I.A 26 04 17	Cherisy Rd East Cem, Heninel.
57946	Plose. F.W.	Forest Gate, London.	Pte.	17th	K.I.A 26 04 17	Arras Memorial.
26633	Robinson. W.H.	Prescot, Lancs.	Pte.	18th	D.O.W 26 04 17	Warlincourt Haite Brit Cem.
36661	Seddon. J.	Farnworth, Lancs.	Pte.	20th	K.I.A 26 04 17	Wancourt Brit Cem.
48269	Strawson. J.	Liverpool.	Pte.	18th	K.I.A 26 04 17	Cherisy Rd East Cem, Heninel.
51640	Stretton. J.J.	Liverpool.	Pte.	17th	K.I.A 26 04 17	Arras Memorial.
56701	Webster. W.	Liverpool.	Pte.	18th	K.I.A 26 04 17	Cherisy Rd East Cem, Henijnel.
29625	James. J.G.	Liverpool.	L/Cpl.	19th	D.O.W 27 04 17	St Sever Cem Ext, Rouen.
	Band. L.	Rock Ferry, Cheshire.	2nd Lieut.	17th	K.I.A 28 04 17	Cherisy Rd East Cem, Heninel.

49033	Brown. J.	Pendleton, Manchester.	L/Cpl.	17th	K.I.A	28 04 17	Cherisy Rd East Cem, Heninel.
34316	Comley. F.	Liverpool.	Pte.	17th	K.I.A	28 04 17	Cherisy Rd East Cem, Heninel.
48411	Fitzgerald. M.	Liverpool.	Pte.	17th	K.I.A	28 04 17	Cherisy Rd East Cem, Heninel.
58019	Heath. E.	Wandsworth, London.	Pte.	17th	K.I.A	28 04 17	Cherisy Rd East Cem, Heninel.
49052	Johns. P.	Aarberth, Pembroke.	Pte.	17th	K.I.A	28 04 17	Cherisy Rd East Cem, Heninel.
24605	Knight. J.	Liverpool.	Pte.	17th	K.I.A	28 04 17	Arras Memorial.
49109	Norton. J.	Stalybridge, Lancs.	Pte.	17th	K.I.A	28 04 17	Cherisy Rd East Cem, Heninel.
49050	Orrell. J.S.	Norden, Lancs.	L/Cpl.	17th	K.I.A	28 04 17	Cherisy Rd East Cem, Heninel.
51721	Speet. A.A.	Wavertree, Liverpool.	Pte.	20th	D.O.W	28 04 17	Warlincourt Halte Brit Cem.
52950	Walton. W.H.	Clacton-on-Sea, Essex.	A/Cpl.	17th	K.I.A	28 04 17	Arras Memorial.
51937	Caldwell. R.	Liverpool.	Cpl.	17th	D.O.W	29 04 17	Arras Memorial.
33195	Allen. G.	Liverpool.	Pte.	17th	D.O.W	01 05 17	St Sever Cem, Rouen.
17823	Fletcher. H.C.	Wickham Market, Suffolk.	L/Cpl.	19th	D.O.W	01 05 17	Hacheston,(All Saints)Suffolk
57566	Robertson. J.	St.Andrews, Fife.	Pte.	18th	D.O.W	02 05 17	Wancourt Brit Cem.
42625	McCormack. J.C.	Crosby, Liverpool.	Pte.	17th	D.O.W	03 05 17	Bucquoy Rd Cem, Ficheux.
	Milliken. F.S.	Anfield, Liverpool.	2nd Lieut	19th	D.O.W	04 05 17	Mont Huon Cem, Le Treport.
22804	Palmer. M.J.	Garston, Liverpool.	Cpl.	20th	Died	12 05 17	Allerton Cem, Liverpool.
59030	Cleminson. P.C.	Edinburgh.	Pte.	19th	Died	17 05 17	Bishop Auckland Cem, Durham.
241768	McLoughlin. J.	Islington, Liverpool.	Pte.	17th	K.I.A	20 05 17	Heninel-Croisilles Rd Cem.
57903	Smith. G.F.	London.	Pte.	19th	D.O.W	21 05 17	Mont Huon Cem, Le Treport.
	Mackie. F.J.	Peterhead, Aberdeen.	2nd Lieut	19th	K.I.A	29 05 17	Vlamertinghe Mil Cem.
29718	Stretch. T.	Liverpool.	Pte.	20th	D.O.W	02 06 17	Hop Store Cem, Vlamertinghe.
269686	Bradshaw. W.	Manchester.	Pte.	19th	K.I.A	04 06 17	Menin Gate Memorial.
23113	Mellor. J.	Neston, Cheshire.	Pte.	20th	D.O.W	05 06 17	Railway Dugouts B.G Zillebeke
52098	Jackson. A.	Manchester.	Pte.	19th	K.I.A	07 06 17	Railway Dugouts B.G Zillebeke
23106	McIlroy. S.	Liverpool.	Pte.	19th	K.I.A	07 06 17	Railway Dugouts B.G Zillebeke
	Sharpies. G.W.	Preston, Lancs.	2nd Lieut	19th	K.I.A	07 06 17	Railway Dugouts B.G Zillebeke
204299	Wiseman G.	Liverpool.	Pte.	19th	K.I.A	07 06 17	Railway Dugouts B.G Zillebeke
49099	Brown. H.A.	Bebington, Cheshire.	Pte.	17th	K.I.A	08 06 17	Menin Gate Memorial.
49180	Walley J E	Stockport, Cheshire.	Pte.	17th	K.I.A	08 06 17	New Irish Farm Cem, Ypres
49490	Bradbury. J.	St.Helens, Lancs.	Pte.	20th	D.O.W	09 06 17	Hop Store Cem, Vlamertinghe.
24432	Davies. E.G.	Liverpool.	Pte.	20th	K.I.A	09 06 17	Hop Store Cem, Vlamertinghe.
49202	Holding. H.	Accrington, Lancs.	Pte.	20th	D.O.W	09 06 17	Pozieres Memorial.
49201	Holman. V.C.	Burnley, Lancs.	L/Cpl.	20th	K.I.A	09 06 17	Hop Store Cem, Vlamertinghe.
73593	Lomas. J.W.	Ashton-under-Lyne, Lancs.	Pte.	18th	Died	09 06 17	Ashton-u-Lyne, Lancs.
23857	Weston J	Liverpool.	L/Cpl.	17th	K.I.A	09 06 17	Menin Gate Memorial.
58057	Wood. H.	Keighley. Yorks.	Pte.	17th	K.I.A	09 06 17	Menin Gate Memorial.
51645	Mercer. G.W.	St.Helens, Lancs.	L/Cpl.	17th	D.O.W	12 06 17	Wimereux CC
16642	Caulfield. J.	Liverpool.	Pte.	18th	K.I.A	13 06 17	Railway Dugout B.G Zillebeke
57636	Crowther. W.	Mirfield, Yorks	L/Cpl.	18th	D.O.W	14 06 17	Lijssenthoek M C Poperinghe
44251	Davies. J.	Llanarth, Cards.	Pte.	18th	D.O.W	15 06 17	Lijssenthoek M C Poperinghe
61793	Millard. W.	London	Pte.	18th	K.I.A	15 06 17	Menin Gate Memorial.
16668	Morgan. S.	Reading, Berks.	Pte.	18th	K.I.A	20 06 17	Dickebusch New Mil Cem.
58561	Parker. S.	Warrington, Cheshire.	Pte.	18th	K.I.A	20 06 17	Railway Dugout B.G Zillebeke
57660	Skevington. G.E.	Brough, Yorks.	Pte.	18th	K.I.A	20 06 17	Railway Dugout B.G Zillebeke
26020	Watts. J.W.	Wigan, Lancs.	Pte.	18th	K.I.A	20 06 17	Dickebusch New Mil Cem.
17033	Yates. J.	Liverpool.	Pte.	18th	K.I.A	20 06 17	Dickebusch New Mil Cem.
11813	Ashbrook. G.	Warrington, Cheshire.	Pte.	19th	K.I.A	23 06 17	Birr Cross Rds Cem, Zillebeke
48421	Collock. M	Liverpool.	Pte.	19th	K.I.A	23 06 17	Menin Gate Memorial.
52121	Garside. N.	Manchester.	Pte.	19th	K.I.A	23 06 17	Menin Gate Memorial.
38434	Harron. A.	Liverpool.	Pte.	19th	K.I.A	23 06 17	Perth Cem, Zillebeke.
52219	Hart. H.	Cloughfold, Lancs.	Pte.	19th	K.I.A	23 06 17	Perth Cem, Zillebeke.
269214	Mathison. H.	Wrexham.	Pte.	19th	K.I.A	23 06 17	Menin Gate Memorial.
52117	Owen. F.	Hulme, Manchester.	Pte.	19th	K.I.A	23 06 17	Birr Cross Rds Cem, Zillebeke.
50058	Spencer. J.	Blackpool, Lancs.	Pte.	19th	K.I.A	23 06 17	Menin Gate Memorial.
36216	Russell. W.	Nelson, Lancs.	Pte.	18th	D.O.W	24 06 17	Menin Gate Memorial.
57605	Wright. R.	Bradford, Yorks.	Cpl.	19th	D.O.W	24 06 17	Railway Dugout B.G Zillebeke
22200	Owens. A.	Liverpool.	Cpl.	20th	D.O.W	25 06 17	Wimereux CC

53087	Baker. L.H.	Rayleigh, Essex.	Pte.	18th	K.I.A	26 06 17	Railway Dugout B.G Zillebeke
27547	Smith. J.	Liverpool.	L/Cpl.	17th	K.I.A	27 06 17	Cerisy-Gailly Mil Cem.
50088	Blozard. G.	Darwen, Lancs.	Pte.	20th	K.I.A	28 06 17	Perth Cem, Zillebeke.
50220	Ward. J.	Kirbymoorside, Yorks.	Pte.	19th	K.I.A	28 06 17	Menin Gate Memorial.
50122	Curry. J.R.	Easington, Durham.	Pte.	17th	K.I.A	30 06 17	Menin Gate Memorial.
25796	Davies. W.F.	Liscard, Cheshire	Pte.	18th	K.I.A	30 06 17	Dickebusch New Mil Cem.
49083	Ellison. J.	Liverpool.	Pte.	20th	D.O.W	30 06 17	Railway Dugout B.G Zillebeke
49156	Goodwin. G.W.	Macclesfield, Ches.	Pte.	17th	K.I.A	30 06 17	Perth Cem, Zillebeke.
49541	Thornton. L.	Oldham, Lancs.	Pte.	20th	K.I.A	30 06 17	Perth Cem, Zillebeke.
15172	Willacy. T.B.	Birkenhead, Cheshire.	Pte.	17th	Died.	30 06 17	Dozingham M C, Westvleteren
57397	Brown. J.	Partick, Glasgow.	Pte.	20th	D.O.W	01 07 17	Railway Dugout B.G Zillebeke
50285	Wood. G.	Blackburn, Lancs.	Pte.	20th	D.O.W	01 07 17	Railway Dugout B.G Zillebeke
57999	Birks. F.W.	Nottingham.	Pte.	17th	K.I.A	02 07 17	Menin Gate Memorial.
47163	Cooil. E.	Arbory, I.O.M.	Pte.	17th	K.I.A	02 07 17	Perth Cem, Zillebeke.
10630	Farrell. J.	Liverpool.	Pte.	20th	K.I.A	02 07 17	Menin Gate Memorial.
49091	Garside. P.L.P.	Oldham, Lancs.	Pte.	20th	D.O.W	02 07 17	Railway Dugout B.G Zillebeke
49573	Hughes. G.H.	Birkenhead, Cheshire.	Pte.	20th	K.I.A	02 07 17	Perth Cem, Zillebeke.
51583	Johnson. H.V.	Liverpool.	Pte.	17th	K.I.A	02 07 17	Perth Cem, Zillebeke.
52872	Bottomley. B.	Nelson, Lancs.	Pte.	17th	D.O.W	03 07 17	Railway Dugout B.G Zillebeke
49013	Brooks. H.	Tottington, Lancs.	Pte.	17th	D.O.W	03 07 17	Perth Cem, Zillebeke.
	Chavasse. A.	Liverpool.	Lieut.	17th	K.I.A	04 07 17	Menin Gate Memorial.
	Peters. C.A.	Acton, London.	2nd Lieut.	17th	K.I.A	04 07 17	Perth Cem, Zillebeke.
33188	Cook. L.	Liverpool.	Pte.	17th	D.O.W	05 07 17	Dickebusch New Mil Cem.
48127	Gouldbourne. J.	Liverpool.	Pte.	19th	K.I.A	05 07 17	Railway Dugout B.G Zillebeke
202845	Roberts. E.	Liverpool.	Pte.	19th	K.I.A	05 07 17	Railway Dugout B.G Zillebeke
49498	Birkhead. N.	Liscard, Cheshire.	Pte.	20th	D.O.W	06 07 17	Railway Dugout B.G Zillebeke
59428	Hampson. R.W.	Liverpool.	Pte.	18th	Died	07 07 17	Longuenesse Cem, St.Omer.
56519	Edwards. J.		Pte.	20th	D.O.W	09 07 17	Brandhoek Mil Cem
235326	Rump. C.	Aylsham, Norfolk.	Pte.	20th	D.O.W	12 07 17	Recques Cem, (Nr Calais).
52110	Dooley. J.P.	Manchester.	L/Cpl.	19th	K.I.A	15 07 17	Tyne Cot Memorial.
24772	Leach. F.	Liverpool.	Pte.	18th	K.I.A	20 07 17	Menin Gate Memorial.
53125	Todd. H.O.	London.	Pte.	18th	Died	20 07 17	Harwicke, Bucks.
19793	Harrison. J. M.M.	Everton, Liverpool.	Sgt.	19th	K.I.A	21 07 17	Menin Gate Memorial.
51720	Meredew. F.	Liverpool.	Pte.	20th	K.I.A	21 07 17	Menin Gate Memorial.
32227	Mellor. G.W.	Manchester.	Pte.	19th	K.I.A	22 07 17	Menin Gate Memorial.
331008	Bowyer. W.	Liverpool.	Pte.	19th	K.I.A	23 07 17	Menin Gate Memorial.
53130	Boycott. J.A.	Shawbury, Salop.	L/Cpl.	19th	K.I.A	23 07 17	Hooge Crater Cem, Zillebeke.
332672	Harrington. F.	Liverpool.	Pte.	19th	K.I.A	23 07 17	Menin Gate Memorial.
31560	Lewis. F.R.	Seacombe, Ches.	Pte.	19th	K.I.A	23 07 17	Menin Gate Memorial.
53134	Morris. A.	Wellington, Salop.	L/Cpl.	19th	K.I.A	23 07 17	Menin Gate Memorial.
57824	Pyburn. T.B.	Carville, Durham.	Pte.	19th	K.I.A	23 07 17	Menin Gate Memorial.
25467	Ryder. J.W.	Nelson, Lancs.	Pte.	19th	K.I.A	23 07 17	Menin Gate Memorial.
235213	Bailey. J.	Burnley, Lancs.	Pte.	19th	K.I.A	24 07 17	Menin Gate Memorial.
38820	Holden. F.	Wigan, Lancs.	Pte.	18th	K.I.A	24 07 17	Menin Gate Memorial.
73301	Monday. I.	High Wycombe, Bucks.	Pte.	17th	Died.	24 07 17	Calais Southern Cem
11439	Stapleton. R.	Waterford, Ireland.	Pte.	18th	K.I.A	24 07 17	Menin Gate Memorial.
235257	Tait. C.A.	Stockton-on-Tees, Durham.	Pte.	18th	K.I.A	24 07 17	Menin Gate Memorial.
25732	Gaskell. T.P.	Wigan, Lancs.	Pte.	18th	K.I.A	25 07 17	Tyne Cot Cem.
54142	Foxley. R.T.	Liverpool.	Pte.	19th	K.I.A	27 07 17	Menin Gate Memorial.
405703	Gibson. G.	Liverpool.	Pte.	19th	K.I.A	27 07 17	Menin Gate Memorial.
57896	Giles. G.	Newcastle-on-Tyne	Pte.	19th	K.I.A	27 07 17	Menin Gate Memorial.
52140	Jones. W.	Manchester.	Pte.	19th	K.I.A	27 07 17	Menin Gate Memorial.
17412	Lightfoot. A.	Runcorn, Ches.	Cpl.	19th	K.I.A	27 07 17	Menin Gate Memorial.
406610	Seacombe. T.G.	Swansea, S.Wales.	Pte.	19th	K.I.A	27 07 17	Menin Gate Memorial.
405666	Wynn. G.T.	Liverpool.	Pte.	19th	K.I.A	27 07 17	Menin Gate Memorial.
235240	Leeming. J.T.	Littleborough, Lancs.	Pte.	19th	K.I.A	28 07 17	Menin Gate Memorial.
50206	Peebles. W.	Musselburgh, Edinburgh.	Pte.	19th	D.O.W	28 07 17	Lijssenthoek M C Poperinghe
57476	Wilson. A.Mc.	Hamilton, Lanarks.	Pte.	19th	Died	28 07 17	Etaples Mil Cem

243873	Lyon. T.H.	West Kirby, Cheshire.	Pte.	19th	K.I.A	29 07 17	Menin Gate Memorial.
22219	Povey. G.F.D.	Birkenhead, Cheshire.	Pte.	20th	D.O.W	29 07 17	Wimereux CC
50211	Smith. E.	Belchford, Lines.	Pte.	19th	K.I.A	29 07 17	Menin Gate Memorial.
17283	Briscoe. H.R.	Liverpool.	Sgt.	19th	K.I.A	30 07 17	Menin Gate Memorial.
202908	Allister. T.H.	Garston, Liverpool.	Pte.	18th	K.I.A	31 07 17	Menin Gate Memorial.
35601	Armstrong. J.T. M.M.	Liverpool.	Cpl.	20th	K.I.A	31 07 17	Menin Gate Memorial.
56662	Ashcroft. H.	Lathom, Lancs.	Pte.	20th	K.I.A	31 07 17	Menin Gate Memorial.
16793	Badderiey. H.	Egremont, Cheshire.	Cpl.	18th	K.I.A	31 07 17	Hooge Crater Cem, Zillebeke.
202814	Ball. A.	Liverpool.	Pte.	18th	K.I.A	31 07 17	Hooge Crater Cem, Zillebeke.
57940	Barker. F.	New Barnet, London	Pte.	18th	K.I.A	31 07 17	Menin Gate Memorial.
50091	Barlow. W.	Barrowford, Lancs.	Pte.	20th	K.I.A	31 07 17	Menin Gate Memorial.
308710	Bateson.J.	Heywood, Lancs.	Pte.	18th	K.I.A	31 07 17	Menin Gate Memorial.
57773	Batey. I.	Dawdon, Durham	Pte.	18th	K.I.A	31 07 17	Menin Gate Memorial.
42705	Baxter. J.	Wavertree, Liverpool.	Pte.	18th	K.I.A	31 07 17	Menin Gate Memorial.
41279	Beaman. J.	Pendleton, Manchester.	Pte.	17th	K.I.A	31 07 17	Menin Gate Memorial.
27638	Beardsworth. R.	Leyland, Lancs.	Pte.	17th	K.I.A	31 07 17	Menin Gate Memorial.
21928	Bell. R.A.	Liverpool.	L/Cpl.	18th	K.I.A	31 07 17	Menin Gate Memorial.
	Bigg. A.C.		2nd Lieut.	18th	K.I.A	31 07 17	Menin Gate Memorial.
14015	Blackwood. D.	Liverpool.	Pte.	19th	K.I.A	31 07 17	Menin Gate Memorial.
36701	Bolton.J.E.	Preston, Lancs.	Pte.	20th	K.I.A	31 07 17	Menin Gate Memorial.
269812	Bowyer. E.B.	Liverpool.	Pte.	18th	K.I.A	31 07 17	Menin Gate Memorial.
59465	Boyer. J.	Liverpool.	Pte.	18th	K.I.A	31 07 17	Menin Gate Memorial.
235283	Brown. A.	Southacre, Norfolk.	Pte.	18th	K.I.A	31 07 17	Menin Gate Memorial.
57775	Brown. T.H.	Fulwell, Sunderland	Pte.	18th	K.I.A	31 07 17	Menin Gate Memorial.
59438	Brown. T.W.	Liverpool.	Pte.	18th	K.I.A	31 07 17	Menin Gate Memorial.
23917	Brown. W.H.	Birkenhead, Cheshire.	Sgt.	19th	K.I.A	31 07 17	Menin Gate Memorial.
52112	Burke. F.	Manchester.	Pte.	19th	K.I.A	31 07 17	Menin Gate Memorial.
57502	Burt. A.E.	Richmond, Surrey	Sgt.	18th	K.I.A	31 07 17	Menin Gate Memorial.
242488	Caldwell. A.W.	Rock Ferry, Cheshire	Pte.	18th	K.I.A	31 07 17	Tyne Cot Cem.
59571	Chandler. F.	Everton, Liverpool.	Pte.	17th	K.I.A	31 07 17	Menin Gate Memorial.
41723	Conroy. P.W.	Liverpool.	Pte.	18th	K.I.A	31 07 17	Menin Gate Memorial.
	Copland. G.H.	New Brighton, Cheshire.	2nd Lieut.	18th	K.I.A	31 07 17	Menin Gate Memorial.
50118	Copley. P.	Hull, Yorks.	Pte.	17th	K.I.A	31 07 17	Menin Gate Memorial.
270090	Cowin. R.	Douglas, I.O.M.	Pte.	18th	K.I.A	31 07 17	Menin Gate Memorial.
64780	Craine. R.W.	Liverpool.	Pte.	17th	K.I.A	31 07 17	Bedford House Cem End 4
57607	Crane. S.A.	Sheffield, Yorks.	Cpl.	18th	K.I.A	31 07 17	Menin Gate Memorial.
57779	Crawford. G.	Philadelphia F.H. Durham	Pte.	18th	K.I.A	31 07 17	Menin Gate Memorial.
59155	Crosby. J.H.	Liverpool.	Pte.	18th	K.I.A	31 07 17	Menin Gate Memorial.
36251	Cross. W.W.	Fulwood, Preston, Lancs.	Pte.	17th	K.I.A	31 07 17	Menin Gate Memorial.
20196	Dale. H.	Manchester	Pte.	18th	K.I.A	31 07 17	Menin Gate Memorial.
16053	Dalton. P.F.	Liverpool.	Sgt.	17th	K.I.A	31 07 17	Zantvoorde Brit Cem.
50234	Davidson. H.	Haslingden, Lancs.	Pte.	17th	K.I.A	31 07 17	Menin Gate Memorial.
200232	Davies. A.	Everton, Liverpool.	Pte.	17th	K.I.A	31 07 17	Menin Gate Memorial.
59224	Davies. J.	Liverpool.	Pte.	18th	K.I.A	31 07 17	Menin Gate Memorial.
22667	Davies. J.F.	Kirkdale, Liverpool.	Pte.	20th	K.I.A	31 07 17	Peronne CC Ext.
23706	Deaville. A.	Manchester.	L/Cpl.	17th	K.I.A	31 07 17	Hooge Crater Cem, Zillebeke.
44530	Dickinson. C.	Stockport, Cheshire.	Pte.	18th	K.I.A	31 07 17	Menin Gate Memorial.
17314	Dicks. A.W.	Liverpool.	Pte.	19th	K.I.A	31 07 17	Menin Gate Memorial.
	Dimond. F.R.		Lieut.	17th	K.I.A	31 07 17	Menin Gate Memorial.
56458	Docherty. J.	Ballycroy, Co.Mayo	Pte.	17th	K.I.A	31 07 17	Menin Gate Memorial.
242545	Dodd. T.S.	Liverpool.	Pte.	18th	K.I.A	31 07 17	Hooge Crater Cem, Zillebeke.
57638	Dodworth. C.G.	Sheffield, Yorks.	Pte.	18th	K.I.A	31 07 17	Menin Gate Memorial.
270033	Downey. W.	Liverpool.	Pte.	18th	K.I.A	31 07 17	Hooge Crater Cem, Zillebeke.
201303	Dutton. W.	Lathom, Lancs.	Pte.	17th	K.I.A	31 07 17	Menin Gate Memorial.
25727	Floyd. E.	Liverpool.	Pte.	17th	K.I.A	31 07 17	Menin Gate Memorial.
59579	Forrest. W.	Liverpool.	Pte.	17th	K.I.A	31 07 17	Ypres Reservoir Cem.
41925	Fox. G.H.	Liverpool.	Pte.	18th	K.I.A	31 07 17	Menin Gate Memorial.
270079	Francis. A.	Garston, Liverpool.	Pte.	18th	K.I.A	31 07 17	Tyne Cot Cem.

57410	Gault. T.	Possilpark, Glasgow.	Pte.	18th	K.I.A	31 07 17	Menin Gate Memorial.
	Goldspink. E.N.		2nd Lieut.	17th	K.I.A	31 07 17	Menin Gate Memorial.
	Graham. W.J.	Warrington, Cheshire.	2nd Lieut.	18th	K.I.A	31 07 17	Menin Gate Memorial.
30332	Griffin. J.	Everton, Liverpool.	Pte.	20th	K.I.A	31 07 17	Zantvoorde Brit Cem.
266905	Hall. W.	Southport, Lancs.	L/Cpl.	18th	K.I.A	31 07 17	Menin Gate Memorial.
202901	Hammond. J.H.	Liverpool.	Pte.	18th	K.I.A	31 07 17	Menin Gate Memorial.
23945	Harrington.!.	Bootle, Liverpool.	Cpl.	20th	K.I.A	31 07 17	Menin Gate Memorial.
38470	Harrison. H.R.	Douglas, I.O.M.	Pte.	20th	K.I.A	31 07 17	Menin Gate Memorial.
32316	Harvey. A.J.	Liverpool.	Pte.	18th	K.I.A	31 07 17	Menin Gate Memorial.
235231	Harwood. A.	Darwen, Lancs.	Pte.	19th	K.I.A	31 07 17	Menin Gate Memorial.
50246	Haworth. J.W.	Haslingden, Lancs.	Pte.	17th	K.I.A	31 07 17	Menin Gate Memorial.
59464	Higginson. F.	Liverpool.	Pte.	18th	K.I.A	31 07 17	Menin Gate Memorial. «
52291	Hircock. J.	Liverpool.	Pte.	17th	D.O.W	31 07 17	Menin Gate Memorial.
57790	Hird. R.	Sunderland	Pte.	18th	K.I.A	31 07 17	Menin Gate Memorial.
11360	Hobrough. F	Liverpool.	Pte.	19th	K.I.A	31 07 17	Hooge Crater Cem, Zillebeke.
269834	Holcroft. P.	Garston, Liverpool.	Pte.	17th	K.I.A	31 07 17	Menin Gate Memorial.
37026	Hornby. J.	Lathom, Lancs.	Pte.	18th	K.I.A	31 07 17	Menin Gate Memorial.
242550	Hudson. A.M.	Market Drayton, Salop	Pte.	18th	K.I.A	31 07 17	Menin Gate Memorial.
52072	Hudson. J.H.	Liverpool.	Pte.	19th	K.I.A	31 07 17	Menin Gate Memorial.
235230	Hulme. J.	Stockport, Ches.	Pte.	19th	D.O.W	31 07 17	Menin Gate Memorial.
26092	Hutton. F.R.	Liverpool.	Pte.	18th	K.I.A	31 07 17	Menin Gate Memorial.
53112	Jarman. J.H.	Cricklewood, London	Pte.	18th	K.I.A	31 07 17	Hooge Crater Cem, Zillebeke.
29175	Jones. A.	Liverpool.	L/Cpl.	20th	K.I.A	31 07 17	Menin Gate Memorial.
204132	Jones. C.	Liverpool.	Pte.	17th	K.I.A	31 07 17	Menin Gate Memorial.
32710	Jones. F.J.	Liverpool.	Pte.	18th	K.I.A	31 07 17	Menin Gate Memorial.
59540	Jubey. S.A.	Liverpool.	Pte.	17th	K.I.A	31 07 17	Menin Gate Memorial.
58602	Kavanagh. L.M.		Pte.	17th		31 07 17	Menin Gate Memorial.
204632	Kelly. P.	Birkenhead, Cheshire.	Pte.	18th	K.I.A	31 07 17	Menin Gate Memorial.
59309	Kneale. W.P.	Liverpool.	Pte.	17th	K.I.A	31 07 17	Menin Gate Memorial.
	Lane. F.A.	Clapham Common, London.	2nd Lieut.	18th	K.I.A	31 07 17	Menin Gate Memorial.
13624	Lanigan. E.J.	Liverpool.	L/Cpl.	18th	K.I.A	31 07 17	Menin Gate Memorial.
406638	Leeke. L.	Nantwich, Cheshire.	Pte.	19th	K.I.A	31 07 17	Oosttaverne Wood Cem.Whytschaete.
46700	Leighton. C.R.	Liverpool.	Pte.	18th	K.I.A	31 07 17	Ypres Reservoir Cem.
50069	Lester. J.	Bolton, Lancs.	Pte.	19th	K.I.A	31 07 17	Menin Gate Memorial.
405620	Llewellyn. E.	Liverpool.	Pte.	19th	K.I.A	31 07 17	Menin Gate Memorial.
202854	Lloyd. R.	Everton, Liverpool.	Pte.	18th	K.I.A	31 07 17	Menin Gate Memorial.
26225	Lomas. J.A.	Altrincham, Cheshire	Pte.	18th	K.I.A	31 07 17	Menin Gate Memorial.
52398	Lyne. S.	Ulverston.	Pte.	18th	K.I.A	31 07 17	Menin Gate Memorial.
269725	MacDonald. J.	Manchester.	Pte.	19th	D.O.W	3107 1 7	Menin Gate Memorial.
32511	Manning. T.	Ardle, Co.Louth.	Pte.	19th	K.I.A	31 07 17	Menin Gate Memorial.
57980	Martland. J.	Appleby Bridge, Lancs.	Pte.	18th	K.I.A	31 07 17	Menin Gate Memorial.
332341	McCann. O.	Liverpool.	Pte.	18th	K.I.A	31 07 17	Menin Gate Memorial.
57435	McGregor. D.	Glasgow.	Pte.	18th	K.I.A	31 07 17	Menin Gate Memorial.
25159	McNiven.W.	Liverpool.	Pte.	18th	K.I.A	31 07 17	Menin Gate Memorial.
30783	Melia. P.	Ormskirk, Lancs.	Pte.	18th	K.I.A	3107 17	Menin Gate Memorial.
50032	Milne. R.	London.	Pte.	20th	K.I.A	31 07 17	Hooge Crater Cem, Zillebeke.
235243	Monks C.	Bolton, Lancs.	Pte.	19th	D.O.W	31 07 17	Menin Gate Memorial.
59562	Moore. J.	Peel, I.O.M.	Pte.	17th	D.O.W	31 07 17	Lijssenthoek M C Poperinghe
15675	Muir L	Liverpool.	Pte.	17th	K.I.A	31 07 17	Zantvoorde Brit Cem.
18487	Murphy. W.	Liverpool.	Pte.	18th	K.I.A	31 07 17	Menin Gate Memorial.
25066	Mylchreest. D.	Union Mills, I.O.M.	Pte.	20th	K.I.A	31 07 17	Menin Gate Memorial.
39038	Mylchreest. H.	Malew, I.O.M.	Pte.	18th	K.I.A	31 07 17	Menin Gate Memorial.
	Nickel. G.G.	Liverpool.	2nd Lieut.	20th	K.I.A	31 07 17	Menin Gate Memorial.
41726	O'Connor. P.	Kirkdale, Liverpool.	Pte.	17th	K.I.A	31 07 17	Menin Gate Memorial.
	Orme. A.L.	Liverpool.	2nd Lieut.	18th	K.I.A.	31 07 17	Menin Gate Memorial.
235117	Parker. C.G.	Harringay, London.	Pte.	19th	K.I.A	31 07 17	Menin Gate Memorial.
49532	Parker. J.	Kirkdale, Liverpool.	L/Cpl.	20th	K.I.A	31 07 17	Menin Gate Memorial.

48658	Parkinson. T.	Beswick, Manchester	Pte.	18th	K.I.A	31 07 17	Menin Gate Memorial.
17158	Parry. R.E.	Bootle, Liverpool.	Sgt.	18th	K.I.A	31 07 17	Menin Gate Memorial.
22808	Peters. J.	Parkgate, Ches.	Cpl.	19th	K.I.A	31 07 17	Menin Gate Memorial.
59572	Phillips. W.	Liverpool.	Pte.	18th	K.I.A	31 07 17	Menin Gate Memorial.
57729	Phillis. F.	Radford, Notts	L/Cpl.	18th	K.I.A	31 07 17	Menin Gate Memorial.
48276	Plummer. F.	Bootle, Liverpool.	Pte.	18th	K.I.A	31 07 17	Menin Gate Memorial.
	Prendiville. LA.	Birkenhead, Cheshire.	2nd Lieut.	18th	K.I.A	31 07 17	Menin Gate Memorial.
58040	Preston. R.	Slough, Bucks.	Pte.	17th	K.I.A	31 07 17	Menin Gate Memorial.
16265	Pugh. H.J.	Willaston, Cheshire.	Sgt.	18th	K.I.A	31 07 17	Menin Gate Memorial.
58553	Rawlings. J.W.	Liverpool.	Pte.	18th	K.I.A	31 07 17	Menin Gate Memorial.
51921	Redfern. A.M.	Liverpool.	Pte.	17th	D.O.W	31 07 17	Huts Cem, Dikebusch
50623	Richardson. J.	Liverpool.	Pte.	20th	K.I.A	31 07 17	Zantvoorde Brit Cem.
15780	Rigby. J.D. M.M.	Runcorn, Cheshire.	Sgt.	17th	K.I.A	31 07 17	Menin Gate Memorial.
32730	Rogers. W.J.D.	Liverpool.	Pte.	17th	K.I.A	31 07 17	Menin Gate Memorial.
49177	Roome. F.	Macclesfield, Ches.	Pte.	17th	K.I.A	31 07 17	Menin Gate Memorial.
42879	Rothwell. T.	Liverpool.	Pte.	18th	K.I.A	31 07 17	Menin Gate Memorial.
51920	Rugen F.	Liverpool.	Pte.	17th	K.I.A	31 07 17	Menin Gate Memorial.
24788	Russell. W.	Liverpool.	Pte.	17th	K.I.A	31 07 17	Menin Gate Memorial.
235254	Schofield. J.	Litherland, Liverpool.	Pte.	18th	K.I.A	31 07 17	Menin Gate Memorial.
57732	Scothon. W.	Nottingham.	Pte.	17th	K.I.A	31 07 17	Zantvoorde Brit Cem.
50277	Sheader. W.W.	Colne, Lancs.	Pte.	20th	K.I.A	31 07 17	Menin Gate Memorial.
29183	Sixsmith. A.	Liverpool.	Pte.	17th	K.I.A	31 07 17	Menin Gate Memorial.
	Stacey. H.L.	Liverpool.	2nd Lieut.	18th	K.I.A	31 07 17	Tyne Cot Cem.
235250	Stafford. R	Blackpool, Lancs.	Pte.	18th	K.I.A	31 07 17	Menin Gate Memorial.
27540	Stanton. P.	Liverpool.	Cpl.	18th	K.I.A	31 07 17	Menin Gate Memorial.
57957	Stephan. E.F.	Islington, London	Pte.	18th	K.I.A	31 07 17	Hagle Dump Cem, Elveringhe.
16051	Stewart. R.A.	Wallasey, Cheshire.	Pte.	17th	K.I.A	31 07 17	Zantvoorde Brit Cem.
50057	Stone. C.	Bridgewater, Somerset.	Pte.	19th	K.I.A	31 07 17	Menin Gate Memorial.
50072	Tanner. H.	Stroud, Glos.	Pte.	19th	K.I.A	31 07 17	Menin Gate Memorial.
58775	Taylor. H.	Middleton, Lancs.	Pte.	18th	K.I.A	31 07 17	Menin Gate Memorial.
235258	Taylor. P.	Monton, Manchester.	Pte.	18th	K.I.A	31 07 17	Menin Gate Memorial.
58708	Thompson. R.B.	Oswaldtwistle, Lancs.	Pte.	20th	K.I.A	31 07 17	Menin Gate Memorial.
29643	Thornton. E.	Liverpool.	Pte.	17th	D.O.W	31 07 17	Huts Cem, Dikebusch.
48129	Trafford. I.	Liverpool.	Pte.	19th	K.I.A	31 07 17	Menin Gate Memorial.
14501	Travis. J.	Thatto Heath, St. Helens Lancs.	Pte.	18th	K.I.A	31 07 17	Menin Gate Memorial
17094	Trow. H.	Liverpool.	Pte.	18th	K.I.A	31 07 17	Ypres Reservoir Cem.
58564	Walsworth F.	Gomersal, Yorks.	Pte.	18th	K.I.A	31 07 17	Menin Gate Memorial.
49315	Walters. W.H.	Chester.	Pte.	20th	K.I.A	31 07 17	Bedford House Cem End 4
22294	Weatherspoon. P.	Liverpool.	Pte.	20th	K.I.A	31 07 17	Menin Gate Memorial.
38880	Welsh. J.	Liverpool.	Pte.	17th	K.I.A	31 07 17	Menin Gate Memorial.
15097	West. LT.	Liverpool.	L/Cpl.	17th	K.I.A	31 07 17	Ypres Reservoir Cem
50113	Westby. P.	Hull, Yorks.	Pte.	17th	K.I.A	31 07 17	Menin Gate Memorial.
308909	Whelan. P.	Liverpool.	L/Cpl.	17th	K.I.A	31 07 17	Menin Gate Memorial.
50173	Whinhan. G.O.	Felling, Durham.	Pte.	17th	K.I.A	31 07 17	Bedford House Cem End 4
51927	Whitehouse. R.A.	Seacombe, Cheshire.	Pte.	17th	K.I.A	31 07 17	Menin Gate Memorial.
270104	Wickstead. I.	Liverpool.	Pte.	18th	K.I.A	31 07 17	Menin Gate Memorial.
17749	Wilcock. C.	Liverpool.	Pte.	18th	K.I.A	31 07 17	Menin Gate Memorial.
235340	Wollaston. P.	Hexham, Suffolk.	Pte.	20th	K.I.A	31 07 17	Menin Gate Memorial.
42762	Wright. W.	Southport, Lancs.	Pte.	17th	K.I.A	31 07 17	Menin Gate Memorial.
42391	Young. W.H.	Liverpool.	Pte.	17th	K.I.A	31 07 17	Menin Gate Memorial.
49328	Anderson. T.	Manchester.	Pte.	18th	D.O.W	01 08 17	Mendinghem B.G,, Proven.
51960	Baker. G.	Eastham, Cheshire.	Pte.	17th	K.I.A	01 08 17	Hooge Crater Cem. Zillebeke.
59287	Baker. T.	Kirkdale, Liverpool.	Pte.	17th	K.I.A	01 08 17	Railway Dugout B.G Zillebeke
49138	Bull. H.	Macclesfield, Cheshire.	L/Cpl.	17th	K.I.A	01 08 17	Menin Gate Memorial.
22656	Connor. H.	Liverpool.	Pte.	20th	K.I.A	01 08 17	Tyne Cot Cem.
235221	Davies. H.	Manchester.	Pte.	19th	K.I.A	01 08 17	Menin Gate Memorial.
57020	Fulcher. W.E.	Croydon, Surrey.	Pte.	17th	K.I.A	01 08 17	Menin Gate Memorial.

22089	Gibson. L.	Widnes, Lancs.	Pte.	20th	K.I.A	01 08 17	Menin Gate Memorial.
33102	Gidney. F.	Liverpool.	Pte.	20th	K.I.A	01 08 17	Menin Gate Memorial.
31507	Hargreaves. H.	Liverpool.	L/Cpl.	17th	K.I.A	01 08 17	Menin Gate Memorial.
50098	Hayhurst. J.	Great Harwood, Lancs.	Pte.	20th	K.I.A	01 08 17	Menin Gate Memorial.
49081	Hazlewood. J.	Bootle, Liverpool.	Pte.	20th	K.I.A	01 08 17	Menin Gate Memorial.
13645	Hill. S.	Warrington, Cheshire.	Pte.	18th	D.O.W	01 08 17	Lijssenthoek M C Poperinghe
	Joseph. J.H.	South Wales.	Capt.	17th	K.I.A	01 08 17	Zantvoorde Brit Cem.
36040	Marshall. W.	Southport, Lancs.	Pte.	20th	K.I.A	01 08 17	Menin Gate Memorial.
41606	McCaffery. H.	Liverpool.	Pte.	17th	K.I.A	01 08 17	Zantvoorde Brit Cem.
27440	Rockcliffe. R.	Birkdale, Southport.	Pte.	18th	D.O.W	01 08 17	Lijssenthoek M C Poperinghe
51966	Sayle. CO.	Fleetwood, Lancs.	Cpl.	17th	K.I.A	01 08 17	Menin Gate Memorial.
58646	Seddon.T.E.	Liverpool.	Cpl.	20th	K.I.A	01 08 17	Bedford House Cem Encl 4
235253	Shaw. F.	Bingley, Lancs.	Pte.	18th	D.O.W	01 08 17	Lijssenthoek M C Poperinghe
	Simpson. J.W. M.C.	New Brighton, Cheshire.	2nd Lieut.	20th	K.I.A	01 08 17	Menin Gate Memorial.
22840	Simpson. P.T.	Liverpool.	Cpl.	20th	D.O.W	01 08 17	Lijssenthoek M C Poperinghe
57971	Winn. F.J.	London.	Pte.	17th	K.I.A	01 08 17	Menin Gate Memorial.
49056	Fell. F.	Manchester.	Pte.	20th	K.I.A	02 08 17	Menin Gate Memorial.
22397	Hatch. C.	Southport, Lancs.	Sgt.	20th	K.I.A	02 08 17	Menin Gate Memorial.
330854	Holmes. W.	Liverpool.	Pte.	19th	D.O.W	02 08 17	Etaples Mil Cem
50031	Hope. C.R.	Maidstone, Kent.	Pte.	20th	K.I.A	02 08 17	Menin Gate Memorial.
22116	Houghton. T.	Litherland, Liverpool.	Pte.	20th	K.I.A	02 08 17	Menin Gate Memorial.
57541	Irvine. D.J.	Dundee.	Pte.	20th	K.I.A	02 08 17	Menin Gate Memorial.
42689	Jackson. D.	Liverpool.	Pte.	17th	D.O.W	02 08 17	Brandhoek New Mil Cem.
53163	Lamb. H.	Little Totham, Essex.	L/Cpl.	20th	K.I.A	02 08 17	Tyne Cot Cem.
25788	Newall. F.C.	Poulton, Cheshire.	Sgt.	18th	D.O.W	02 08 17	Lijssenthoek M C Poperinghe
34416	O'Hagan. C.	Liverpool.	Pte.	20th	K.I.A	02 08 17	Menin Gate Memorial.
50627	Riding. J.	Lancaster.	Pte.	20th	K.I.A	02 08 17	Menin Gate Memorial.
50273	Sculley. T.	Blackburn, Lancs.	Pte.	20th	K.I.A	02 08 17	Menin Gate Memorial.
57463	Smith. A.	Glasgow.	Pte.	20th	K.I.A	02 08 17	Menin Gate Memorial.
22850	Sturrock. W.	Crosby, Liverpool.	A/Sgt.	20th	K.I.A	02 08 17	Menin Gate Memorial.
52927	Vinall. T.A.	Tunbridge, Kent.	Pte.	17th	D.O.W	02 08 17	Lijssenthoek M C Poperinghe
235337	Whitehead. H.	Oldham, Lancs.	Pte.	20th	K.I.A	02 08 17	Menin Gate Memorial.
52058	Aston. G.	Liverpool.	Pte.	19th	K.I.A	03 08 17	Menin Gate Memorial.
50181	Bagley. D.	Smethwick, Staffs.	Pte.	19th	K.I.A	03 08 17	Menin Gate Memorial.
47440	Blakeman. J.	Liverpool.	Pte.	19th	K.I.A	03 08 17	Menin Gate Memorial.
235304	Gregory. W.S.	Blackpool, Lancs.	Pte.	20th	D.O.W	03 08 17	Mendinghem B.C., Proven.
335227	Grossett. A.B.	Chorlton-o-M, Manchester.	Pte.	19th	K.I.A	03 08 17	Menin Gate Memorial.
57591	Horan. M.L.	Dundee.	Pte.	19th	K.I.A	03 08 17	Menin Gate Memorial.
50138	Howden. C.F.	Sunderland.	Pte.	17th	D.O.W	03 08 17	Mendinghem B.C., Proven.
36802	Lowens. J.	Liverpool.	Pte.	20th	K.I.A	03 08 17	Menin Gate Memorial.
50636	Salthouse. G.A.	Out Rawcliffe, Lancs.	Pte.	20th	K.I.A	03 08 17	Hooge Crater Cem, Zillebeke.
50210	Sketchley. H.	Hinckley, Leics.	Pte.	19th	D.O.W	03 08 17	Mendinghem B.C., Proven.
35865	Bailey. T.	Manchester.	Pte.	17th	K.I.A	04 08 17	Menin Gate Memorial.
57995	Behr. W.E.	Camberwell, Surrey.	Pte.	17th	K.I.A	04 08 17	Menin Gate Memorial.
51676	Broster. E.H.	Heswell, Cheshire.	Pte.	17th	D.O.W	04 08 17	Huts Cem, Dickebusch.
42606	Brown. G.	Liverpool.	Pte.	17th	K.I.A	04 08 17	Menin Gate Memorial.
51898	Carlisle. J.	Liverpool.	Pte.	17th	K.I.A	04 08 17	Menin Gate Memorial.
52097	Chesters. T.H.	Chorlton o M, Manchester	Pte.	17th	K.I.A	04 08 17	Menin Gate Memorial.
50228	Cottam. E.	Blackburn, Lancs.	Pte.	17th	K.I.A	04 08 17	Zantvoorde Brit Cem.
15598	Davies. E.H.	Liverpool.	L/Cpl.	17th	K.I.A	04 08 17	Menin Gate Memorial.
50233	Dixon. J.	Blackburn, Lancs.	Pte.	17th	K.I.A	04 08 17	Menin Gate Memorial.
58015	Francis. H.S.	Hoddesdon, Herts.	Pte.	17th	K.I.A	04 08 17	Menin Gate Memorial.
48219	Freestone. C.	Liverpool.	Pte.	17th	K.I.A	04 08 17	Menin Gate Memorial.
38789	Gaskell. T.	Seaforth, Liverpool.	Pte.	17th	K.I.A	04 08 17	Hagle Dump Cem, Elveringhe.
58017	Gentle. S.G.	Tottenham, London.	Pte.	17th	K.I.A	04 08 17	Menin Gate Memorial.
29623	Harvey. R.H.	Woolton, Liverpool.	Pte.	17th	K.I.A	04 08 17	Poelcapelle Brit Cem.
54280	Hetherington. T.	Liverpool.	Pte.	17th	K.I.A	04 08 17	Menin Gate Memorial.
50137	Hogg. J.	Lesbury, Northumberland.	Pte.	17th	K.I.A	04 08 17	Ypres Reservoir Cem.

42974	Jackson. J.	Bootle, Liverpool.	Pte.	17th	K.I.A	04 08 17	Hagle Dump Cem, Elveringhe.
57445	Mair. C.	Glasgow.	Pte.	17th	K.I.A	04 08 17	Menin Gate Memorial.
41821	Marsden. T.	Scarisbrick, Lancs.	L/Cpl.	17th	K.I.A	04 08 17	Menin Gate Memorial.
35696	Martin. J.A.	Liverpool.	Pte.	18th	D.O.W	04 08 17	Brandhoek New Mil Cem.
269882	McCabe. P.	Seaforth, Lancs.	L/Cpl.	17th	K.I.A	04 08 17	Menin Gate Memorial.
48423	McDonald. A.	Liverpool.	L/Cpl.	20th	Died.	04 08 17	Reninghelst New Mil Cem
49047	McLaren. J.	Accrington, Lancs.	Pte.	17th	K.I.A	04 08 17	Menin Gate Memorial.
49106	Murray. E.	Stockport, Cheshire.	Pte.	17th	K.I.A	04 08 17	Menin Gate Memorial.
50178	Neilson. J.A.	South Shields.	Pte.	17th	K.I.A	04 08 17	Menin Gate Memorial.
51653	Neligan. W.D.	Liverpool.	Pte.	17th	K.I.A	04 08 17	Menin Gate Memorial.
15340	Ollason. A.	Liverpool.	Cpl.	17th	K.I.A	04 08 17	Menin Gate Memorial.
41843	Robinson. T.	Liverpool.	L/Cpl.	17th	K.I.A	04 08 17	Menin Gate Memorial.
202266	Simmons. J.A.	Fairfield, Liverpool.	Pte.	20th	Died.	04 08 17	Bedford House Cem Encl 2
41806	Spencer. G.	Liverpool.	Pte.	17th	K.I.A	04 08 17	Zantvoorde Brit Cem.
32751	Travis. F.	Oldham, Lancs.	L/Cpl.	17th	K.I.A	04 08 17	Hooge Crater Cem, Zillebeke.
42688	Tweedy. G.	Liverpool.	Pte.	20th	D.O.W	04 08 17	Lijssenthoek M C Poperinghe
204777	Johnston. G.	Cockermouth, Cumberland.	Pte.	19th	D.O.W	05 08 17	Boulogne East Cem.
202865	Swift. R.	Liverpool.	Pte.	19th	D.O.W	05 08 17	Brandhoek New Mil Cem.
50174	Wilkinson. J.	Washington, Co.Durham.	Pte.	17th	D.O.W	05 08 17	Lijssenthoek M C Poperinghe
235111	Clements. R.	Dover, Kent.	Pte.	20th	D.O.W	07 08 17	Brandhoek New Mil Cem.
52905	Harrison. A.	Bootle, Liverpool.	Pte.	19th	D.O.W	08 08 17	Mendingham B.C., Proven.
23513	Mobbs. H.	Norwich, Norfolk.	Pte.	20th	D.O.W	13 08 17	Boulogne East Cem.
16529	King. R.	Liverpool.	Pte.	18th	D.O.W	16 08 17	St Sever Cem Ext, Rouen.
49203	Hargreaves. A.	Burnley, Lancs.	Pte.	20th	D.O.W	17 08 17	Etaples Mil Cem
33155	Wilkinson. F.	Liverpool.	Pte.	18th	Died	17 08 17	Vlamertinghe Mil Cem
16260	Parr. G.H.	Liverpool.	Pte.	18th	D.O.W	22 08 17	Anfield Cem, Liverpool.
16289	Brough. G.	Warrington, Cheshire.	Pte.	18th	K.I.A	25 08 17	Derry House Cem No 2, Whytschaete.
24765	Burrows. W.	Ellesmere Port, Cheshire	Pte.	18th	K.I.A	25 08 17	Derry House Cem No 2, Whytschaete.
16466	Lindsay. T.	Liverpool.	Cpl.	18th	K.I.A	25 08 17	Derry House Cem No 2, Whytschaete.
52437	Ridgeway. W.	Salford, Manchester.	Pte.	18th	K.I.A	25 08 17	Derry House Cem No 2, Whytschaete.
16486	Spencer. H.	Derby.	Sgt.	18th	D.O.W	25 08 17	Huts Cem, Dikebusch
	Averill. T.H.	Great Witley, Worcs.	2nd Lieut.	17th	K.I.A	30 08 17	Dranoutre Mil Cem.
57673	Johnson. A.J.	Nottingham.	L/Sgt.	17th	K.I.A	30 08 17	Dranoutre Mil Cem.
22868	Ennis. W. (Turton)	Liverpool.	Pte.	18th	K.I.A	03 09 17	Torreken Farm Cem No1 Whytschaete.
57563	Pairman. A.A.	Dundee	Pte.	18th	D.O.W	03 09 17	Montrose,(Rosehill)Scotland.
265312	Smith. J.	Bootle, Liverpool.	L/Cpl.	18th	K.I.A	03 09 17	Torreken Farm Cem No1 Whytschaete.
25798	Elton. F.J.	Liscard, Cheshire	L/Cpl.	18th	K.I.A	04 09 17	Neuville-Vitasse Rd Cem.
203305	Lamb. J.	Burscough, Lancs.	Pte.	18th	D.O.W	04 09 17	Bailleul CC Ext.
24603	Richardson. R.	Liverpool.	Pte.	18th	K.I.A	04 09 17	Neuville-Vitasse Rd Cem.
52929	Smith. J.	Bolton, Lancs.	Pte.	17th	EXECUTED	05 09 17	Kemmel Chateau Mil Cem
35736	French. H.	Litherland, Liverpool.	Pte.	20th	D.O.W	06 09 17	Bailleul CC Ext.
235302	Galley. G.	Norwich.	Pte.	20th	K.I.A	06 09 17	Voormezeele Cem Encl No.3
49581	Gallaghan. J.	Seaforth, Liverpool	Pte.	20th	K.I.A	08 09 17	Torreken Farm Cem No1 Whytschaete.
38824	Loughran. H.A.	Liverpool.	Pte.	20th	K.I.A	08 09 17	Torreken Farm Cem No1 Whytschaete.
23109	McLaren. E. M.M.	Liverpool.	L/Cpl.	20th	K.I.A	08 09 17	Torreken Farm Cem No1 Whytschaete.
54137	Murray. R.	Liverpool.	Pte.	20th	K.I.A	08 09 17	Torreken Farm Cem No1 Whytschaete.
200827	Silvey. R.M.	Liverpool.	Cpl.	20th	D.O.W	10 09 17	Outtersteene CC Ext, Bailleul.
15236	Terry. D.	Oxton, Birkenhead, Ches.	Pte.	17th	D.O.W	11 09 17	Birkenhead, Flaybrick Cem.
12067	Dandy. J.	Aintree, Liverpool.	L/Cpl.	19th	K.I.A	12 09 17	Tyne Cot Memorial.

25315	Boyes. W.E.	Liverpool.	Pte.	20th	D.O.W	15 09 17	Voormezeele Cem Encl No.1 & 2
27360	Fell. H.	Birkenhead, Cheshire.	Pte.	17th	K.I.A	16 09 17	Kemmel Chateau Mil Cem
57943	Phillips. A.	Missenden, Bucks.	Pte.	17th	K.I.A	18 09 17	Kemmel Chateau Mil Cem
52142	Armstrong. R.	Manchester.	Pte.	19th	K.I.A	20 09 17	Tyne Cot Memorial.
50591	Birchall. W.	Earlstown, Lancs.	Pte.	19th	K.I.A	20 09 17	Tyne Cot Memorial.
50193	Fotherby. J.H.	Bakewell, Derbys.	Pte.	19th	K.I.A	20 09 17	Tyne Cot Memorial.
265006	Gilduff. J.	Liverpool.	Cpl.	19th	K.I.A	20 09 17	Bridge House Cem, Langemarck.
57377	Hodge. D.R.	Muirkirk, Ayr.	Sgt.	19th	K.I.A	20 09 17	Tyne Cot Memorial.
57421	Hunter. J.F.	Paisley, Renfrew.	Pte.	19th	K.I.A	20 09 17	Tyne Cot Memorial.
	Laird. C.	Chester.	Capt.	19th	K.I.A	20 09 17	Tyne Cot Memorial.
17888	Madeley. B.	Warrington, Cheshire.	Pte.	19th	K.I.A	20 09 17	Tyne Cot Memorial.
27377	Vaughan. D.W.	Liverpool.	L/Cpl.	19th	K.I.A	20 09 17	Tyne Cot Memorial.
235281	Bitten. C.H.	Pidley, Hunts	Pte.	18th	K.I.A	21 09 17	Torreken Farm Cem No1 Whytschaete.
300040	Boon. J.	Bury, Lancs.	Pte.	18th	K.I.A	21 09 17	Torreken Farm Cem No1 Whytschaete.
207277	Clifford. H.	Liverpool.	Pte.	18th	K.I.A	21 09 17	Torreken Farm Cem No1 Whytschaete.
57859	Dodd. H.A.	Newcastle on Tyne.	Pte.	18th	K.I.A	21 09 17	Torreken Farm Cem No1 Whytschaete.
29693	Lyon. J.	Prescot, Lancs.	Pte.	18th	K.I.A	21 09 17	Torreken Farm Cem No1 Whytschaete.
300042	Reed. W.H.	Liverpool.	Pte.	18th	K.I.A	21 09 17	Torreken Farm Cem No1 Whytschaete.
57694	Brearley. B.S.	Nottingham.	Pte.	18th	K.I.A	22 09 17	Torreken Farm Cem No1 Whytschaete.
202830	Conway. J. (McCabe)	Liverpool.	Pte.	18th	K.I.A	22 09 17	Torreken Farm Cem No1 Whytschaete.
57784	Elliot. W. M.M.	Bebside, Northumberland	Pte.	18th	K.I.A	22 09 17	Torreken Farm Cem No1 Whytschaete.
203094	Fearson. G.	Liverpool.	Pte.	18th	K.I.A	22 09 17	Torreken Farm Cem No1 Whytschaete.
58710	Johnson. H.H.	Shaw, Lancs.	Pte.	20th	K.I.A	22 09 17	Torreken Farm Cem No1 Whytschaete.
17076	Malcolm. D.	Liverpool.	Sgt.	18th	K.I.A	22 09 17	Torreken Farm Cem No1 Whytschaete.
44339	Perry. H.J.	Aberbargoed, Glamorgan	Pte.	18th	D.O.W	22 09 17	Pond Farm Cem, Wulverghem
203179	Williams. N.	Bootle, Liverpool.	Pte.	18th	K.I.A	22 09 17	Torreken Farm Cem No1 Whytschaete.
42757	Davies. R.	Liverpool.	Pte.	18th	K.I.A	23 09 17	Torreken Farm Cem No1 Whytschaete.
34538	Lee. P.	Tankardstown, Co.Wicklow	Pte.	18th	D.O.W	23 09 17	Outtersteene CC Ext, Bailleul.
16593	McElnea. R.H.	Liverpool.	Pte.	18th	D.O.W	23 09 17	Outtersteene CC Ext, Bailleul.
	Riddell. D.M.	Belfast, Ireland.	2nd Lieut.	17th	Died.	23 09 17	Belfast City Cem, Ireland.
17070	Lucas. A.F.	Liverpool.	Cpl.	18th	D.O.W	25 09 17	Mont Huon Cem, Le Treport.
21729	Wells. F.S.	Liverpool.	Pte.	18th	D.O.W	25 09 17	Longuenesse Cem, St.Omer.
46129	Conroy. B.	Liverpool.	Pte.	19th	K.I.A	26 09 17	Oosttaverne Wood Cem.Whytschaete.
16271	Roberts. W.E.	Liverpool.	Pte.	19th	K.I.A	26 09 17	Oosttaverne Wood Cem.Whytschaete.
57820	Osborne. H.	Sunderland.	Pte.	19th	D.O.W	27 09 17	Trois-Arbres Cem, Steenwerck.
17408	Lloyd. G.R.	Wallasey, Cheshire.	Pte.	19th	D.O.W	28 09 17	Trois-Arbres Cem, Steenwerck.
	Chaning-Pearce.W.T. (MC)	Ramsgate, Kent.	Capt.	18th RAMC.	K.I.A	01 10 17	Derry House Cem No. 2, Whytschaete.
300035	Jones. J.	Anfield, Liverpool.	Pte.	18th	K.I.A	01 10 17	Tyne Cot Memorial.
24682	Wooldridge. J.H.	Liverpool.	Pte.	18th	K.I.A	01 10 17	Derry House Cem No 2, Whytschaete.
300412	Wyman. A.	Widnes, Lancs.	Pte.	18th	K.I.A	01 10 17	Tyne Cot Memorial.
57466	Innes. A.	Fraserburgh, Aberdeen	L/Cpl.	18th	D.O.W	03 10 17	Torreken Farm Cem No1 Whytschaete.
59557	Jones. J.T.	Liverpool.	Pte.	18th	D.O.W	03 10 17	Etaples Mil Cem

300029	Craddock. G.R.	Clapham, London	Pte.	18th	D.O.W	04 10 17	Bailleul CC Ext.
331166	Aird. J.M.	Edge Hill, Liverpool.	Pte.	18th	K.I.A	06 10 17	Tyne Cot Memorial.
300190	Bergin. T.	Ashton-in-Makerfield, Lancs.	Pte.	18th	K.I.A	06 10 17	Tyne Cot Memorial.
300306	Lowe. W.J.	Rainhill, Lancs.	Sgt.	18th	K.I.A	06 10 17	Tyne Cot Memorial.
300409	Worthington. T.E.	Warrington, Cheshire.	Cpl.	18th	K.I.A	06 10 17	Tyne Cot Memorial.
300482	Harrison. R.	Aigburth, Liverpool.	Pte.	18th	K.I.A	07 10 17	Tyne Cot Memorial.
300283	Hayes. J.	Earlstown, Lancs.	Sgt.	18th	K.I.A	07 10 17	Tyne Cot Memorial.
300518	Webster. V.	Wigan, Lancs.	Pte.	18th	K.I.A	07 10 17	Tyne Cot Memorial.
52101	Davies. P.	Bristol.	Pte.	19th	K.I.A	08 10 17	Tyne Cot Memorial.
300253	Gilvray. F.	St.Helens, Lancs.	Sgt.	18th	D.O.W	08 10 17	St Sever Cem Ext, Rouen.
300038	Wilson. J.	Penrith, Cumberland,	Pte.	18th	K.I.A	08 10 17	Tyne Cot Memorial.
235201	Arrowsmith. H.	Droylsden, Manchester.	Pte.	19th	K.I.A	09 10 17	Tyne Cot Memorial.
47735	Sweeney. C.	Liverpool.	Pte.	18th	D.O.W	09 10 17	St Sever Cem Ext, Rouen.
58663	Adams. C.	Hyde, Cheshire.	Sgt.	20th	D.O.W	11 10 17	Greenwich Cem, London.
49038	Yardley. E.	Manchester.	L/Cpl.	17th	K.I.A	6 10 17	Kemmel Chateau Mil Cem
52899	Cregeen. E.	Castletown, I.O.M.	Pte.	19th	K.I.A	8 10 17	Tyne Cot Memorial.
17371	Hughes. H.L	Birkenhead, Cheshire.	Pte.	19th	K.I.A	8 10 17	Tyne Cot Memorial.
25941	Leevers. F.	Southport, Lancs.	Pte.	19th	K.I.A	8 10 17	Tyne Cot Memorial.
50697	Nuttall. J.	Middleton, Lancs.	Pte.	17th	D.O.W	8 10 17	Outtersteene CC Ext, Bailleul.
49197	Dugdale. R.W.	Great Harwood, Lancs.	L/Cpl.	20th	D.O.W	19 10 17	Kemmel Chateau Mil Cem
32328	Stewart. G.A.	Liverpool.	Pte.	18th	D.O.W	19 10 17	St Sever Cem Ext, Rouen.
235264	Ward. E.M.	Darlington.	Pte.	17th	D.O.W	20 10 17	Trois-Arbres Cem, Steenwerck.
	Draper. A.I.	Bebington, Cheshire.	Major.	17th	K.I.A	21 10 17	Kemmel Chateau Mil Cem
17690	Phipps. H.	Liverpool.	Pte.	19th	D.O.W	24 10 17	Bailleul CC Ext.
202001	Thomas. F.A.	Bootle, Liverpool.	Pte.	19th	D.O.W	25 10 17	Bailleul CC Ext.
202839	Sanderson. H.J.	Liverpool.	Pte.	19th	K.I.A	27 10 17	Kemmel Chateau Mil Cem
21428	Rouse. F.	Liverpool.	Pte.	19th	D.O.W	29 10 17	Dozingham M C, Westvleteren
34410	Clarkson. T.	Liverpool.	Pte.	19th	D.O.W	31 10 17	Outtersteene CC Ext, Bailleul.
72265	Smith. J.	Mossley, Lancs.	Pte.	20th	K.I.A	31 10 17	Kemmel Chateau Mil Cem
203034	Godding. A.J.	Liverpool.	Pte.	19th	K.I.A	04 11 17	Tyne Cot Memorial.
51744	Greenwood. E.	Burnley, Lancs.	L/Cpl.	20th	K.I.A	04 11 17	Tyne Cot Memorial.
63809	Lane. R.	Newcastle-on-Tyne.	Pte.	17th	D.O.W	05 11 17	Outtersteene CC Ext, Bailleul.
46697	Hancock. T.C.	Liverpool.	Pte.	17th	K.I.A	07 11 17	Kemmel Chateau Mil Cem
86440	Nicholls. A.E.	Ebrington, Glos.	Pte.	17th	K.I.A	07 11 17	Kemmel Chateau Mil Cem
17992	Walberg. R.S.	Liverpool.	L/Cpl.	19th	Died	07 11 17	Bailleul CC Ext.
51659	Williams. R.D.	Liverpool.	Pte.	17th	K.I.A	07 11 17	Kemmel Chateau Mil Cem
24359	Ledger. A.	Liverpool.	Pte.	20th	D.O.W	08 11 17	Toxteth Park Cem, Liverpool.
86364	Tomlinson. L.D.	Leeds.	Pte.	17th	K.I.A	08 11 17	Kemmel Chateau Mil Cem
17315	Dowling. E.H.	Bristol.	Sgt.	19th	D.O.W	11 11 17	Bailleul CC Ext.
46416	Dangerfield. T.	Liverpool.	Pte.	19th	D.O.W	12 11 17	Lijssenthoek M C Poperinghe
	Barlow. P.	Manchester.	2nd Lieut	19th	K.I.A	15 11 17	Ypres Reservoir Cem.
17286	Blackstone. L.W.	Liverpool.	Sgt.	19th	K.I.A	16 11 17	Tyne Cot Memorial.
22407	Hill. E.	Liverpool.	Sgt.	20th	Died.	16 11 17	Llandudno/Llanros,(St E & M)
86428	Copp. H.	L. Broughton, Manchester.	Pte.	17th	K.I.A	17 11 17	Kemmel Chateau Mil Cem
50286	Walkden. J.	Darwen, Lancs.	Pte.	20th	Died.	19 11 17	Wimereux CC
55689	Williams. J.B.	Liverpool.	Pte.	18th	D.O.W	20 11 17	Bootle Cem, Liverpool.
50189	Dewhurst. R.T.	Blackburn, Lancs.	Pte.	19th	D.O.W	22 11 17	Nine Elms Brit Cem, Poperinghe
26007	Shone. J.	Liverpool.	Pte.	18th	D.O.W	28 11 17	Outtersteene CC Ext, Bailleul.
27333	Clarke. W.G.F.	Liverpool.	Pte.	17th	K.I.A	03 12 17	Tyne Cot Memorial.
269183	Howard. H.	Blackburn, Lancs.	Pte.	19th	K.I.A	03 12 17	Tyne Cot Memorial.
38868	Skillicorn. S.	Douglas, I.O.M.	Pte.	17th	K.I.A	03 12 17	Bedford House Cem Encl 4
80807	Barker. J.	Salford, Manchester.	Pte.	17th	K.I.A	04 12 17	Bedford House Cem Encl 4
300522	Baron. J.G.	St.Helens, Lancs.	Sgt.	18th	K.I.A	04 12 17	Hooge Crater Cem, Zillebeke.
16131	Carroll. H.E.	Egremont, Cheshire.	L/Cpl.	17th	K.I.A	04 12 17	Bedford House Cem Encl 4
35227	Greenwood. C.	Todmorden, Yorks.	Pte.	17th	K.I.A	04 12 17	Tyne Cot Memorial.
57718	Haywood. J.G.	Newark, Notts.	Pte.	18th	K.I.A	04 12 17	Hooge Crater Cem, Zillebeke.
202836	Hennesy. T.J.	Liverpool.	Pte.	18th	K.I.A	04 12 17	Hooge Crater Cem, Zillebeke.
300388	Palfrey. C.	St.Helens, Lancs.	Pte.	18th	K.I.A	04 12 17	Hooge Crater Cem, Zillebeke.

26617	Roberts. E.E.	Liverpool.	Pte.	18th	K.I.A	04 12 17	Hooge Crater Cem, Zillebeke.
300413	Walsh. P.	Widnes, Lancs.	Pte.	18th	K.I.A	04 12 17	Hooge Crater Cem, Zillebeke.
17536	Baxter. J.H.	Eastham, Cheshire.	Pte.	19th	K.I.A	05 12 17	Hooge Crater Cem, Zillebeke.
52198	Smith. C.	Hulme, Manchester.	Pte.	19th	K.I.A	05 12 17	Tyne Cot Memorial.
203219	Swinburn. J.	Liverpool.	Pte.	19th	K.I.A	05 12 17	Hooge Crater Cem, Zillebeke.
300182	Baker. J.	St.Helens, Lancs.	Pte.	18th	K.I.A	06 12 17	Bedford House Cem Encl 4
15776	Evans. A.L.	Pendre, Flint.	Pte.	17th	D.O.W	06 12 17	Bailleul CC Ext.
300177	Atherton. P.	St.Helens, Lancs.	Pte.	18th	K.I.A	07 12 17	Tyne Cot Memorial.
58565	Moore. T.	Seacombe, Cheshire.	Pte.	18th	K.I.A	07 12 17	Hooge Crater Cem, Zillebeke.
49253	Moss. T.	Carnforth. Lancs.	Pte.	18th	K.I.A	07 12 17	Tyne Cot Memorial.
52178	Thompson. W.	Manchester.	Pte.	19th	D.O.W	07 12 17	Bailleul CC Ext.
17254	Moore. A.J.	Liverpool.	Cpl.	18th	K.I.A	08 12 17	Bedford House Cem End 4
63587	Sulivan. J.P.	Patricroft, Manchester.	Pte.	17th	K.I.A	08 12 17	Bedford House Cem End 4
52158	Sutcliffe. J.	Salford, Manchester.	L/Cpl.	19th	D.O.W	08 12 17	Outtersteene CC Ext, Bailleul.
51725	Mitchell. S.P.	Liverpool.	Pte.	20th	D.O.W	12 12 17	Wimereux CC
300365	Silcock. R.	Downall Green, Lancs.	Pte.	18th	D.O.W	12 12 17	Outtersteene CC Ext, Bailleul.
15647	Keiffer. A.H.	Liverpool.	L/Cpl.	17th	K.I.A	15 12 17	Tyne Cot Memorial.
50595	Beament. A.W.	Broadstone, Dorset.	Pte.	19th	K.I.A	19 12 17	Railway Dugout B.G Zillebeke
48113	Cliffe. O.	Liverpool.	Pte.	19th	K.I.A	19 12 17	Tyne Cot Memorial.
300470	Cuddy. A.G.	Liverpool.	Pte.	18th	D.O.W	19 12 17	Lijssenthoek M C Poperinghe
50057	Goodman. T.	Paddington, London.	Pte.	19th	K.I.A	19 12 17	Railway Dugout B.G Zillebeke
202819	Martindale. G.D.	Liverpool.	Pte.	19th	K.I.A	19 12 17	Railway Dugout B.G Zillebeke
50266	Spiller. W.H.	Burnley, Lancs.	Pte.	20th	K.I.A	19 12 17	Railway Dugout B.G Zillebeke
52138	Bushell. H.H.	Manchester.	L/Sgt.	19th	K.I.A	20 12 17	Tyne Cot Memorial.
35103	Ellison. J.	Woolton, Liverpool.	Pte.	17th	K.I.A	20 12 17	Hooge Crater Cem, Zillebeke.
21247	Hughes. J.E.	Wigan, Lancs.	Pte.	19th	K.I.A	20 12 17	Tyne Cot Memorial.
52009	Leach. J.H.M.	Stockport, Cheshire.	Pte.	17th	K.I.A	20 12 17	Hooge Crater Cem, Zillebeke.
57398	Caldwell. A.A.	Glasgow	Pte.	18th	K.I.A	21 12 17	Bedford House Cem End 4
52909	Smith. E.F.	Bootle, Liverpool.	Pte.	19th	D.O.W	21 12 17	Menin Road South, Mil Cem.
50262	Milsom.J.	Haslingden, Lancs.	Pte.	20th	D.O.W	23 12 17	Hooge Crater Cem, Zillebeke.
235305	Hine. G.E.	London.	Pte.	20th	D.O.W	24 12 17	Bailleul CC Ext.
26438	Dukes. E.	Birmingham.	Pte.	17th	K.I.A	31 12 17	Hooge Crater Cem, Zillebeke.
34646	Hunter. H.	Fazackerley, Liverpool.	Pte.	17th	D.O.W	31 12 17	Outtersteene CC Ext, Bailleul.
50683	Williams. W.	Liverpool.	Pte.	17th	D.O.W	31 12 17	Bailleul CC Ext.
300199	Bickley. H.G.	Birkenhead, Cheshire.	Pte.	18th	K.I.A	02 01 18	Hooge Crater Cem, Zillebeke.
300010	Hilton. S.	Ashton-in-Makerfield, Lancs.	Pte.	18th	K.I.A	02 01 18	Hooge Crater Cem, Zillebeke.
56928	Rigby. W.	Southport, Lancs.	Pte.	18th	K.I.A	02 01 18	Hooge Crater Cem, Zillebeke.
300359	Seddon. J.	St.Helens, Lancs.	L/Cpl.	18th	K.I.A	02 01 18	Hooge Crater Cem, Zillebeke.
59569	Colville. A.	Liverpool.	Pte.	20th	K.I.A	03 01 18	Tyne Cot Memorial.
33103	Cottier. W.D.	German, I.O.M.	Pte.	18th	D.O.W	03 01 18	Outtersteene CC Ext, Bailleul.
57575	Suttie. H.	Dunshalt, Fife.	Pte.	20th	K.I.A	03 01 18	Railway Dugout B.G Zillebeke
	Moore. E.N. M.C.	St.Johns Wood, London.	Rev RACD	20th	K.I.A	05 01 18	Railway Dugout B.G Zillebeke
14928	Wright. H.	Liverpool.	L/Cpl.	20th	K.I.A	05 01 18	Tyne Cot Memorial.
50289	Shorrock. R.	Darwen, Lancs.	Pte.	20th	D.O.W	06 01 18	Outtersteene CC Ext, Bailleul.
57755	Trewhett. R.C.	Sunderland.	Pte.	20th	K.I.A	06 01 18	Tyne Cot Memorial.
266894	Hatton. W.	Southport, Lancs.	Pte.	19th	D.O.W	08 01 18	Lijssenthoek M C Poperinghe
13452	Roberts. A.H.	Liverpool.	Pte.	18th	K.I.A	09 01 18	Ferme Olivier Cem,Elverdinghe
235328	Shrimpston. H.A.	Berkhamstead, Herts.	Pte.	20th	D.O.W	13 01 18	Outtersteene CC Ext, Bailleul.
15618	Jardine. D.	Liverpool.	Pte.	17th	Drowned	14 01 18	Cerisy-Gailly Mil Cem.
34298	Armstrong. S.	Carlisle.	L/Sgt.	20th	Died	27 01 18	Longuenesse Cem, St.Omer.
50144	Lambert. G.W.	Cockfield, Durham.	Pte.	17th	Died	30 01 18	Ham British Cem.
52669	Millett. J.	Oldham, Lancs.	A/L/Cpl	17th	K.I.A	05 03 18	Grand Seraucourt Brit Cem
16800	Donal. R.	Edge Hill, Liverpool.	Pte.	18th	Died.	07 03 18	Toxteth Park Cem, Liverpool.
406053	Roberts. A.E.	Rochdale.	Pte.	18th	Died	09 03 18	Healey,(Christchurch)Rochdale
325219	Houghton. W.T.	Upper Holloway, London.	Pte.	17th	Died.	12 03 18	Ham British Cem.
240953	Hughes. A.	Liverpool.	L/Cpl.	17th	K.I.A	17 03 18	Chapelle Brit Cem, Holnon.
30295	Doyle. J.	Liverpool.	Pte.	19th	K.I.A	22 03 18	Pozieres Memorial.
235119	Charles. W.W.	London.	Pte.	17th	Died	21 03 18	Pozieres Memorial.

406581	Connell. J.	Manchester	Pte.	18th	K.I.A	21 03 18	Pozieres Memorial.
300216	Critchley. G.	St.Helens, Lancs.	Pte.	18th	K.I.A	21 03 18	Pozieres Memorial.
300026	Frame. J.	Henley, Oxford	Pte.	18th	K.I.A	21 03 18	Roye New Brit Cem
300434	Guile. H.W.	Liverpool.	Pte.	18th	D.O.W	21 03 18	Ham British Cem.
57388	Alexander. D	Glasgow.	Pte.	19th	K.I.A	22 03 18	Pozieres Memorial.
	Ashcroft. W.	Birkenhead, Cheshire.	Lieut.	19th	K.I.A	22 03 18	Savy Brit Cem.
	Barnes. A.	St.Helens, Lancs.	2nd Lieut.	17th	K.I.A	22 03 18	Pozieres Memorial.
17276	Barnes. W.H. M.M.	Liverpool.	Sgt.	19th	K.I.A	22 03 18	Pozieres Memorial.
203049	Bishop. C.H.	Kendal.	Pte.	19th	K.I.A	22 03 18	Savy Brit Cem.
	Booth. A.W.	Cheshire.	2nd Lieut	19th	K.I.A	22 03 18	Pozieres Memorial.
13587	Burns. E.	Kirkdale, Liverpool.	Pte.	19th	K.I.A	22 03 18	Pozieres Memorial.
17290	Clayton. J.T.	Liverpool.	L/Cpl.	19th	K.I.A	22 03 18	Pozieres Memorial.
241890	Conlan. R.	Manchester.	Pte.	19th	K.I.A	22 03 18	Pozieres Memorial.
57495	Cowie. A.E.	Grimsby, Yorks.	Cpl.	19th	K.I.A	22 03 18	Pozieres Memorial.
20755	Crank. W.	Liverpool.	Pte.	19th	K.I.A	22 03 18	Pozieres Memorial.
42657	Douglas. J.	Liverpool.	Pte.	17th	K.I.A	22 03 18	Chapelle Brit Cem, Holnon.
53741	Driver. R.	Colne, Lancs.	Pte.	18th	K.I.A	22 03 18	Pozieres Memorial.
57531	Fairfull. R.	Fife.	Pte.	19th	K.I.A	22 03 18	Pozieres Memorial.
300249	Farquhar. N.H.	Tuebrook, Liverpool.	Pte.	18th	K.I.A	22 03 18	Pozieres Memorial.
11243	Fisher. W.	Liverpool.	Cpl.	19th	K.I.A	22 03 18	Pozieres Memorial.
57408	Foster. A.G.	Glasgow.	Pte.	19th	K.I.A	22 03 18	Pozieres Memorial.
12733	Fowler. J.	Liverpool.	Pte.	17th	D.O.W	22 03 18	Ham British Cem.
22380	Frodsham. C.	Liverpool.	L/Cpl.	19th	K.I.A	22 03 18	Pozieres Memorial.
29228	Griffiths. J.M.	Liverpool.	Pte.	19th	K.I.A	22 03 18	Savy Brit Cem.
57418	Henderson. J.	Glasgow.	Pte.	19th	K.I.A	22 03 18	Pozieres Memorial.
30839	Highcock. A.E.	Liverpool.	L/Cpl.	19th	K.I.A	22 03 18	Pozieres Memorial.
22409	Hindley. W.	Liverpool.	Cpl.	19th	K.I.A	22 03 18	Grand Seraucourt Brit Cem
53190	Hooper. LW.	Staines, Middlesex.	Pte.	19th	K.I.A	22 03 18	Pozieres Memorial.
17357	Hughes. O.	Bootle, Liverpool.	L/Cpl.	19th	K.I.A	22 03 18	Savy Brit Cem.
51685	Jacobson. S.	Liverpool.	Pte.	19th	K.I.A	22 03 18	Pozieres Memorial.
50602	James. E.	Ealing, Middlesex.	Pte.	19th	K.I.A	22 03 18	Pozieres Memorial.
235307	Jinman. E.	Eastbourne, Sussex.	Pte.	19th	K.I.A	22 03 18	Savy Brit Cem.
17387	Jones. J.H.	Liverpool.	Pte.	19th	K.I.A	22 03 18	Savy Brit Cem.
330175	Kenny. J.	Ormskirk, Lancs.	Pte.	19th	K.I.A	22 03 18	Pozieres Memorial.
21850	Kewley. J.H.	Liverpool.	Pte.	19th	K.I.A	22 03 18	Pozieres Memorial.
	Lawson. J.	Waterloo, Liverpool.	Capt.	18th	K.I.A	22 03 18	Pozieres Memorial.
52242	Under. P.	Manchester.	Pte.	19th	K.I.A	22 03 18	Savy Brit Cem.
32696	Lloyd. G.F.	Liverpool.	Pte.	19th	K.I.A	22 03 18	Savy Brit Cem.
203512	Lowe. G.J.H.	Litherland, Liverpool.	Pte.	19th	K.I.A	22 03 18	Pozieres Memorial.
57428	Luke. F.A.	Dunoon, Argyll.	Pte.	19th	K.I.A	22 03 18	Pozieres Memorial.
23098	Martin. J.	Liverpool.	Pte.	19th	K.I.A	22 03 18	Pozieres Memorial.
28345	Metcalfe. W.	Port Talbot, Pembroke.	Cpl.	17th	K.I.A	22 03 18	Chapelle Brit Cem, Holnon.
58566	Milsom. A.	Warrington, Cheshire.	Pte.	19th	K.I.A	22 03 18	Pozieres Memorial.
241648	Moses. W.H.	Liverpool.	Pte.	17th	K.I.A	22 03 18	Pozieres Memorial.
17437	Nail. T.	Liverpool.	Sgt.	19th	K.I.A	22 03 18	Pozieres Memorial.
12820	Nolan. N.	Manchester.	Pte.	19th	K.I.A	22 03 18	Pozieres Memorial.
12744	Parker. W.	Liverpool.	Pte.	19th	K.I.A	22 03 18	Pozieres Memorial.
39412	Quilliam. R.W.	Liverpool.	Pte.	19th	K.I.A	22 03 18	Pozieres Memorial.
38819	Quine. A.R.	Douglas, I.O.M.	Pte.	19th	K.I.A	22 03 18	Pozieres Memorial.
21431	Reid. D.	Liverpool.	Pte.	19th	K.I.A	22 03 18	Pozieres Memorial.
57832	Rock. J.	Sunderland.	Pte.	19th	K.I.A	22 03 18	Pozieres Memorial.
90920	Rothwell. W.	Wigan, Lancs.	Pte.	17th	K.I.A	22 03 18	Pozieres Memorial.
51730	Rowley. J.W.	Freshfield, Liverpool.	Pte.	19th	Died.	22 03 18	Roye New Brit Gem
	Salisbury. R.C.	Liverpool.	Capt.	19th	K.I.A	22 03 18	Pozieres Memorial.
200469	Smith. L.A.	Huyton, Liverpool.	L/Cpl.	19th	K.I.A	22 03 18	Savy Brit Cem.
57465	Stewart. R.	Airdrie, Lanarks.	L/Cpl.	19th	K.I.A	22 03 18	Pozieres Memorial.
27321	Stuart. E.	Liverpool.	Pte.	19th	K.I.A	22 03 18	Pozieres Memorial.
21862	Sumner. B.	Birkenhead. Cheshire.	Cpl.	19th	K.I.A	22 03 18	Savy Brit Cem.

307010	Tamber. J.	Liverpool.	Pte.	19th	K.I.A	22 03 18	St.Souplet Brit Cem
51585	Taylor. E.	Liverpool.	Pte.	17th	K.I.A	22 03 18	Pozieres Memorial.
52930	Taylor. H.	Derby.	Pte.	17th	K.I.A	22 03 18	Pozieres Memorial.
17487	Topping. J.T.	Liverpool.	L/Sgt.	19th	K.I.A	22 03 18	Pozieres Memorial.
12942	Turner. J.J.	Liverpool.	Pte.	17th	K.I.A	22 03 18	Pozieres Memorial.
	Villar. R.P.	Taunton, Somerset.	Capt	18th	K.I.A	22 03 18	Pozieres Memorial.
17499	Webster. R.	Birkenhead. Ches.	Cpl.	19th	K.I.A	22 03 18	Savy Brit Cem.
22896	Williams. R.A. D.C.M.	Liverpool.	Cpl.	19th	K.I.A	22 03 18	Pozieres Memorial.
47140	Callister. J.E.	Douglas, I.O.M.	Pte.	19th	K.I.A	23 03 18	Pozieres Memorial.
30955	Dickinson. J.B.	Altcar, Lancs.	L/Cpl.	17th	K.I.A	23 03 18	Pozieres Memorial.
72325	Dutch. W.E.	Liverpool.	Pte.	18th	K.I.A	23 03 18	Ham British Cem.
33081	Hill. J.	Liverpool.	Pte.	19th	D.O.W	23 03 18	St.Souplet Brit Cem
51941	Moore. J.T.	New Brighton, Cheshire.	L/Cpl.	17th	K.I.A	23 03 18	Pozieres Memorial.
49556	Parry. F.	Liverpool.	Pte.	17th	K.I.A	23 03 18	Pozieres Memorial.
15716	Proctor. E. D.C.M.	Milnthorpe, Westmoreland	L/Sgt.	17th	D.O.W	23 03 18	Ham British Cem.
	Sutton. G.S. M.C.	Southport, Lancs.	Capt.	19th	K.I.A	23 03 18	Savy Brit Cem.
49220	Bailey. E.	Morecambe, Lancs.	Pte.	17th	K.I.A	24 03 18	Pozieres Memorial.
57978	Finnamore. A.T.	London.	Sgt.	17th	K.I.A	24 03 18	Pozieres Memorial.
58014	Fowle. P.J.	Cheam, Surrey.	Pte.	17th	K.I.A	24 03 18	Pozieres Memorial.
57928	Gilbert. H.R.	Bermondsey, London.	Pte.	19th	K.I.A	24 03 18	Roye New Brit Cem
	MacHale. J.R.J.	Balham, London.	Lieut.	19th	K.I.A	24 03 18	Ham British Cem.
49116	Pope. W.E.	Chester.	L/Cpl.	17th	K.I.A	24 03 18	Pozieres Memorial.
33013	Radcliffe. J.	Liverpool.	Pte.	19th	K.I.A	24 03 18	Pozieres Memorial.
	Smith. H.L	Newton Abbot, Devon.	Capt.	19th	K.I.A	24 03 18	Pozieres Memorial.
36714	Tatlock. J.	Bolton, Lancs.	Pte.	17th	K.I.A	24 03 18	Pozieres Memorial.
15729	Turner. C.	Southport, Lancs.	Pte.	17th	K.I.A	24 03 18	Pozieres Memorial.
	Watson. W.E.	Leitholm, Scotland.	2nd Lieut.	17th	K.I.A	24 03 18	Pozieres Memorial.
36254	Fazackerley. H.	Whittingham, Preston, Lancs.	L/Cpl.	17th	K.I.A	25 03 18	Pozieres Memorial.
31193	Wilson. W.M.	Liverpool.	Pte.	17th	K.I.A	25 03 18	Ham British Cem.
51899	Cook. W.	Everton, Liverpool.	Pte.	17th	K.I.A	26 03 18	Mezieres CC Ext
22258	Smith. T.A. DCM, CdeG.	Liverpool.	C.S.M.	20th	K.I.A	26 03 18	Bouchoir New Brit Cem
34115	Thornton. F.H.	Appleby, Yorks.	L/Cpl.	17th	K.I.A	26 03 18	Ham British Cem.
26192	Williamson. P.J.	Runcorn, Cheshire.	Pte.	17th	K.I.A	26 03 18	Pozieres Memorial.
22324	Bellingham. R.	Ellesmere Port, Ches.	Pte.	19th	K.I.A	27 03 18	Pozieres Memorial.
12422	Danby. J.	Liverpool.	Pte.	17th	K.I.A	27 03 18	Pozieres Memorial.
31887	Doyle. P.	Aspull, Lancs.	Pte.	19th	K.I.A	27 03 18	Pozieres Memorial.
25301	Earle. G.	New Ferry, Cheshire.	Pte.	19th	K.I.A	27 03 18	Pozieres Memorial.
16538	Lello. J.	Liverpool.	Pte.	18th	D.O.W	27 03 18	Namps-Au-Val Brit Cen
	Mather. R.		2nd Lieut.	20th	K.I.A	27 03 18	Pozieres Memorial.
52015	Mitchell. A.	London.	L/Cpl.	17th	K.I.A	27 03 18	Pozieres Memorial.
22194	Norris. G.	Liverpool.	Pte.	17th	K.I.A	27 03 18	Pozieres Memorial.
22543	Stansfield. H.N.	Liverpool.	Cpl.	19th	D.O.W	27 03 18	St.Souplet Brit Cem
90939	Turner. G.	Bolton, Lancs.	Pte.	17th	K.I.A	27 03 18	Roye New Brit Cem
57579	Williams. J.	Dundee	Pte.	18th	D.O.W	27 03 18	Pozieres Memorial.
37887	Ashcroft. W.	Wigan, Lancs.	L/Cpl.	18th	K.I.A	28 03 18	Pozieres Memorial,
25340	Baker. T. M.M.	Liverpool.	Pte.	19th	K.I.A	28 03 18	Pozieres Memorial.
41320	Barlow. A.E.	Salford, Manchester.	Pte.	17th	K.I.A	28 03 18	Pozieres Memorial.
57772	Barrett. G.	Sunderland.	Pte.	18th	K.I.A	28 03 18	Pozieres Memorial.
22615	Bethell. C.E.	Liverpool.	Pte.	18th	K.I.A	28 03 18	Pozieres Memorial.
21849	Blyde. H.	Liverpool.	Pte.	19th	K.I.A	28 03 18	Pozieres Memorial.
27340	Boswell. H.	Liverpool.	Pte.	18th	K.I.A	28 03 18	Pozieres Memorial.
29558	Bretherton. J.	Liverpool.	Pte.	18th	K.I.A	28 03 18	Pozieres Memorial.
35988	Butterworth. J.W.	Royston, Lancs.	Pte.	18th	K.I.A	28 03 18	Pozieres Memorial.
22948	Caldwell. A.	Liverpool.	Pte.	18th	K.I.A	28 03 18	Savy Brit Cem.
37054	Campion. T.	Liverpool.	Pte.	18th	K.I.A	28 03 18	Pozieres Memorial.
64769	Cheetham. J.	Ormskirk, Lancs.	Pte.	18th	K.I.A	28 03 18	Pozieres Memorial.
51785	Clarke. S.W.	Manchester	Pte.	18th	K.I.A	28 03 18	Pozieres Memorial.
24344	Congdon. F.C.	Knotty Ash, Liverpool.	Sgt.	19th	K.I.A	28 03 18	Pozieres Memorial.

21469	Cornish. A.J.	Stanley, Liverpool.	Cpl.	19th	K.I.A	28 03 18	Pozieres Memorial.
300539	Davenport. H.	Newbridge-on-Wye, Radnor.	Pte.	18th	K.I.A	28 03 18	Pozieres Memorial.
300244	Digby. J.J.W.	Tranmere, Cheshire.	Pte.	18th	K.I.A	28 03 18	Pozieres Memorial.
22981	Dohren. J.	Liverpool.	Pte.	18th	K.I.A	28 03 18	Pozieres Memorial.
57527	Doig. W.	Newport-on-Tay, Fife.	Cpl.	18th	K.I.A	28 03 18	Pozieres Memorial.
94237	Driver. J.	Rawtenstall, Lancs.	Pte.	19th	K.I.A	28 03 18	Pozieres Memorial.
300540	Eccleston. F.	Widnes, Lancs.	Pte.	18th	K.I.A	28 03 18	Pozieres Memorial.
300247	Eckersley. J.	Ashton-in-Makerfield, Lancs.	Pte.	18th	K.I.A	28 03 18	Roye New Brit Cem
22066	Ellison. J.	Bootle, Liverpool.	Cpl.	17th	K.I.A	28 03 18	Ham British Cem.
332470	Ensor. T.	Liverpool.	Pte.	18th	K.I.A	28 03 18	Pozieres Memorial.
300246	Evans. H.	Birkenhead, Cheshire.	Pte.	18th	K.I.A	28 03 18	Pozieres Memorial.
305724	Fahey. P.	Preston, Lancs.	Cpl.	18th	K.I.A	28 03 18	Pozieres Memorial.
35362	Foy. J.	Liverpool.	Pte.	18th	K.I.A	28 03 18	Pozieres Memorial.
300259	Gibbons. A.	Liverpool.	Pte.	18th	K.I.A	28 03 18	Pozieres Memorial.
17347	Gulland. W.	Aintree, Liverpool.	Pte.	18th	K.I.A	28 03 18	Pozieres Memorial.
241632	Guy. S.	Litherland, Liverpool.	L/Cpl.	19th	K.I.A	28 03 18	Pozieres Memorial.
34528	Hackett. T.	Camborne, Cornwall.	Pte.	17th	K.I.A	28 03 18	Pozieres Memorial.
57416	Hanton. J.A.	Glasgow.	Pte.	17th	K.I.A	28 03 18	Caix Brit Cem
17613	Harker. G.J.	Liverpool.	Cpl.	19th	K.I.A	28 03 18	Pozieres Memorial.
48257	Harland. J.	Liverpool.	Pte.	19th	K.I.A	28 03 18	Pozieres Memorial.
17358	Haydn. P.	Liverpool.	Pte.	19th	K.I.A	28 03 18	Pozieres Memorial.
	Hewitt. E.A.F.	Plaistow, Essex.	2nd Lieut.	18th	K.I.A	28 03 18	Pozieres Memorial.
406047	Kay. J.	Salford, Manchester	Pte.	18th	K.I.A	28 03 18	Pozieres Memorial.
12446	Kelly. J.	Manchester	Pte.	18th	K.I.A	28 03 18	Pozieres Memorial.
33134	Kewley. W.G.	Douglas, I.O.M.	Pte.	18th	K.I.A	28 03 18	Pozieres Memorial.
46928	Kilday. S.	Stockport, Cheshire.	Pte.	19th	D.O.W	28 03 18	Namps-Au-Val Brit Cem.
22442	Kirk. L.	Birkenhead, Cheshire.	Cpl.	17th	K.I.A	28 03 18	Bouchoir New Brit Cem.
12090	Kirkpatrick. A.	Bootle, Liverpool.	Pte.	18th	K.I.A	28 03 18	Pargny Brit Cem.
22443	Knowles. J.R.	Birkenhead, Cheshire.	Pte.	19th	K.I.A	28 03 18	Pozieres Memorial.
36935	Lynch. M.	Inneskean, Cork, Ireland.	Pte.	17th	K.I.A	28 03 18	Pozieres Memorial.
27382	Malley. W.N.	Liverpool.	Pte.	17th	K.I.A	28 03 18	Pozieres Memorial.
22466	March. T.	Liverpool.	Pte.	19th	K.I.A	28 03 18	Pozieres Memorial.
23100	Marshall. G.	Walsall, Staffs.	Pte.	19th	K.I.A	28 03 18	Pozieres Memorial.
94178	Mathews. J.H.	Northwich, Ches.	Pte.	17th	K.I.A	28 03 18	Pozieres Memorial.
57558	Millar. B.	Dundee	L/Cpl.	18th	K.I.A	28 03 18	Pozieres Memorial.
23796	Morris. M.E.	Birkenhead, Cheshire.	Pte.	19th	K.I.A	28 03 18	Pozieres Memorial.
31656	Muirhead. J.	Liverpool.	Pte.	18th	K.I.A	28 03 18	Pozieres Memorial.
16473	Osterfield. HJ.	Eastham, Cheshire.	Cpl.	18th	K.I.A	28 03 18	Savy Brit Cem.
17919	Perry. C.W.	Liverpool.	Pte.	19th	K.I.A	28 03 18	Pozieres Memorial.
13188	Redpath. T.L.	Ramsey, I.O.M.	Pte.	17th	K.I.A	28 03 18	Pozieres Memorial.
14812	Rigby. J.T.	Southport, Lancs.	Pte.	17th	K.I.A	28 03 18	Pozieres Memorial.
300354	Roberts. D.	Whiston, Lancs.	Pte.	18th	K.I.A	28 03 18	Pozieres Memorial.
300372	Sale. G.D.	Liverpool.	Pte.	18th	K.I.A	28 03 18	Pozieres Memorial.
238084	Smith. A.S.	Stockport, Cheshire.	Sgt.	17th	K.I.A	28 03 18	Pozieres Memorial.
16484	Smith. J.H.	Liverpool.	Pte.	18th	Died	28 03 18	Pozieres Memorial.
50103	Smith. W.	Liverpool.	Pte.	18th	K.I.A	28 03 18	Pozieres Memorial.
42956	Stothart. W.H.	Liverpool.	Pte.	18th	K.I.A	28 03 18	Pozieres Memorial.
57468	Taylor. A.	Glasgow.	Pte.	18th	K.I.A	28 03 18	Pozieres Memorial.
300558	Thomson. W.	Birkenhead, Cheshire.	Pte.	18th	K.I.A	28 03 18	Pozieres Memorial.
51747	Treacy. D.F.	Widnes, Lancs.	Pte.	18th	K.I.A	28 03 18	Pozieres Memorial.
57964	Trueman. A.W.	Highgate, London.	Pte.	18th	K.I.A	28 03 18	Pargny Brit Cem.
300406	Watkinson. A.	Prescot, Lancs.	Pte.	18th	K.I.A	28 03 18	Pozieres Memorial.
18053	White. J.	Liverpool.	Cpl.	19th	K.I.A	28 03 18	Pozieres Memorial.
50281	Whittaker. J.E.	Rawtenstall, Lancs.	Pte.	18th	K.I.A	28 03 18	Pozieres Memorial.
52258	Wild. E.	Manchester.	Pte.	19th	K.I.A	28 03 18	Pozieres Memorial.
300031	Williams. H.E.	Kentmere, Westmorland.	Pte.	18th	K.I.A	28 03 18	Pozieres Memorial.
27304	Willshaw. W.	Liverpool.	Pte.	19th	K.I.A	28 03 18	Pozieres Memorial.
17744	Wilson. H.J.	Liverpool.	Cpl.	19th	K.I.A	28 03 18	Pozieres Memorial.

57973	Wren. J.S.	London.	Pte.	19th	K.I.A	28	03	18	Pozieres Memorial.
300520	Young. T.	Wallasey, Cheshire.	Pte.	18th	K.I.A	28	03	18	Pozieres Memorial.
300239	Dunn. J.W.	Newton-le-Willows, Lancs.	Pte.	18th	D.O.W	29	03	18	Etretat C.Y. Ext.
94251	Fitton. W.	Bury, Lancs.	Pte.	19th	K.I.A	29	03	18	Pozieres Memorial.
23131	Ollerenshaw. G.W.	Liverpool.	L/Cpl.	19th	D.O.W	29	03	18	Namps-Au-Val Brit Cem.
25015	Andrews-Jones. H. (Jones)	W.Ealing, Middlesex.	Pte.	19th	K.I.A	30	03	18	Pozieres Memorial.
52250	Bethell. A.	Manchester.	L/Cpl.	19th	K.I.A	30	03	18	Pozieres Memorial.
50094	Blinkhorn. T.	Langroyd, Kent.	Pte.	19th	K.I.A	30	03	18	Ham British Cem.
307837	Booth. W.T.	Southport, Lancs.	L/Cpl.	19th	K.I.A	30	03	18	Savy Brit Cem.
57521	Bruce. J.F.	Dundee.	Pte.	19th	K.I.A	30	03	18	Savy Brit Cem.
30535	Burgess. A.E.	Manchester.	Pte.	19th	K.I.A	30	03	18	Savy Brit Cem.
21713	Campbell. I.Mc. M.M.	Liverpool.	Cpl.	19th	K.I.A	30	03	18	Savy Brit Cem.
49064	Collier. P	Liverpool.	Pte.	19th	K.I.A	30	03	18	Savy Brit Cem.
50605	Coulton. T.	Nelson, Lancs.	Pte.	19th	K.I.A	30	03	18	Savy Brit Cem.
330704	Cross. W.	Liverpool.	L/Cpl.	19th	K.I.A	30	03	18	Savy Brit Cem.
47186	Dalton. N.J.	Southport, Lancs.	Pte.	19th	K.I.A	30	03	18	Savy Brit Cem.
86424	Davidson. J.I.	Coventry.	Pte.	19th	K.I.A	30	03	18	Pozieres Memorial.
12098	Devon. E.	Liverpool.	Pte.	19th	K.I.A	30	03	18	Savy Brit Cem.
86423	Dyer. A.E.	Whitney, Oxford.	Pte.	19th	K.I.A	30	03	18	Pozieres Memorial.
17334	Fitzsimmons. G	Liverpool.	L/Sgt.	19th	K.I.A	30	03	18	Pozieres Memorial.
52195	Fletcher. W.	Manchester.	L/Cpl.	19th	K.I.A	30	03	18	Savy Brit Cem.
53066	Herring. W.G.	Leicester.	Pte.	19th	K.I.A	30	03	18	Pozieres Memorial.
23434	Hodgson. J.M.	Flixton, Manchester.	Pte.	19th	K.I.A	30	03	18	Savy Brit Cem.
331120	Holden. E.	Liverpool.	Pte.	19th	K.I.A	30	03	18	Savy Brit Cem.
17615	Hopkins. P.	Liverpool.	Pte.	19th	K.I.A	30	03	18	Pozieres Memorial.
58671	Jackson. P.	Manchester.	Pte.	19th	K.I.A	30	03	18	Savy Brit Cem.
22434	Jones. J.A.	Bootle, Liverpool.	Pte.	19th	K.I.A	30	03	18	Pozieres Memorial.
17873	Kempsey. W.J.	Liverpool.	Pte.	19th	K.I.A	30	03	18	Pozieres Memorial.
50609	Larner. B.T.	Fairford, Glos.	Pte.	19th	K.I.A	30	03	18	Pozieres Memorial.
52032	Lawrence. J.R.	Liverpool.	Pte.	19th	K.I.A	30	03	18	Pozieres Memorial.
265744	Leedale. F.	Southport, Lancs.	Pte.	19th	K.I.A	30	03	18	Pozieres Memorial.
22452	Lees. J.	Kendal.	Sgt.	19th	K.I.A	30	03	18	Pozieres Memorial.
52111	Little. R.	Manchester.	L/Cpl.	19th	K.I.A	30	03	18	Savy Brit Cem.
204394	McKenzie. T.	Liverpool.	Pte.	19th	K.I.A	30	03	18	Pozieres Memorial.
23107	McKibbin. G.	Liverpool.	Sgt.	19th	K.I.A	30	03	18	Pozieres Memorial.
57817	Moore. T.	Newcastle-on-Tyne.	Pte.	19th	K.I.A	30	03	18	Savy Brit Cem.
51777	Plunkett. J.E.	Liverpool.	Pte.	19th	K.I.A	30	03	18	Savy Brit Cem.
58067	Poncia. F.H.	Boreham Wood, Herts.	Pte.	19th	K.I.A	30	03	18	Savy Brit Cem.
37849	Roberts. W.H.	Liverpool.	Pte.	19th	K.I.A	30	03	18	Savy Brit Cem.
57830	Robson. A.E.	New Shildon, Durham.	Pte.	19th	K.I.A	30	03	18	Savy Brit Cem.
203700	Smith. A.P.	Manchester.	Pte.	19th	K.I.A	30	03	18	St. Souplet Brit Cem
52122	Thomas. S.	Manchester.	Pte.	19th	K.I.A	30	03	18	Pozieres Memorial.
202905	Whittaker. G.H.	Liverpool.	Pte.	19th	K.I.A	30	03	18	Pozieres Memorial.
52041	Williams. W.H.	Liverpool.	Pte.	19th	K.I.A	30	03	18	Pozieres Memorial.
41897	Jones. A.	Liverpool.	Pte.	19th	D.O.W	31	03	18	St Souplet Brit Cem
21559	Lake. G.H.	Liverpool.	Pte.	19th	D.O.W	01	04	18	St Sever Cem Ext, Rouen.
	Beaumont. E.P. M.C.	Rock Ferry, Cheshire.	Lieut.	17th	D.O.W	02	04	18	St Sever Cem, Rouen.
	Sheard. F.M. M.C.	Great Crosby, Liverpool.	Capt.	18th	D.O.W	02	04	18	Premont Brit Cem.
22702	Harper. J.L.	Liverpool.	Pte.	20th	Died.	07	04	18	Anfield Cem, Liverpool.
51687	Smith. A.F.	Liscard, Cheshire.	L/Cpl.	19th	Died	08	04	18	Wallasey, Rake Lane Cem.
91288	Fowden. W.F.	Manchester.	Pte.	17th	D.O.W	09	04	18	St Sever Cem Ext, Rouen.
86416	Hayward. A.	Manchester.	Pte.	18th	Died	09	04	18	Chambiere French Nat Cem Metz
57452	Niven. R.	Bridge-of-Weir, Renfrew	Pte.	18th	Died	09	04	18	Ham British Cem.
88135	Tall. F.J.	London.	Pte.	18th	K.I.A	09	04	18	Cement House Cem, Langemarck
86359	Thompson. T.	Frodsham, Cheshire.	Pte.	17th	D.O.W	10	04	18	Foreste CC (Aisne)
51602	Blackburn. J.E.	Liverpool.	Pte.	17th	D.O.W	11	04	18	St.Souplet Brit Cem
57462	Smith. J.	Glasgow.	Pte.	17th	D.O.W	12	04	18	Glasgow Western Cem, Scotland

24348	Keenan. A.	Rock Ferry, Cheshire.	Cpl.	17th	Died	15 04 18	St.Hilaire CC Ext, Frevant
270085	Holmes. J.	Everton, Liverpool.	Pte.	18th	K.I.A	16 04 18	Cement Hse Cem, Langemarck.
52884	McNicoll. C.F.	Southport, Lancs.	Pte.	19th	D.O.W	16 04 18	St Sever Cem Ext, Rouen.
235259	Thomas. J.	Henllan, Cardigan.	Pte.	17th	K.I.A	17 04 18	Westourtre Brit Cem..
57698	Burton. H.	Nottingham.	Cpl.	19th	K.I.A	18 04 18	Mont-Noir M.C.St.Jans-Cappel
59347	Eves. W.	Liverpool.	Pte.	18th	K.I.A	18 04 18	Tyne Cot Memorial.
202844	Farcy. L.	Bootle, Liverpool.	Pte.	19th	K.I.A	18 04 18	Bailleul CC Ext.
49547	Gaskell. S.J.	Moss Side, Manchester.	Pte.	19th	K.I.A	18 04 18	Tyne Cot Memorial.
203773	Green. L.		Pte.	19th	K.I.A	18 04 18	Mont-Noir M.C.St.Jans-Cappel
50020	Hollidge. T.	Croyden, Surrey.	Pte.	19th	K.I.A	18 04 18	Tyne Cot Memorial.
27143	Jones. R.G.W.	Newton-le-Willows, Lancs.	Pte.	19th	K.I.A	18 04 18	Tyne Cot Memorial.
54155	Kavanagh. J.	Liverpool.	Pte.	19th	K.I.A	18 04 18	Tyne Cot Memorial.
31798	Lawler. P.	Liverpool.	Pte.	19th	K.I.A	18 04 18	Mont-Noir M.C.St.Jans-Cappel
52907	Merrick. F.	Manchester.	Pte.	19th	K.I.A	18 04 18	Mont-Noir M.C.St.Jans-Cappel
49558	Parker. A.W.	Litherland, Liverpool.	Pte.	17th	Died.	18 04 18	Bootle Cem, Liverpool.
50620	Parker. T.	Morecambe, Lancs.	Pte.	17th	K.I.A	18 04 18	Bailleul CC Ext.
	Gill. R. M.C. M.M.	Worthing, Sussex.	2nd Lieut.	17th	K.I.A	19 04 18	Tyne Cot Memorial.
	Crook. H.	Liverpool.	2nd Lieut.	17th	D.O.W	20 04 18	Tyne Cot Memorial.
86268	Fasham. G.J.	Camberwell, London.	Pte.	17th	D.O.W	20 04 18	Haringhe Mil Cem
20781	Green. J	Wigan, Lancs.	Pte.	19th	D.O.W	20 04 18	Mendingham B.C., Proven.
57856	Shotton. H.	Acklington, Northumberland	Sgt.	18th	K.I.A	20 04 18	Tyne Cot Memorial.
	Sparks. R.W.	Minchinhampton, Glos.	2nd Lieut.	18th	K.I.A	20 04 18	Tyne Cot Memorial.
90940	Tomlinson. F.W.	Droylsden, Manchester.	Pte.	17th	K.I.A	20 04 18	Tyne Cot Memorial.
23876	Young. H.	Liverpool.	Sgt.	18th	K.I.A	20 04 18	Tyne Cot Memorial.
49222	Charles. W.	Burnley, Lancs.	Pte.	18th	D.O.W	21 04 18	Avesnes-sur-Helpe CC
94199	Evans. A.Z.	Caersws, Montgomery.	Pte.	17th	D.O.W	23 04 18	Boulogne East Cem.
300299	Jones. W.H.	Birkenhead, Cheshire.	Pte.	18th	D.O.W	23 04 18	Tourgeville Mil Cem.
	Edwards. U.S. M.C.	Liverpool.	Capt.	18th	D.O.W	24 04 18	Haringhe Mil Cem
51675	Relph. E.G.	Penrith, Cumberland.	Pte.	17th	K.I.A	24 04 18	Tyne Cot Memorial.
15967	Webb. J.L	Liverpool.	Pte.	17th	D.O.W	24 04 18	Toxteth Park Cem, Liverpool.
49217	Baines. E.	Dalton, Lancs.	Pte.	18th	K.I.A	26 04 18	Perth Cem, Zillebeke.
30573	Alcock. A	Liverpool.	Pte.	19th	D.O.W	27 04 18	Guise French Nat Cem.
59280	Moodie. W.	Liverpool.	Pte.	17th	D.O.W	27 04 18	Boulogne East Cem.
57378	Robertson. R.	Partick, Glasgow.	Sgt.	18th	Died	27 04 18	Grand Seraucourt Brit Cem
50064	Astle. T	Birmingham.	Pte.	19th	K.I.A	28 04 18	Tyne Cot Memorial.
	Bloore. R.H.	Walton, Liverpool.	Lieut.	17th	K.I.A	28 04 18	Voormezeele Cem Encl No.3
50654	Byron. J.	Beswick, Manchester.	Pte.	17th	K.I.A	28 04 18	Tyne Cot Memorial.
15632	Crail. W.D.	Liscard, Cheshire.	Cpl.	17th	K.I.A	28 04 18	Tyne Cot Memorial.
50701	Gainford. H.	Frizington, Cumberland	L/Cpl.	17th	K.I.A	28 04 18	Tyne Cot Memorial.
50661	Hankin. C.	Liverpool.	Pte.	17th	K.I.A	28 04 18	Tyne Cot Memorial.
	Harrop. T.	Morley, Yorks.	2nd Lieut.	17th	K.I.A	28 04 18	Tyne Cot Memorial.
94260	Hetherington. H.A.	Ulverston.	Pte.	19th	K.I.A	28 04 18	Tyne Cot Memorial.
42758	Kermode. G.W.	Ballaugh, I.O.M.	Pte.	17th	K.I.A	28 04 18	Tyne Cot Memorial.
31631	Macdonald. A.	Liverpool.	Pte.	17th	K.I.A	28 04 18	Tyne Cot Memorial.
59341	Millward. J.	Liverpool.	Pte.	17th	K.I.A	28 04 18	Tyne Cot Memorial.
21348	Shyman. W.	Manchester.	Pte.	19th	K.I.A	28 04 18	Tyne Cot Memorial.
57464	Somers. G.	Glasgow.	Pte.	17th	K.I.A	28 04 18	Tyne Cot Memorial.
49041	Taylor. L	Brinscall, Lancs.	Cpl.	17th	K.I.A	28 04 18	Tyne Cot Memorial.
15307	Thompson. J.E.	Liverpool.	Cpl.	17th	K.I.A	28 04 18	Voormezeele Cem Encl No.3
50028	Valentine. W.T.	Wrexham, N.Wales.	Pte.	17th	D.O.W	28 04 18	Perth Cem, Zillebeke.
94184	Aspey. J.G.	Liverpool.	Pte.	17th	K.I.A	29 04 18	Tyne Cot Memorial.
50225	Ball. B.	Haslingden, Lancs.	Pte.	17th	K.I.A	29 04 18	Tyne Cot Memorial.
29106	Barlow. J.B.	Liverpool.	Pte.	18th	K.I.A	29 04 18	Tyne Cot Memorial.
24318	Bellis. A.J.	Liverpool.	Pte.	19th	K.I.A	29 04 18	Voormezeele Cem Encl No.3
50025	Benstead. A.	St. Mary Cray, Kent.	Pte.	19th	K.I.A	29 04 18	Voormezeele Cem Encl No.3
42949	Birch. A.	Liverpool.	L/Cpl.	19th	K.I.A	29 04 18	Tyne Cot Memorial.
50587	Bollom. P.	Sandford, Hants.	Pte.	19th	K.I.A	29 04 18	Tyne Cot Memorial.
94195	Bowman. A.	Moreton, Cheshire.	Pte.	17th	K.I.A	29 04 18	Tyne Cot Memorial.

48259	Brennan. P.	Liverpool.	Pte.	19th	K.I.A	29 04 18	Voormezeele Cem Encl No.3
406704	Bridson. T.H.	Douglas. I.O.M.	Pte.	18th	K.I.A	29 04 18	Tyne Cot Memorial. "
50186	Chamberlain. W.R.	Old Overton, Leicester.	Pte.	19th	K.I.A	29 04 18	Tyne Cot Memorial.
300366	Clayton. H.G.V.	Liverpool.	Pte.	18th	K.I.A	29 04 18	Tyne Cot Memorial.
300229	Coulthard. J.	Carlisle	Pte.	18th	K.I.A	29 04 18	Voormezeele Cem Encl No.3
300219	Cox. F.	Wigan, Lancs.	Pte.	18th	K.I.A	29 04 18	Tyne Cot Memorial.
16361	Cuddy. G.	Walton, Liverpool.	L/Sgt.	17th	K.I.A	29 04 18	Tyne Cot Memorial.
23927	Davies. A.G.	Liverpool.	Pte.	17th	K.I.A	29 04 18	Tyne Cot Memorial.
94238	Dean. F.	Liverpool.	Pte.	19th	K.I.A	29 04 18	Tyne Cot Memorial.
94239	Declaire. H.	Penarth, Glam.	Pte.	19th	K.I.A	29 04 18	Tyne Cot Memorial.
29766	Dentith. R.	Crewe, Cheshire.	Cpl.	17th	D.O.W	29 04 18	Arneke British Cem.
41175	Dillon. J.E.	Cardiff, Glam.	Pte.	17th	K.I.A	29 04 18	Voormezeele Cem Encl No.3
50123	Dinning. R.	Gateshead, Durham.	Pte.	17th	K.I.A	29 04 18	Tyne Cot Memorial.
50044	Dougherty. H	Burnley, Lancs.	Pte.	19th	K.I.A	29 04 18	Whytschaete Mil Cem
300445	Eaves. H.	Earlestown, Lancs.	Sgt.	18th	K.I.A	29 04 18	Tyne Cot Memorial.
94245	Entwistle. J.	L. Broughton, Manchester.	Pte.	19th	K.I.A	29 04 18	Tyne Cot Memorial.
94247	Evans. T.A.	Bromborough, Cheshire.	Pte.	17th	K.I.A	29 04 18	Tyne Cot Memorial.
300254	Gaskell. W.	St. Helens, Lancs.	L/Cpl.	18th	K.I.A	29 04 18	Tyne Cot Memorial.
202812	Gill. J.H.	Liverpool.	Pte.	18th	K.I.A	29 04 18	Tyne Cot Memorial.
300596	Graham. W.H.	Sale, Cheshire.	Pte.	17th	K.I.A	29 04 18	Tyne Cot Memorial.
35928	Hargreaves. H.	Bolton, Lancs.	Pte.	18th	K.I.A	29 04 18	Tyne Cot Memorial.
300266	Heaton. G.	St. Helens, Lancs.	Pte.	18th	K.I.A	29 04 18	Tyne Cot Memorial.
41800	Heery. J.	Liverpool.	Pte.	18th	K.I.A	29 04 18	Tyne Cot Memorial.
91289	Hides. A.W.	Stacksteads, Lancs.	Pte.	17th	K.I.A	29 04 18	Tyne Cot Memorial.
300273	Hitchin. W.	Liverpool.	Pte.	18th	K.I.A	29 04 18	Tyne Cot Memorial.
	Hough. E.B. M.C.	Birkenhead, Cheshire.	Capt.	19th	K.I.A	29 04 18	Tyne Cot Memorial.
22418	Houlgrave. P.S.	Liverpool.	Pte.	19th	D.O.W	29 04 18	White House Cem, Ypres
300598	Jewsbury. H.	Stockport, Cheshire.	Pte.	17th	K.I.A	29 04 18	Perth Cem, Zillebeke.
91297	Kinsey. N.D.	Newchurch, Lancs.	Pte.	17th	K.I.A	29 04 18	Tyne Cot Memorial.
57434	Macartney. R.	Wallacetown, Ayr.	Pte.	17th	K.I.A	29 04 18	Tyne Cot Memorial.
265827	Martin. L.	Southport, Lancs.	Pte.	18th	K.I.A	29 04 18	Tyne Cot Memorial.
300597	Martin. P.	Macclesfield, Ches.	Pte.	17th	K.I.A	29 04 18	Tyne Cot Memorial.
194185	Massey. R.	Dukinfield, Cheshire.	Pte.	17th	K.I.A	29 04 18	Tyne Cot Memorial.
59289	McCambridge. H.P.	Liverpool.	Pte.	17th	K.I.A	29 04 18	Tyne Cot Memorial.
300321	McGinley. J.	Liverpool.	Pte.	18th	K.I.A	29 04 18	Tyne Cot Memorial.
21963	Meek. F.	Liverpool.	Pte.	18th	K.I.A	29 04 18	Tyne Cot Memorial.
52130	Millhouse. C.	Manchester.	L/Cpl.	19th	K.I.A	29 04 18	Tyne Cot Memorial.
	Munro. W.	Liverpool.	2nd Lieut	19th	K.I.A	29 04 18	Voormezeele Cem Encl No.3
48207	O'Hare. W.	Liverpool.	Pte.	18th	K.I.A	29 04 18	St.Souplet Brit Cem
31142	Orme. R.J.	Liverpool.	Pte.	17th	K.I.A	29 04 18	Tyne Cot Memorial.
353008	Patterson. E.	Everton, Liverpool.	Pte.	18th	K.I.A	29 04 18	Tyne Cot Memorial.
29610	Richardson. J.E.	Liverpool.	Pte.	18th	K.I.A	29 04 18	Tyne Cot Memorial.
23221	Roberts. T.	Chorley, Lancs.	Pte.	17th	K.I.A	29 04 18	Tyne Cot Memorial.
235265	Rodgers. J.	Manchester.	Pte.	17th	K.I.A	29 04 18	Tyne Cot Memorial.
15291	Rymill. A.	Liverpool.	Pte.	17th	K.I.A	29 04 18	Tyne Cot Memorial.
64834	Shortall. J.	Liverpool.	Pte.	18th	K.I.A	29 04 18	Tyne Cot Memorial.
90932	Smith. E.W.	Ashton-under-Lyne, Lancs.	Pte.	17th	K.I.A	29 04 18	Tyne Cot Memorial.
51706	Smith. T.	Liverpool.	Pte.	19th	K.I.A	29 04 18	Tyne Cot Memorial.
49204	Spiers. J.D.	Liverpool.	Pte.	17th	K.I.A	29 04 18	Tyne Cot Memorial.
24634	Thoel. W.	Liverpool.	L/Cpl.	18th	K.I.A	29 04 18	Tyne Cot Memorial.
50217	Townsend. J.J.	Gate Helmsley, Yorks.	Pte.	19th	K.I.A	29 04 18	Tyne Cot Memorial.
57666	Willey. J.W.	Leeds, Yorks.	Pte.	18th	K.I.A	29 04 18	Tyne Cot Memorial.
50991	Williams. G.H.	Liverpool.	A/CSM	19th	K.I.A	29 04 18	Voormezeele Cem Encl No.3
300032	Woodburn. S.	Liscard, Cheshire.	Pte.	18th	K.I.A	29 04 18	Tyne Cot Memorial.
204723	Woods. J.	Liverpool.	Pte.	19th	K.I.A	29 04 18	Voormezeele Cem Encl No.3
51749	Woodward. W.	Liverpool.	L/Cpl.	18th	K.I.A	29 04 18	Tyne Cot Memorial.
29203	Alexander. V.	Liverpool.	L/Cpl.	18th	D.O.W	30 04 18	Lijssenthoek M C Poperinghe
27948	Atherton. J.	Liverpool.	Pte.	18th	D.O.W	30 04 18	Esquelbecq Mil Cem.

51896	Banks. R.E.	Scarisbrick, Lancs.	Pte.	17th	K.I.A	30	04	18	Voormezeele Cem Encl No.3
46630	Bell. W.	Liverpool.	Pte.	17th	D.O.W	30	04	18	Arneke British Cem.
	Brewerton. R.H.	Marple, Cheshire.	Lieut.	19th		30	04	18	Tyne Cot Memorial.
51708	Dod. C.N.	Wallasey, Cheshire.	Pte.	19th	K.I.A	30	04	18	Tyne Cot Memorial.
94246	Emison. F.	Haslingden, Lancs.	L/Cpl.	19th	D.O.W	30	04	18	Esquelbecq Mil Cem.
23056	Jones. J.W.	Northwich, Cheshire.	Pte.	18th	D.O.W	30	04	18	Esquelbecq Mil Cem.
49572	Leadbeater. J.H.	Knottingley, Yorks.	Pte.	17th	D.O.W	30	04	18	Esquelbecq Mil Cem.
22462	Lowe. J.	Birkenhead, Cheshire.	Sgt.	17th	K.I.A	30	04	18	Tyne Cot Memorial.
51712	Murray. E.D.	Liverpool.	Pte.	17th	D.O.W	30	04	18	Esquelbecq Mil Cem.
269899	Pitts. H.	St.John's, I.O.M.	Pte.	17th	D.O.W	30	04	18	Esquelbecq Mil Cem.
42874	Slifkin. I.	Liverpool.	Pte.	17th	D.O.W	30	04	18	Arneke British Cem.
50692	Williams. S.B.R.	Potters Bar, Middlesex.	Pte.	17th	K.I.A	30	04	18	Tyne Cot Memorial.
300188	Bansor. A.C.	Chelmsford, Essex.	Pte.	18th	D.O.W	01	05	18	Esquelbecq Mil Cem.
64776	Blank. G.	Liverpool.	Pte.	17th	D.O.W	01	05	18	Harlebeke New Brit Cem
57831	Conolly. L.	Clapham, London.	Sgt.	17th	K.I.A	01	05	18	Tyne Cot Memorial.
94253	Firth. E.	Haslingden, Lancs.	L/Cpl.	19th	D.O.W	01	05	18	Longuenesse Cem, St.Omer.
49533	Rigby. H.	Bootle, Liverpool.	Pte.	18th	D.O.W	01	05	18	Boulogne East Cem.
17721	Squirrell. C.H.	Bootle, Liverpool.	Pte.	19th	K.I.A	01	05	18	Tyne Cot Memorial.
53165	Corke. J.T.	Burwash, Sussex.	Pte.	17th	D.O.W	02	05	18	Esquelbecq Mil Cem.
57549	Lennon. R.	Forfar	Pte.	18th	K.I.A	02	05	18	Klein-Vierstraat Brit Cem.
23145	Poole. E.	Ellesmere Port, Ches.	Pte.	17th	D.O.W	02	05	18	Arneke British Cem.
235256	Schlank. J.	Congleton, Yorks.	Pte.	18th	K.I.A	02	05	18	Klein-Vierstraat Brit Cem.
17895	Mervyn. A.	Liverpool.	Cpl.	19th	D.O.W	03	05	18	Haringhe Mil Cem
52714	Robinson. W.	Oldham, Lancs.	Pte.	18th	Died	03	05	18	Grand Seraucourt Brit Cem
59339	Anders. T.	Liverpool.	Pte.	18th	D.O.W	04	05	18	Arneke British Cem.
36398	Greenwood. D.	Bolton, Lancs.	Pte.	18th	K.I.A	06	05	18	La Clytte Mil Cem.
17470	Stone. LE.	Bootle, Liverpool.	Pte.	19th	K.I.A	06	05	18	La Clytte Mil Cem.
30804	Warburton. J.E.	Liverpool.	Pte.	18th	K.I.A	06	05	18	La Clytte Mil Cem.
23858	Whitney. G.	Runcorn, Cheshire.	Cpl.	17th	K.I.A	06	05	18	La Clytte Mil Cem.
235275	Airey. M.C.	South Bank, Yorks.	Pte.	18th	K.I.A	07	05	18	Tyne Cot Memorial.
16698	Anderson. W.	Birkenhead, Cheshire.	Pte.	17th	K.I.A	07	05	18	La Clytte Mil Cem.
22701	Gyte. H.	Kirkdale, Liverpool.	Cpl.	18th	Died	07	05	18	Tyne Cot Memorial
53111	Hill. A.E.	Enfield, London	Pte.	18th	K.I.A	07	05	18	Tyne Cot Memorial.
406176	Stafford. H.K.	Middleton, Lancs.	Pte.	18th	K.I.A	07	05	18	Tyne Cot Memorial.
28125	Allen. T.E. M.M	Radcliffe, Lancs.	Sgt.	18th	K.I.A	08	05	18	Tyne Cot Memorial.
57546	Amess. J.	Stanley, Perthshire.	Pte.	19th	D.O.W	08	05	18	Esquelbecq Mil Cem.
35125	Bampton. C.	Liverpool.	Pte.	18th	K.I.A	08	05	18	Tyne Cot Memorial.
	Black. D.H.	Ventnor, Isle of Wight.	2nd Lieut.	17th	K.I.A	08	05	18	Tyne Cot Memorial.
47371	Bulfield. R.	Liverpool.	Pte.	17th	K.I.A	08	05	18	Tyne Cot Memorial.
58002	Butler. A.	Peterborough.	Pte.	17th	K.I.A	08	05	18	Tyne Cot Memorial.
235112	Clarke. W.J.	Pluckley, Kent.	Pte.	17th	K.I.A	08	05	18	Tyne Cot Memorial.
235202	Davies. E.L	North Ormesby, Yorks.	Pte.	17th	K.I.A	08	05	18	Tyne Cot Memorial.
	Dean. T.A.W.	Sculcoates, Yorks.	2nd Lieut.	18th	D.O.W	08	05	18	Tyne Cot Memorial.
54111	Forshaw. A.E.	Liverpool.	Pte.	18th	K.I.A	08	05	18	Tyne Cot Memorial.
34368	Greenway. F.B.	West Hartlepool.	Pte.	20th	D.O.W	08	05	18	Arneke British Cem.
16074	Griffiths. F.N.	Liverpool.	L/Cpl.	17th	K.I.A	08	05	18	Nine Elms Brit Cem., Poperinghe.
300264	Guppy. R.J.	Sherbourne, Dorset.	Pte.	18th	K.I.A	08	05	18	Tyne Cot Memorial.
300479	Hankinson. E.S.	Ashton-in-Makerfield, Lancs.	Pte.	18th	K.I.A	08	05	18	Tyne Cot Memorial.
	Henry. N. M.C.	Liverpool.	Capt.	17th	K.I.A	08	05	18	Tyne Cot Memorial.
22135	Jones. J.M.	Montgomery, M.Wales.	Pte.	17th	D.O.W	08	05	18	Arneke British Cem.
17073	MacDonald. A.	Liverpool.	L/Cpl.	18th	K.I.A	08	05	18	Tyne Cot Memorial.
38436	Maddocks. R.	Formby, Lancs.	Pte.	17th	K.I.A	08	05	18	Klein-Vierstraat Brit Cem.
50669	McDermott. W.	Liverpool.	Pte.	17th	K.I.A	08	05	18	Tyne Cot Memorial.
269819	McKay. R.G.	Liverpool.	Pte.	17th	K.I.A	08	05	18	Tyne Cot Memorial.
50151	Moffat. R.	Stamfordham, Northumberland	Pte.	17th	K.I.A	08	05	18	Tyne Cot Memorial.
50670	Nicholson. E.	Manchester.	Pte.	17th	K.I.A	08	05	18	Tyne Cot Memorial.
22812	Pittman. W.	Liverpool.	Pte.	17th	K.I.A	08	05	18	Tyne Cot Memorial.
57984	Russell. A.R.	London.	L/Sgt.	17th	D.O.W	08	05	18	Tyne Cot Memorial.

34358	Sefton. J.T.H.	Birkenhead, Cheshire.	Pte.	18th	D.O.W	08 05 18	Boulogne East Cem.
50650	Stuart. W.H.	Burnley, Lancs.	Pte.	17th	K.I.A	08 05 18	Tyne Cot Memorial.
15678	Symonds. H.	Liverpool.	Cpl.	17th	K.I.A	08 05 18	Tyne Cot Memorial.
49181	Walsh. J.H.	Hyde, Cheshire.	Pte.	17th	K.I.A	08 05 18	Tyne Cot Memorial.
51658	Waywell. E.	Garston, Liverpool.	Pte.	17th	K.I.A	08 05 18	Tyne Cot Memorial.
94254	Griffiths. J.	Manchester	Pte.	19th	D.O.W	09 05 18	Lijssenthoek M C Poperinghe
50049	Holley. H.J.	Wellington, Somerset.	Pte.	19th	K.I.A	09 05 18	Tyne Cot Memorial.
29854	Keating. H.	Hollinwood, Lancs.	Pte.	17th	D.O.W	09 05 18	Arneke British Cem.
300498	O'Laverty. J.	Anfield, Liverpool.	Pte.	18th	D.O.W	09 05 18	Boulogne East Cem.
50646	Wilson. W.	Lancaster.	Pte.	17th	K.I.A	09 05 18	Tyne Cot Memorial.
203014	Barker. C.	Liverpool.	Pte.	19th	K.I.A	10 05 18	Tyne Cot Memorial.
36708	Leigh. C.R.	Kearsley, Lancs.	Pte.	19th	K.I.A	10 05 18	Tyne Cot Memorial.
	Threlfall. J.A.	Birkenhead, Cheshire.	Lieut.	18th	D.O.W	10 05 18	Arneke British Cem.
49029	Whittam. R.	Pendleton, Manchester.	L/Cpl.	17th	D.O.W	10 05 18	Esquelbecq Mil Cem.
51838	Kerr. L.	Manchester.	Pte.	17th	D.O.W	11 05 18	Ghent City Cem.
235239	Lockett. A.	Crewe, Cheshire.	Pte.	19th	K.I.A	11 05 18	La Clytte Mil Cem.
	Ashcroft. E.S.	Birkenhead, Cheshire.	Lieut.	17th	D.O.W	12 05 18	Harlebeke New Brit Cem
21687	Williams. E.M.	Liverpool.	Sgt.	19th	Died	12 05 18	Boulogne East Cem.
202833	Donelan. L.	Everton, Liverpool.	Pte.	19th	D.O.W	16 05 18	Esquelbecq Mil Cem.
37751	McCormick. M.	Liverpool.	Pte.	19th	Died	17 05 18	Kirkdale Cem, Liverpool.
30068	Devereux. R.	Liverpool.	Pte.	19th	Died	19 05 18	Epsom Cem, Surrey.
94248	Eastwood. G.B.	Southport, Lancs.	L/Cpl.	19th	D.O.W	20 05 18	Etaples Mil Cem
50614	McCray. W.	Lancaster.	Pte.	17th	D.O.W	20 05 18	Harlebeke New Brit Cem
269622	Boyd. E.	Manchester.	Pte.	17th	D.O.W	22 05 18	Tyne Cot Memorial.
44402	Viner. J.	London.	Pte.	17th	D.O.W	24 05 18	Islington Cem, Finchley.
17591	Hargreaves. E.	Burnley, Lancs.	Pte.	19th	D.O.W	26 05 18	Cheltenham Cem, Glos.
16427	Kewley. G.	Liverpool.	Pte.	18th	D.O.W	27 05 18	Boulogne East Cem.
50652	Barnes. A.	Blackpool, Lancs.	Pte.	17th	D.O.W	01 06 18	Wimereux CC
21547	Kehoe. P.	Liverpool.	Pte.	19th	D.O.W	04 06 18	Hautmont CC
49078	Edwards. G.	Liverpool.	Pte.	18th	Died	06 06 18	Ham British Cem.
35028	Tomlinson. T.	Wheelton, Lancs.	Pte.	18th	Died	06 06 18	Ham British Cem.
21591	Perryman. W.T.	Liverpool.	Sgt.	19th	D.O.W	09 06 18	Wimereux CC
21718	Hargreaves. F.	Liverpool.	L/Cpl.	19th	Died	13 06 18	Premont Brit Cem.
57615	Preddy. F.W.	Rotherham, Yorks.	Pte.	18th	D.O.W	16 06 18	La Capelle-en-Thierache
50634	Skirrow. T.	Lancaster.	Pte.	18th	Died	22 06 18	Kronenburg Mil Cem Strasbourg
24596	Madden. W.A.	Bootle, Liverpool.	Pte.	17th	D.O.W	25 06 18	Niederzwehren Cem, Cassel, Germany.
18323	Byrne. S.	Liverpool.	Pte.	19th	Died.	27 06 18	Berlin South West Cem.
58764	Cocker. H.	Burnley, Lancs.	Pte.	17th	Died	27 06 18	Berlin South West Cem.
24661	Alker. H.	Wigan, Lancs.	Pte.	18th	Died	03 07 18	Berlin South West Cem.
300380	Smith. E.	Widnes, Lancs.	Pte.	18th	K.I.A	05 07 18	Tyne Cot Memorial.
23828	Shaw. F.	Liverpool.	L/Cpl.	20th	D.O.W	11 07 18	St Sever Cem, Rouen.
26112	Parker. W.	Liverpool.	Pte.	17th	Died.	16 07 18	Levallois-Perret CC
42267	Smethurst. W.	Liverpool.	Pte.	17th	K.I.A	24 07 18	Wancourt Brit Cem.
17861	Jones. H.A.	Liverpool.	Pte.	19th	Died	29 07 18	Berlin South West Cem.
21663	Youde. J.	Liverpool.	Pte.	19th	K.I.A	30 07 18	Guillemont Rd Cem.
43483	Parry. T.H.	Bootle, Liverpool.	Pte.	17th	Died.	02 08 18	Cologne South Cem, Germany.
58682	Jones. H.	Manchester.	Pte.	17th	K.I.A	05 08 18	Tyne Cot Memorial.
57902	Mowlam. N.G.	Kingston-on-Thames.	L/Cpl.	17th	D.O.W	08 08 18	Harlebeke New Brit Cem
50689	Chambers. F.S.	Canning Town, London.	L/Cpl.	17th	K.I.A	15 08 18	Ancre Brit Cem. B-Hamel.
269892	Rooney. R.	Liverpool.	L/Cpl.	17th	D.O.W	15 08 18	Berlin South West Cem.
64787	Cheater. E.	Bootle, Liverpool.	Pte.	18th	K.I.A	21 08 18	Warry Copse Cem, Courcelles-le-Compte
	Friend. J.B.	Dover, Kent.	2nd Lieut.	17th	K.I.A	21 08 18	Vis-en-Artois Memorial.
21655	Webster. J.S.	Liverpool.	Cpl.	19th	Died	21 08 18	Etaples Mil Cem
50180	Alliott. A.H.	Nottingham.	Cpl.	19th	D.O.W	25 08 18	St.Hilaire CC Ext, Frevant
52869	Gregson. J.	Southport, Lancs.	Pte.	17th	D.O.W	26 08 18	Bucquoy CC Ext, Ficheux.
300473	Evans. J.	Bootle, Liverpool.	Pte.	18th	K.I.A	27 08 18	Tyne Cot Memorial.

52419	Hollinghurst. A.E.	Bootle, Liverpool.	L/Cpl.	18th	K.I.A	31 08 18	HAC Cem, Ecourt-Saint Mein
307968	Myland. A.S.	Longsight, Manchester.	Pte.	18th	K.I.A	31 08 18	Vis-en-Artois Memorial.
48499	Lyon. J.	Liverpool.	Pte.	19th	Died	04 09 18	Berlin South West Cem.
88128	Rimmer. J.W.	Southport, Lancs.	Pte.	17th	Died.	19 09 18	Cologne South Cem, Germany.
94181	Townsley. A.	Swansea.	Pte.	17th	Died.	20 09 18	Cologne South Cem, Germany.
17308	Dean. J.H.	Liverpool.	Pte.	19th	D.O.W	29 09 18	Villers Hill Brit Cem.
	Carline. T.	Port Sunlight, Wirral.	2nd Lieut.	18th	K.I.A	30 09 18	Uplands Cem, Magny-le-Fosse.
22657	Cooper. J.	Wavertree, Liverpool.	Pte.	18th	Died	02 10 18	Wavertree, H.T.CY Liverpool.
406080	Jackson. W.	Eccles, Lancs.	Pte.	18th	K.I.A	02 10 18	Honnechy Brit Cem
24247	Furnival. J.E.	Farnworth, Lancs.	L/Cpl.	18th	K.I.A	08 10 18	Serain CC Ext.
17943	Roughley. W. M.M.	Ormskirk, Lancs.	Pte.	18th	K.I.A	08 10 18	Busigny CC Ext.
18764	Stafford. G.	Liverpool.	Pte.	18th	K.I.A	08 10 18	Busigny CC Ext.
306287	Strickland. G.	Warton, Lancs.	Pte.	18th	K.I.A	08 10 18	Prospect Hill Cem, Gouy
18892	Wilson. W.G.	Liverpool.	Pte.	18th	D.O.W	08 10 18	Busigny CC Ext.
21646	Wright. C.J. M.M.	Chester	L/Cpl.	18th	K.I.A	08 10 18	Busigny CC Ext.
13516	McHugh. W.	Liverpool.	Pte.	18th	K.I.A	09 10 18	Vis-en-Artois Memotial.
12369	Wood. J.	Liverpool.	Pte.	18th	D.O.W	09 10 18	Serain CC Ext.
56356	Archer. A.	Manchester.	Pte.	18th	K.I.A	10 10 18	Le Gateau Mil Cem.
23477	Beresford. G.	Manchester	Pte.	18th	K.I.A	10 10 18	Highland Cem, Le Cateau.
30297	Blackburn. H.	Liverpool.	Pte.	18th	K.I.A	10 10 18	Vis-en-Artois Memorial.
23777	Cartwright. W.	Manchester	Pte.	18th	K.I.A	10 10 18	Montay Brit Cem
41489	Consterdine. F.W.	Whitefield, Manchester	Pte.	18th	K.I.A	10 10 18	Highland Cem, Le Cateau.
20141	Cottrell. W.E.	Manchester	Pte.	18th	K.I.A	10 10 18	Highland Cem, Le Cateau.
84988	Davies. W.H.	Much Wenlock, Salop.	Pte.	18th	K.I.A	10 10 18	Vis-en-Artois Memorial.
85267	Gloyne. H.	Flint, Flintshire.	Pte.	18th	K.I.A	10 10 18	Montay Brit Cem
	Hayes. L.J.	Bootle, Liverpool.	2nd Lieut.	18th	K.I.A	10 10 18	Highland Cem, Le Cateau.
18398	Holland. W.	Patricroft, Manchester	Pte.	18th	K.I.A	10 10 18	Vis-en-Artois Memorial.
25412	Holt. F.	Skelmersdale, Lancs.	Pte.	18th	K.I.A	10 10 18	Le Cateau Mil Cem.
201530	Hornby. W.R.	Wavertree, Liverpool.	Pte.	18th	K.I.A	10 10 18	Vis-en-Artois Memorial.
202164	Jowett. H.	Liverpool.	Pte.	18th	K.I.A	10 10 18	Highland Cem, Le Cateau.
84856	Leece. W.E.	Peel, I.O.M.	Pte.	18th	K.I.A	10 10 18	Highland Cem, Le Cateau.
56298	Lock. W.	Liverpool.	Pte.	18th	K.I.A	10 10 18	Le Cateau Mil Cem.
14520	Robbins. G.	Liverpool.	Sgt.	18th	K.I.A	10 10 18	Le Cateau Mil Cem.
	Sanders. F.E. M.C.	Great Crosby, Liverpool.	Lieut.	18th	K.I.A	10 10 18	Highland Cem, Le Cateau.
41833	Simpson. J.W.	Castletown, I.O.M.	Pte.	18th	K.I.A	10 10 18	Le Cateau Mil Cem.
24039	Thompson. S.	Manchester.	Sgt.	18th	K.I.A	10 10 18	Highland Cem, Le Cateau.
24250	Tinsley. C.	Bootle, Liverpool.	L/Cpl.	18th	K.I.A	10 10 18	Montay Brit Cem
56255	Walker. R.	High Wheatley Hill, Durham	Pte.	18th	K.I.A	10 10 18	Highland Cem, Le Cateau.
56257	Waterworth. B.	Huttons Sambo, Malton.Yorks	L/Cpl.	18th	K.I.A	10 10 18	Vis-en-Artois Memorial.
20467	Whittaker. S.	Salford, Manchester.	Sgt.	18th	D.O.W	10 10 18	Roisel CC Ext
84868	Wilcock. T.	Appleby Bridge, Lancs.	Pte.	18th	K.I.A	10 10 18	Vis-en-Artois Memorial.
18759	Woods. H.	Liverpool.	Pte.	18th	K.I.A	10 10 18	Montay Brit Cem
49045	Hampson. A.F.	Boothstown, Lancs.	Pte.	17th	D.O.W	11 10 18	Cologne South Cem, Germany
332160	Hanley. J.	Everton, Liverpool.	Pte.	18th	D.O.W	11 10 18	Roisel CC Ext
21268	Kirkham. A.	Darwen, Lancs.	Pte.	18th	K.I.A	11 10 18	Vis-en-Artois Memorial.
27259	Carter. E.	West Didsbury.Manchester	Pte.	19th	D.O.W	12 10 18	Bailleul CC Ext.
243457	Cronin. W.	Wavertree, Liverpool.	Pte.	18th	D.O.W	13 10 18	Rocqigny-Equancourt Rd B.C.
85187	Hoy. J.	Bootle, Liverpool.	Pte.	18th	D.O.W	16 10 18	Roisel CC Ext
306003	McFarlane. W.J.	Everton, Liverpool.	Pte.	18th	K.I.A	18 10 18	Maurois.
22846	Smith. S.	Liverpool.	Sgt.	18th	K.I.A	18 10 18	Maurois CC.
23838	Sutton. R.	Kirkdale, Liverpool.	Cpl.	20th	Died	18 10 18	Kirkdale Cem, Liverpool.
21263	Townsend. J.	Darwen, Lancs.	Pte.	18th	D.O.W	18 10 18	Maurois CC.
25398	Townsend. W.J.	Sheldon, Warwicks.	Pte.	18th	K.I.A	18 10 18	Le Gateau Mil Cem.
20506	Yates. E.	Manchester.	Pte.	18th	K.I.A	18 10 18	Highland Cem, Le Gateau.
27167	Howarth. J.	Ashton-under-Lyne, Lancs.	Pte.	18th	D.O.W	19 10 18	Roisel CC Ext
93359	Clancy . S.	Sligo, Co.Roscommon	Pte.	18th	K.I.A	20 10 18	Le Cateau Mil Cem.
56271	Cell. AT.	Everingham, Yorks	Pte.	18th	K.I.A	20 10 18	Le Cateau Mil Cem.
18990	Gill. I.E.	Tuebrook, Liverpool.	Pte.	18th	K.I.A	20 10 18	Vis-en-Artois Memorial.

267375	Rimmer. J.	Southport, Lancs.	Pte.	18th	D.O.W	20 10 18	Southport, Duke St Cem, Lancs.
32555	Taylor. G.	Marple, Cheshire.	Pte.	18th	D.O.W	20 10 18	Maurois CC.
13784	Wilson. J.	Liverpool.	Pte.	18th	D.O.W	21 10 18	Ford Cem, Liverpool.
35226	Russell. A.A.	West Derby, Liverpool.	Pte.	18th		23 10 18	Romeries CC Ext.
308326	Speakman. J.	Wigan, Lancs.	Pte.	18th	K.I.A	24 10 18	Kemmel Chateau Mil Cem
12056	Woods. S.	Liverpool.	Pte.	19th	Died	30 10 18	Berlin South West Cem
19404	Bolton. H.	Kirkby, Liverpool.	Sgt.	18th	K.I.A	04 11 18	Dourlers CC Ext
	Lee. R.C.	Wallasey, Cheshire.	2nd Lieut.	18th	K.I.A	04 11 18	Romeries CC Ext.
241106	McDougall. E.A.	Waterloo, Liverpool.	Cpl.	18th	D.O.W	04 11 18	Roisel CC Ext
114148	Brown. G.H.	York.	Pte.	17th	Died	06 11 18	Archangel Allied Cem.
268610	Acton. G.	Walton, Liverpool.	Pte.	18th	K.I.A	08 11 18	Dourlers CC Ext
332819	Adams. W.	Liverpool.	Pte.	18th	K.I.A	08 11 18	Dourlers CC Ext
	Baker. H.L. M.C.	Derbyshire.	Lieut.	18th	K.I.A	08 11 18	Dourlers CC Ext
202891	Bartley. J.	Liverpool.	Pte.	18th	K.I.A	08 11 18	Dourlers CC Ext
18192	Cartwright. E	Sparkbrook, Birmingham.	Pte.	18th	K.I.A	08 11 18	Dourlers CC Ext
204609	Curran. H.	Liverpool.	Pte.	18th	K.I.A	08 11 18	Pont-sur-Sambre CC
269493	Derry. S.H.	Pendleton, Lancs.	Pte.	18th	K.I.A	08 11 18	Dourlers CC Ext
65291	Grant. N.G.	Plumstead, Kent.	Pte.	18th	K.I.A	08 11 18	Dourlers CC Ext
13453	Henderson. H.	Liverpool.	Pte.	18th	K.I.A	08 11 18	Dourlers CC Ext
62009	Lowe. J.	Hucknall, Notts	Pte.	18th	Died.	08 11 18	Terlincthun Brit Cem.Wimille
85178	Margerison. G.J.	Port Soderick, I.O.M.	Pte.	18th	K.I.A	08 11 18	Dourlers CC Ext
11490	McGanity. W.	Liverpool.	Pte.	18th	K.I.A	08 11 18	Dourlers CC Ext
18474	Nesbit. C.B.	Liverpool.	Pte.	18th	K.I.A	08 11 18	Dourlers CC Ext
88140	O'Brien. P.J.	Portsmouth.	Pte.	18th	K.I.A	08 11 18	Dourlers CC Ext
25356	Pye. E.	Ormskirk, Lancs.	Pte.	18th	D.O.W	08 11 18	Le Cateau Mil Cem.
52479	Carter. G.H.	Kettering, Northants	L/Cpl.	18th	D.O.W	09 11 18	Busigny CC Ext.
51567	McElroy. J.	Liscard, Cheshire.	Pte.	17th	Died.	09 11 18	Berlin South West Cem.
267329	Nichol. T.	Bootle, Liverpool.	Pte.	18th	D.O.W	09 11 18	Avesnes-le-Sec CC Ext.
20423	Kelly. J.	Manchester	Cpl.	18th	D.O.W	10 11 18	Busigny CC Ext.
91275	Bennett. C.	Birmingham.	Pte.	19th	Died	11 11 18	Poznan, Poland.
17378	Hoggarth. H.	Liverpool.	Pte.	19th	Died	11 11 18	Terlincthun Brit Cem,Wimille
33036	Neild. J.	Timperley, Cheshire	Pte.	18th	D.O.W	13 11 18	Timperley Cem, Cheshire.
300536	Cook. J.	Dolphinholme, Lancs.	Pte.	18th	Died	14 11 18	Grenvillers Brit Cem
41506	Brown. J.S.	Manchester	Pte.	18th	Died	15 11 18	Manchester Southern Cem, Lancs.
35551	Kirkman. F.	Bolton, Lancs.	Pte.	17th	Died	17 11 18	Heaton Cem, Bolton, Lancs.
	Berry. J.F.W.	Swinton, Lancs.	Lieut.	17th	Died.	22 11 18	Terlincthun Brit Cem,Wimille
58081	Foden. R.E.	Liverpool.	Pte.	18th	D.O.W	24 11 18	Mont Huon Cem, Le Treport.
21067	McQuiggan. J.	Southport, Lancs.	Pte.	18th	D.O.W	24 11 18	St Sever Cem Ext, Rouen.
	Maddick. S.A.		Lieut.	20th	Died.	02 12 18	Nairobi South Cem, Kenya.
25235	Wood. H.	Darwen, Lancs.	Pte.	18th	D.O.W	02 12 18	St Sever Cem Ext, Rouen.
22202	Orme. T.G.	Liverpool.	Pte.	18th	Died	03 12 18	Etaples Mil Cem
266299	Ainsworth. C.H.	Litherland, Liverpool.	Pte.	17th	K.I.A	07 12 18	Archangel Allied Memorial.
33092	Brown. R.	Dingle, Liverpool.	Pte.	17th	K.I.A	07 12 18	Archangel Allied Memorial.
24938	Greany. P.M. M.M.	Claughton, Birkenhead.	Sgt.	17th	D.O.W	07 12 18	Archangel Allied Memorial.
58354	Houghton. J.	Farnworth, Lancs.	Pte.	17th	K.I.A	07 12 18	Archangel Allied Memorial.
330381	Owens. A.	Liverpool.	Pte.	17th	K.I.A	07 12 18	Archangel Allied Memorial.
114356	Turner. H.J.	Birmingham.	Pte.	17th	K.I.A	07 12 18	Archangel Allied Memorial.
114275	Midgley. P.B.	Nelson, Lancs.	Pte.	17th	Died.	15 12 18	Archangel Allied Memorial.
16167	Corlett. W.R.	Port Sunlight, Cheshire	Pte.	18th	Died	17 12 18	Cologne South Cem, Germany.
114187	Sugden. B.R.	Bradford, Yorks.	Pte.	17th	K.I.A	30 12 18	Selenskoe Cem, Russia.
86974	Robson. E.A.	Liverpool.	Pte.	18th	Died	07 01 19	Huy (La Sarte) CC
108580	Graham. W.	Liverpool.	Pte.	17th	K.I.A	07 02 19	Archangel Allied Memorial.
	Stephenson. E.A.	Pocklington, Yorks.	2nd Lieut.	17th	K.I.A	07 02 19	Archangel Allied Memorial.
391327	Corlett. C.E.	Douglas, I.O.M.	Pte.	17th	K.I.A	09 02 19	Seletscoe Cem, Russia.
114116	Kenworthy. J.	Ashton-u-Lyne, Lancs.	Pte.	17th	K.I.A	09 02 19	Seletscoe Cem, Russia.
18619	Maher. C.	Garston, Liverpool.	Pte.	17th	K.I.A	09 02 19	Seletscoe Cem, Russia.
29664	McDonough. W.	Kirkdale, Liverpool.	Pte.	17th	K.I.A	09 02 19	Seletscoe Cem, Russia.
50785	Milton. C.L.	Goosnargh, Preston, Lancs.	L/Cpl.	17th	K.I.A	09 02 19	Seletscoe Cem, Russia.

114144	Roberts. J.W.	Shaw, Lancs.	Pte.	17th	K.I.A	09 02 19	Seletscoe Cem, Russia.
114191	Wright. A.H.	Tulse Hill, London.	Cpl.	17th	K.I.A	09 02 19	Seletscoe Cem, Russia.
22472	McLenahan. J.	Mossley Hill, Liverpool.	Pte.	20th	Died	18 02 19	Allerton Cem, Liverpool.
22669	Clatworthy. H.	Dingle, Liverpool.	CQMS.	18th	Died.	21 02 19	Allerton Cem, Liverpool.
	Kelly. J.A.C.	Edge Hill, Liverpool.	Pte.	19th	D.O.W	01 05 19	West Derby Cem, Liverpool.
108666	Murphy. J.	Wigan, Lancs.	Pte.	17th	K.I.A	05 05 19	Ust-Vaga Burial Ground, Russia.
114307	Stoneley. A.	Handsworth, Birmingham.	Pte.	17th	Died.	30 05 19	Chequevo Cem, Russia.
114100	Murray. B.	Keighley, Yorks.	Pte.	17th	Died.	28 06 19	Archangel Allied Cem.
81814	Jennings. J.	Waterloo, Liverpool.	Pte.	17th	Died	15 09 19	Windlesham Abbey RC Cem.
49121	Robins. E.	Tranmere, Birkenhead.	Pte.	17th	Died	09 01 20	Bebington C. Cheshire
235237	Lee A	Bispham, Lancs.	Pte.	19th	Died	17 01 21	Bispham Churchyard.
47353	Jackson. J.	Rainford.	Pte.	19th	Died	17 03 21	West Derby Cem, Liverpool.

* Also Commemorated on Thiepval Memorial.

Sergeant Harry Keay
17 Bn. Killed in action 27 June 1916. aged 26. The youngest son of Edward and Mary Keay. 25 Chepstow Avenue, Egremont, Wallasey, Cheshire.

Lance Corporal Thomas Hugh Jones
18 Bn. Killed in action 1 July 1916. He lived at 73 Hornby Boulevard, Bootle.

Sergeant John Milne
18 Bn. Killed in action 1 July 1916.

Private F. Morris
Killed in action 30th July 1916, at Guillemont.

Private John Murray
18 Bn. Killed in action 1 July 1916, whilst attending wounded in No Mans Land.

Private Louis Molyneux
19 Bn. Killed in action 30 July 1916.

Private J. W. Smith
18 Bn. Killed in action 1 July 1916.

Private Harold John Lover
19 Bn. Killed in action 30 July 1916, at Guillemont.

Private T. H. Parry
19 Bn. Killed in action 12 October 1916, at Fliers. He came from Liscard, Cheshire and was 23 years old.

Private John Alexander Jones
19 Bn. Killed in action 30 March 1918.

Private Tommy Ismay
18 Bn. Killed in action 2 September 1918.

Trench Raid

Orders issued to No.2 Company, 18th Battalion for the night of 15/16 March, 1917

See Chapter Nine, page 154

Headquarters,
10th March 1917.

Order No.1.

1. A specially detailed party of No.2 Coy. 18th K.L.R
    will carry out a raid on the enemy's trenches during
    the night 15/16th March.   This raid will be covered
    by Artillery and T.M.Fire.
    At approximately the same time the 8/Beford Regt.will
    carry out a "silent" raid about 1000 yards to our
    right and a "feint raid" - action by Artillery only -
    will be made by the Brigade on our left.
    The objects of the raid will be:-
    (a) To secure identifications and intelligence.
    (b) To inflict casualties on the enemy.
    (c) To secure booty and destroy dugouts and emplacement

2.
Objectives.
    The following are the objectives:-
    (a). Enemy Front Line Trench M.16.a.05.35 to M.16.a.40.5
    (b). Enemy Support Line, Trench, M.16.a.20.08 to
                                    M.16.a.46.27.
    (c). The two communication trenches connecting (a) and (b)

3.
Force
Employed. **Headquarters.**   O.C.Raid   Capt R.W.Jones 18th KLR
                           Intell.Off. 2nd.Lt. A.W.Brown.
                           Artillery Liaison Officer &
                           telephonists.
                           2 L.G. teams, 4 miners, 2 Signaller
                           1 bugler, 4 stretcher Bearers,
                           18th K.L.R.

         1st Wave.   O.C.Assault. 2nd.Lt.A.Coupe, 18th
                                                  K.L.R.
                           1 Officer & 40 O.R. 18th K.L.R.
                           2.R.E. personnel and 2 T.M.personnel

         2nd.Wave.   1 Officer & 48 O.R. 18th K.L.R.
                           2 R.E.personnel and 2 T.M.personnel

4. Artillery T.M.   The General Plan of Artillery, T.M. &
   & M.G.Support.   M.G. Support will be as follows (for
                           details see Artillery Orders issued to
                           all concerned.
                           (a) Zero to 0.05. 3 18.pr.batteries
                               barrage hostile front line between
                               M.15.b.82.25 and M.16.a.50.62.
                           1. 4.5" How.Battery barrages selected
                               points round the objectives.
                           5. Light T.M's (Stokes) barrage enemy
                               front trenches on the flanks of
                               the raided portion & also Sap Y.27
                               on the BUCQUOY RD.
                               ....Guns, 21st M.G.Coy.will fir
                               on........
                           (b) 0.05 to 0.10. 3. 18-pr.Batteries
                               lift to German Support line.
                               4.5" Hows. L.T.M. & M.G's
                               continue as before.
                           (c) 0.10 to 0.5. 3 18-pr Batteries
                               lift to Marlboro' Trench and
                               form box barrage. Remainder a.
                               before.
                           (d) Heavy Artillery will bombard road
                               junction M.15.d.65.95 and enemy's
                               support line South of this point
                               during the raid.

5. Time Table.   Raiding party will move from AGNY up
GREY ST. and will assemble in Front
Line Trench opposite gaps in our wire:
this move to be complete with all
stores issued by - 25 (25 minutes
before Zero).

  - 20 to -5. Raiding Party moves through gaps in
our wire and assembles in No Man's
Land under bank between Saps.
Y.30 and Y.32. Lewis Guns move to
positions covering the flanks of raid.

  Zero. Artillery barrage starts.
1st Wave preceded by wire cutters &
followed by 2nd Wave moves through
gaps in wire on objective, closing up
to the barrage.

  + 5. Artillery lifts to support line.
1st Wave enters front line trench.
2nd Wave crosses trench and follows
up our barrage: moving over the top
close to the communication trenches.
connecting German Front and Support
lines as far as the SUNKEN ROAD.
Blocks will be dropped at the SUNKEN
ROAD and then the remainder of the
advance continued underground except
for small parapet parties moving on to

  + 10. Artillery lifts to Marlboro' Trench
and forms box barrage.
2nd Wave enters German support trench.

  + 30. Bugler at Raid H.Q. sounds 1. "G" -
2nd Wave commences to withdraw.

  + 35. Bugler sounds 2 "G"s - 1st Wave
prepares to withdraw and does so
after 2nd Wave has got clear.
Headquarters withdraws to front line
trench when 2nd Wave has passed.
When all are in Lewis Guns will be
withdrawn by O.C. Raid.

  + 50. Artillery, T.M. & M.G's cease fire unless
asked to continue.

6. Zero.   Zero will be communicated by B.A.B. code to
all concerned by 6 p.m. on the 15th inst.
It will be approximately at M.30 p.m. unless
the night is very dark in which case it will
be postponed till after moonrise, i.e. about
2.30 a.m. 16th March.

7. Distinctive   (a) All ranks moving over the parapet will
Marks and   wear a strip of white cloth sewn under the
Dress.   collar of the jacket.
At Zero collars will be turned up, thus
exposing the white cloth.
Officers will in addition wear a white band
2½" wide on the left upper arm. This will be
covered up to Zero, with a piece of Khaki
loosely tacked on. Engineers and T.M.
personnel will wear white shoulder straps,
Similarly covered up to Zero.
(b) The O.C. Raid is held personally
responsible that every sign of identification
is removed from all taking part in the raid
before the party leaves billets. Officers
will wear jackets with universal buttons.
(c) Details of Equipment to be carried are
given in Appendix II of these orders and are
issued to all concerned.

8. Synchronisation
of Watches.

  The Battalion Intelligence Officer will ensure
that the watches of all officers taking part
are synchronised and will check the time with
Artillery, Trench Mortar and M.G. Liaison
Officers.

9. Communication.

    (a) Advanced Batt. H.Q. - RALPH Redoubt
                                (M.9.d.6.2)
            "   Aid Post -   Dugout in GRAINGER ST.
                                (M.9.d.3.5)
    (b) The following Telephone lines are being
        arranged for:-
    (1) Batt.H.Q. in AGNY to RALPH Redoubt (buried
                                cable).
    (2) RALPH Redoubt to Russian Sap in Front Line.
    (3) Russian Sap to Raid H.Q. in No Man's Land.
    (4) Artillery wire from Raid H.Q. in No Man's
        Land to Artillery.Brigade. H.Q. via RALPH
        Redoubt and AGNY H.Q.
    (c) Runners. With O.C.Battn. at RALPH Redoubt.
                                    ....4.
                        With O.C. Raid.............4.
                        Other Officers (1 each)....3.
    (d) The following officers will join Advanced
        Batt.H.Q. at RALPH Redoubt:- Artillery
        Liaison Officer and an Officer from the
        21st L.T.M. Battery and 21st M.G.Coy. each
        with a runner.

10. Withdrawal.   All ranks will finally withdraw down
                    GRAINGER STREET and will be checked at
                    the point where it crosses the tramway.

11. Appendices.   The following appendices are issued to
                    all concerned:-
                    Appendix I.   Field Artillery Orders.
                      " II.   Detail of Equipment carried
                      "III.   Detailed instructions to
                                parties.

                      (SD) W.R. PINWILL, Lieut.Col.
                        Cmdg. 18th Ser.Bn. K.L.R.

Copies issued as under.	Copy No.	Appendix I.	II.	III.
File	1	1	1	1
O.C.No.2 Coy	2-5		4	4
" 1 "	6		1	1
" 3 "	7		1	1
" 4 "	8		1	1
Signals	9		1	1
21st Infy.Brigade	10		1	1
148th Bde.R.F.A.	11-15			
Heavy Artillery	16			
21st.M.G.Coy.	17			
21st L.T.M.Battery	18			
2nd Bn.Wilts Regt.	19			
2nd Bn.Bedfordshire Rgt.	20.			
202nd Field Coy.R.E.	21			
O.C. Kopje	22			

APPENDIX II.

18th Ser.Bn. King's Liverpool Regt.

Details of Equipment - Issued with Order No.1 dated 10th
                          March 1917.
              ----

1. (a) All ranks will wear Service dress with all
        identification marks removed; Caps or Balaclava
        Comforters, care being taken that the latter do
        not impede hearing.   Gas Helmets will be worn;
        box respirators being left at ACHICOURT.
   (b) Officers will wear belt, revolver & knobkerrie
        and will carry a whistle and luminous compass.
   (c) Other Ranks, except throwers of blocking parties
        will carry rifle and fixed bayonet. Throwers
        carry knobkerries.
   (d) All armed with the rifle carry 10 clips in the right
        hand coat pocket, 9 rounds in the magazine and 1
        in the chamber.
   (e) Those armed with the revolver will carry 5 rounds
        in the chambers and 12 in the pocket. The chamber
        opposite the hammer when the hammer is down will
        be left empty.

APPENDIX III.

18th Ser.Bn. King's Liverpool Regt.

Detailed Instructions to Raiding Party issued with
Order No.1 dated 10th March 1917.
----

1. (a) Movement beyond our trenches must be covered by
advance patrols found from the Raiding Party which
must not move till a report, that advance is
possible, has been received.   Those patrols must
act boldly and with the utmost possible speed.

   (b) Generally speaking, every party will proceed straight
to its objective with as little delay and noise as
possible. There will be neither bombing nor firing
unless it is impossible to make headway without it;
the bayonet will be used as far as possible.

   (c) Dugout entrances when first passed will normally
be treated with a Mills Bomb and then left to be
dealt with by the Cleaners.

   (d) Cleaners of the 1st Wave will work outwards from
the point where the trench is entered; those of
the 2nd Wave will work to the centre of the enemy
support trench.

   (e) "P" Bombs will not be used till the final withdrawal
commences, as the smoke is liable to hamper movement.
"M.S.K." (Lachrymatory) Bombs will be thrown into
the trench by Blocking Parties immediately before
they finally withdraw, so as to hinder the enemy
should he attempt to follow up closely.

2. Assembly in No Man's Land.   This will be timed so as
to give the least possible pause before Zero at the
Bank. The advance over No Man's Land to be done by
crawling: should Very Lights go up all must remain
motionless.   Two minutes before Zero a patrol will
commence to move through the gap in the wire.

3. Action by 1st Wave.   The 1st Wave moves through gap
in wire in two columns, each in single file;   and
closes up to the barrage when gap is reported
practicable.
When the first "lift" occurs bombing and blocking
parties enter trench and proceed along it direct
to their objectives.   Cleaners place ladders for
2nd Wave and then commence cleaning from point of
entry outwards.
Should the wire be found insufficiently cut, this
will be reported and the wire will be cut by
bangalore torpedo placed by the R.E.

4. Action by 2nd Wave.   Crosses enemy front line trench
as quickly as possible and moves over the top along
communication trenches as far as the SUNKEN ROAD;
then takes to communication trenches covered by
parties moving over the top.   Road Blocking Parties
are dropped at the SUNKEN ROAD.   Blocking parties
proceed straight to their objectives and cleaners
work to the centre, special care being taken as they
approach one another.   The parapet parties will
remain on top till the withdrawal commences.

5. Prisoners.   Prisoners taken will at once be sent back
under escort to the parapet party on the front line and
thence to dugout 43.   They should be handcuffed and
if obstreperous tied up with a wire necklace or killed.
They are to be rapidly searched for hidden weapons
immediately they surrender.

6. The method of cleaning Dugouts will be as follows:-
Cleaners standing clear of the entrance will call
out to the occupants to come out; if there is no answer
a light will be flashed down and the dugout entered.
Opposition will be dealt with by rifle or bomb as found
necessary. A dugout after being thoroughly cleaned
may be blown up by a Stokes Bomb.
As an alternative to the above, particularly if prisoners
have been already secured, the cleaners may tell the
enemy to come out at once and after a short pause to
enable them to do so, blow up the dugout.
The cleaners with each wave specially detailed to collect
intelligence will search dugouts, dead bodies etc. for
pay books, identification discs, shoulder straps, numeral
copies of orders, etc. They will also note the
construction & state of enemy trenches.

7. Withdrawal. In timing the withdrawal officers will
rely almost entirely on their watches; the bugle call
is to be looked on as subsidiary as it probably will
not be heard.
When the withdrawal commences as many of the 1st Wave
as possible will be sent back to Raid H.Q.: a portion
of the 1st Wave will get on top of the parapet to assist
the 2nd. Wave to climb out of the trench.

8. All ranks are to be warned that in the event of their
being taken prisoner they are when questioned to KNOW
NOTHING except their name and number.
The Germans frequently use English words of command,
particularly the word "Retire", to cause confusion.
Parties must be on the look-out for this.

9. A party of 10 N.C.O's & men with 4 Stretcher Bearers under
2nd.Lt. Ewing will be held in readiness in dugout 43
to act as a search party if necessary.

10. The checking of men down GRAINGER ST. will be done as
expeditiously as possible so that the earliest intimation
may be sent as to the number of missing to 2nd.Lieut.
Ewing (Vide para.9.)

                    (SD) W.R. PINWILL, Lieut.Col.
                    Comdg. 18th Ser.Bn. K.L.R.

21st Inf.Bde.   69/3
K.p.1.
            16 March 1917.
   xxxxxxxx   Report on Raid.

1. The Raid carried out by No.2 Coy, 18 K.L.R last night on
the German Trenches between Sap Y.30 & Y.31 was
successful in so far that the German Front Line was
entered: dugouts blown up, some of them containing
Germans: and the Raiding Party safely withdrawn.
No prisoners or other identifications were however
obtained.

2. The Raid was carried out in accordance with programme,
except that the final barrage was continued for an
extra five minutes as the raiding party was not clear.

3. The wire had been most successfully cut by the Artillery
& 2" T.M. & no difficulty was experienced in gaining the
front line trench.
A few bombs thrown from this trench caused casualties
among our leading men: but after this no enemy were
seen in the open. From sounds heard in certain
dugouts, it is certain that some of the enemy were
hiding there, they wld. not come out when called on
to do so & the dugouts were treated with stokes bombs.
The party detailed to raid the German Support Line did
not get much further than the Sunken Road. The
reason appears to be that considerable delay was
caused in the German Front Line Trench owing to the
presence of an iron obstacle which checked the leading
wave & thus prevented them from clearing the way for
the 2nd wave.
The withdrawal was carried out in an orderly manner.
The enemy made no attempt to follow up.

4. The Artillery barrage arrangements were all that could
be desired & the Stokes Mortar firing on the flanks
gave specially valuable aid. The Heavy Artillery
kept possible T.M's quiet & two guns 21st M.G. Coy.
kept up a well-sustained fire on the ARRAS - BUCQUOY
Road.

286

5. Some minutes after the Raid started, enemy M.Guns began to fire from the right & left of the raided portion. Their fire was intermittent & seemed mostly high. There was practically no artillery retaliation either during or soon after the raid. A few 77s fell in the vicinity of RALPH Redoubt. It was noticed that the enemy sent up fewer Very Lights than usual when the bombardment started.

6. All the arrangements made for the raid worked perfectly & the party moved up to the front line & back again with a minimum of exertion & delay in spite of heavy mud in places. The assistance given by the 20th K.L.R. was largely responsible for this. The Signal & Medical Arrangements were adequate & worked well. The R.E. Sappers attached for blowing a gap in the wire were not made use of but they had all arrangements made in case they were called on.

7. I am indebted to the Liason Officers of Field Artillery, Heavy Artillery, 21st Bde. T.M.B. & 21 Bde. M.G.Coy. for their assistance. Also to the Bde. Major & Signal Officer 21 Inf.Bde. for valuable assistance both before and during the raid.

8. The casualties were

Killed   1 N.C.O.
Wounded   6 O/R.
   "   accidentally   1 O/R.

9. I propose to submit names for immediate Rewards in connection with this Raid shortly.

3.45 pm.   (SD) W.R. PINWILL, Lt.Col.
                                18 K.L.R.

                              VIIth Corps, G.C.R. 693/8.

30th Division.

        The Corps Commander wishes you to convey to the G.O.C. 21st Brigade, to your Artillery, to the O.C. 18th King's Liverpool Regt. and to all ranks engaged in the raid carried out on the 15th., his appreciation of the skilful manner in which the enterprise was carried out.
        He considers that the operation was very creditable to all concerned; though it is to be regretted that no identification was obtained.

                    (SD) J. BURNETT STUART,
17.3.17.   B.G. C.S. VIIth Corps.

        Copy telegram received from 3rd Army 26/3/17.

Army Commander sends his best congratulations to all ranks of 21st Brigade on the successful raid carried out by them on March 16th.

Pte. R. Barrows

24382 Pte G. Beecroft

Pte J. Borrowscale

17109 Pte T. Burns

22638 Pte G. R. Carter

22965 Pte J. Crawford

21867 Pte J. L. Edwards

21868 L/Cpl Edwards

22994 Pte H. O. Evans

21915 Pte W. Heyes

Pte C. Hobbs

Pte W. Holland

17629 Pte D. M. Jones

CSM P. Lyons

15909 Pte T. W. Maybury

Pte S. F. McCready

L/Cpl A. Moore

17458 Pte G. R. Radcliffe

21912 Pte T. Rosbotham

Pte W. Stephenson

22273 Sgt W. Thompson

23611 Pte E. M. Underwood

24548 Sgt/Maj A. Walters

A selection of men who served in the four Pals battalions of the Kings Liverpool Regiment, and survived the war – from photographs kindly supplied by relatives.